Neurocultural Health and Wellbeing

Aim of this devoted book Series in neurology is to highlight the relationship between neuroscience and culture. Nowadays, there is more evidence of how our brain is influenced by the various artistic and cultural disciplines, both in the form of entertainment as well as – and above all – through the emotions and gratification which lead to wellbeing. The emotional mechanisms, the different cultural manifestations provided, are due to the activation of a perceptual and cognitive range that constitute the basis of the social behaviors. All this makes us aware of the benefits the arts have on personal and collective health.

This concept was already known to philosophers in ancient times, who were convinced that inner balance was influenced by culture and in particular by music.

During the centuries, men understood that the *cultivation of the spirit*, or *humanitas*, had a certain role on behavior and from the beginning of modern experimental science, in the fifteenth, the spread of the notions of neuroanatomy allowed artists to get closer to the knowledge of the brain's mechanisms and reveal the emotional and empathic responses at the basis of creativity, and indeed of their own psychophysical wellbeing. A dialogue between the science of the mind and artistic disciplines was born. The results of this meeting made it possible to better define which mental processes are involved when we come into contact with the various humanistic disciplines, and how they can be applied, for instance, to the treatment of mental disorders and neuropsychiatric diseases.

The goal of this Series is both to prove the role of biological brain mechanisms and the influence of various artistic forms on clinical practice, especially in neuro-psychiatric disorders, as well as to trace different therapeutic and psycho-physical well-being applications based on scientific evidence from medical literature.

Volumes of the Series will be edited by experts under the supervision of an international editorial committee. Each book, focused on a specific discipline, will provide knowledge on relationship between the brain activity and different forms of language, communication, art. This close inter-relationship with specific focus on different forms of art will explain the effectiveness of this kind of approaches in neuro-psychiatric diseases.

This Series will allow to understand how the culture is one of the fundamental tools to improve general well-being, quality of life and motivation in neurological diseases.

This Series also find a correlation with SDG3 goal "ensuring healthy lives and promoting wellbeing for all at all ages" for the progress of the health and wellbeing considering that neurological diseases are on the rise worldwide including in the developing countries.

Amy Ione

Neuroscience and Art

The Neurocultural Landscape

Springer

Amy Ione
The Diatrope Institute
Berkeley, CA, USA

ISSN 2731-4464 ISSN 2731-4472 (electronic)
Neurocultural Health and Wellbeing
ISBN 978-3-031-62335-6 ISBN 978-3-031-62336-3 (eBook)
https://doi.org/10.1007/978-3-031-62336-3

This Springer imprint is published by the registered company Springer Nature Switzerland AG
The registered company address is: Gewerbestrasse 11, 6330 Cham, Switzerland

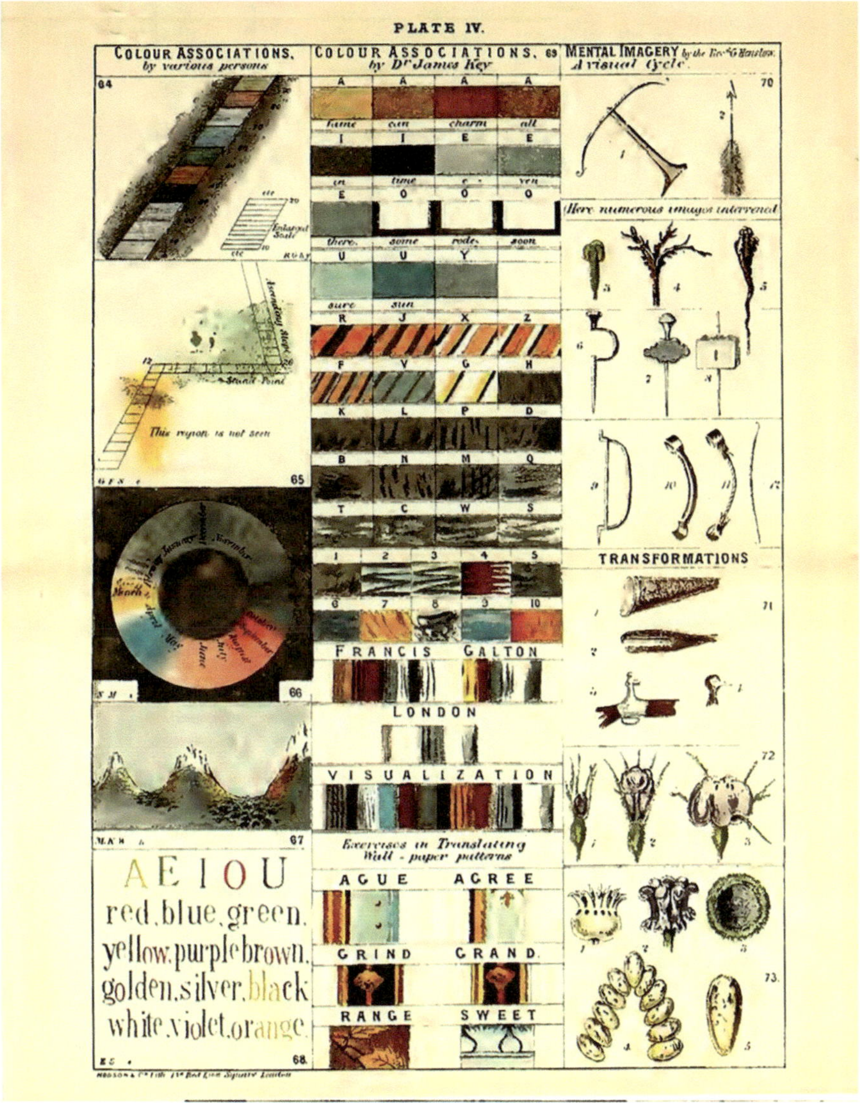

Plate IV from Francis Galton's *Inquiries into Human Faculty and Its Development* (1883). This compilation is an abridged summary of Galton observations on mental imagery. This book brought what we now call synesthesia to the attention of the scientific community

In Memory of Dame [Jean] Iris Murdoch (1919–1999) whose thought-provoking book The Fire and the Sun: Why Plato Banished the Artists first motivated me to think about the complexity of art and culture many years ago.

And

For Christopher Tyler, whose stimulating mind encompasses both an expertise in neuroscience and a knowledge of art that has become an intrinsic part of this book's texture.

Preface

Understanding ourselves as humans is incomplete without considering both biological and cultural aspects. In this book, the neurocultural perspective reflects how everything in the world is filtered back and forth between the brain and the cultural milieu. The thrust of the book, therefore, is to explore the power of art in creating a bridge between cultural and neuroscientific lines of inquiry. Looking at both clinical and non-clinical populations, the text examines historical foundations, distinguishes congenital/developmental conditions from those that are acquired, and emphasizes how the brain constructs our sensory experiences. Topics include the value of case studies, film, war's impact on medicine and art, therapeutics, and education. A range of conditions are integrated, such as prosopagnosia, synesthesia, autism, Alzheimer's disease, and so forth. Artists from all genres are incorporated. Among them are Iris Murdoch, Ludwig van Beethoven, Leonardo da Vinci, Cristoforo de Predis, Rembrandt, Federico Fellini, Chuck Close, and David Hockney. This book will be of interest to a broad range of clinicians and neuroscientists as well as those in the arts and humanities in general.

Berkeley, CA, USA Amy Ione

Acknowledgments

Many thanks to Andrea Ridolfi and Smitha Diveshan at Springer for their help and guidance while I was organizing and writing this book. Thanks are also due to the many museums, organizations, and individuals who provided free access to images in their collections. I am among the many who appreciate this move toward increased image accessibility. Thanks, too, to all those who have contributed thoughts and ideas from the beginning. There are too many to name, but I would like to single out Lorenzo Lorusso and Christopher Tyler for their encouragement.

Contents

About the Author

Amy Ione is the Director of The Diatrope Institute, an organization for the dissemination of interdisciplinary information. She is a writer, artist, and educator and has studied the extensive linkages between art and neuroscience for over 30 years. She has published several books, including *Innovation and Visualization: Trajectories, Strategies, and Myths*; *Art and the Brain: Plasticity, Embodiment, and the Unclosed Circle*; and *Nature Exposed to Our Methods of Questioning*. *Neuroscience and Art: The Neurocultural Landscape* is her sixth book. Amy had also published over 200 articles and book reviews on art, neuroscience, technology, and cultural issues. In addition, Amy Ione is a lifelong artist, with artwork commissioned by the City of San Francisco, exhibited internationally, and found in many collections.

List of Figures

Part I

Overview

Historical Neuroculture

1

1.1 Why a Neurocultural Perspective?

Why write a book about neuroscience and art from a neurocultural perspective? The obvious reason is that understanding who we are as humans is incomplete without considering the interplay between biological and cultural perspectives, or neuroculture. Neuroculture in this book is addressed along the lines mentioned by Oliver Sacks when explaining his approach to neurology. From childhood he gravitated to medicine in part because he learned to see it as the study of stories of people. Each evening at the dinner table his physician parents talked about the patients they had seen during their day. This interest in the patient as an individual followed him to medical school and eventually led him to share the stories of lives impacted by disease or injury, conveying the uniqueness of each person in the process [1].

Neuroculture includes this comprehensive view of human health and well-being as seen through the profile of each individual. On the following pages, this spectrum is examined through neuroscientific research and a range of art forms that demonstrate the value of the arts. The arts are adept in communicating our joys and sorrows. They may tell a personal story or show declining health. They also record family interactions, social matters, and the indescribable. Syntax varies among the arts, just as intentions in making art do. Indeed, art may be expressive, therapeutic, challenging, uplifting, or defy characterization. Because the production, enjoyment, and value of art includes more than "art as beauty," we all benefit from comprehensively acknowledging the degree to which art and the brain interactions encompass all aspects of the complex human experience.

In his book *Musicophilia* [1], Oliver Sacks spoke to this point from another perspective when he noted that music was always one of the first things he looked up in any new neurology or physiology textbook. He adds that he developed this habit because he could scarcely find any mention of the subject until the 1977 publication of *Music and the Brain* [2] edited by Macdonald Critchley and R. A. Henson. Sacks hypothesized that one reason for the earlier scarcity of musical case histories was

A. Ione, *Neuroscience and Art*, Neurocultural Health and Wellbeing,
https://doi.org/10.1007/978-3-031-62336-3_1

that physicians rarely ask their patients about mishaps of musical perception whereas linguistic problems immediately come to light [1]. This chapter explores the history behind this lacuna, as does the next chapter on experiment, theory, and applications.

Music and the Brain [2] documented a symposium Critchley and Henson assembled in 1972. In their introduction to the proceedings they note that participants, to their surprise, realized that this event was the first occasion the subject of music and the brain had been submitted to serious discussion by neurologists, although the sedative and even therapeutic effects of music had been referred to on and off since the pre-Homeric myth of Orpheus and Eurydice [2]. Elaborating on this historical vacuum, Henson acknowledges that musicians in effect filled the gap as they pursued and perfected their craft:

> There is a wealth of distinguished writing by neurologists on philosophical and literary matters, and music has fared ill by comparison.... Even Henry Head limited his work on music to observations on the effects of brain damage on musical functions, although he had a good knowledge of music and the intellect and imagination to make a major contribution.... musicians have rightly made their own studies in practical or applied neurology. Teachers have expounded their ideas on motor activity and voice production as they relate to performance, hitherto with little assistance from neurophysiologists. Others have recorded their views on memory, an area where neurological knowledge is enlarging and increasingly capable of helping the musician (p. 3, [3]).

Recent historiographical work has begun to fill this gap with coverage of a wide array of topics, controversies, and competing perspectives [4–9]. Therefore, at this time there are many resources for those who wish to go beyond the historical summary offered here.[1] In examining this topic neuroculturally, this chapter and the next demonstrate how our understanding of brain function, sensory input evaluation, and higher-order functions has changed over the course of time. For example, looking at older material allows us to identify assumptions that became implicit and how they were initially interwoven into the cultural fabric before they were superseded as new knowledge emerged. This kind of comparative analysis also allows us to better comprehend current understandings of brain functions, brain injury, sensory modalities, brain plasticity, and the brain/culture exchange. A key variable, often underemphasized, is that it was not always understood that the brain constructs our experience. In other words, even with less understanding than the limited understanding we have today about neuroscience and the arts, people understood that the body, the mind, the brain, and culture are connected. So, on an experiential level, it was known that listening to music, for example, is not just auditory or emotional, it is motoric as well—and adds to human experience.

Before I turn to this historical material, let me note that art in this volume is not defined as Fine Art or an aesthetic that operates on some kind of lofty plane. Rather,

[1] In earlier work I have offered detailed historical chronologies on intersections between art and the brain, including collaborative work between artists and medical practitioners that aided research projects (see [10–17]). I have also examined art in relation to the history of psychology, the mind, and the brain elsewhere (see [12, 18]).

art covers a spectrum. It is a form of communication; a tool used to foster health and well-being; a therapeutic tool; a creative and innovative approach to problem-solving or expression; and it is an array of language forms, for each medium uses its own syntax. It is static, like painting and sculpture. It is dynamic, like music and the performing arts. It is found in both popular culture and more refined venues.

Finally, a few words about the book's overall organization. The four chapters that comprise Part I offer an overview that sets the tone for the book. Chapters 1 and 2 introduce historical assumptions and distinguish them from our contemporary knowledge base. Chapters 3 and 4 take this further, distinguishing biological and genetic conditions from those acquired later life. Together these two chapters stress that although an acquired condition is clearly biological and possibly related to genetics or environmental factors, it often creates a rupture that distinguishes it from a condition one knows from birth and infancy. In other words, baseline matters.

Part II presents a series of individual and thematic case studies. It opens with two chapters that contrast the work and conditions of two storytellers, Iris Murdoch and Oliver Sacks. Murdoch, a novelist/philosopher and a cultural icon, invented stories. Sacks, a neurologist, introduced the narratives of his patients to a larger public. Their health challenges and derivatives further frame each in terms of their inspirations, aspirations, and accomplishments. The case studies presented in Chapters 7, 8, and 9 offer thematic neurocultural examinations. Topics include technology, war, and therapeutics. The closing chapters that comprise Part III focus in on practices. The thrust encompasses both how individuals learn about and/or translate specific neuroatypical conditions and also covers how novices, specialists, and the atypical differ.

1.2 Why History Matters

History shows that seeking to comprehend life's complexity biologically and culturally has long roots. People have always asked who am I, what is nature, and how do we best live together? The "answers" have touched on ways of healing the ill, how to teach social mores to the community, and the development of communication tools such as language, art, music, and the performing arts. Within this, the loss of biological function and the preservation of well-being were always of great concern—and loss was interpreted as something that needed understanding, compassion, and cure to the degree possible. Thus, throughout time people have responded to health challenges, body changes, and mishaps by investigating the onset of illness or injury; just as they have found ways to enhance sensory experience and community.

We can identify investigations across the globe, even rudimentary work in ancient times, that included studies of the how the body works, surgeries to heal it, and pharmacological methods that were developed from the testing of crude extracts of plants, animals, and minerals for their medicinal properties [4, 6, 9, 19]. We can also identify experimentation and discussion regarding healthy living and education. We additionally find that in earlier times the neuroatypical and those with congenital anomalies had fewer developmental prospects than those with acquired conditions, a point expanded on throughout this volume.

Art, too, was a part of the human equation. This is shown by a recent discovery of prehistoric bone and ivory flutes from the early Aurignacian period of southwestern Germany. These instruments affirm there was a musical tradition when modern humans colonized Europe, more than 35,000 calendar years ago [20]. Painting, like music, is found historically, as the known Cave Art throughout the world and extending back 40,000 years reminds us. Singing, a musical behavior that leaves no physical trace behind is no doubt a part of this history too for we see it depicted in ancient visual artwork. And surely the universality of beat perception would have given people the urge to move when their instruments were played, even among those listeners who were only minimally musical.

Although non-western lines of thought are largely outside the scope of this study, some are mentioned briefly in this chapter to underscore that not only were physiological functions and healing methodologies debated within cultures, but we can recognize them across cultures as well.[2] What I want to first stress here is that interpreting the earliest known cultural paradigms is challenging for us today due to the difficulty in separating magic and religion from clinical and experimental science in the documents and artifacts. Source materials that have survived do nonetheless offer a wide range of clues.

1.3 Egypt

In Egypt we can find evidence of sophisticated efforts to understand neurological conditions recorded in the *Edwin Smith Surgical Papyrus* and the *Ebers Papyrus* [6, 24]. These date from about 1500–1700 BCE, with the *Smith Papyrus* considered the older of the two. Several of the cases discuss the brain, meninges (coverings of the brain), spinal cord, and cerebrospinal fluid. Some include head injuries. From their descriptions we can see that the Egyptian physicians were aware that symptoms of central nervous system injuries could be felt far from the locus of damage [4, 6]. We also find sensory discussions. For example, in the *Ebers Papyrus* we find the earliest mention of deafness and otology on the list of medical remedies and spells against common ailments, which includes "the ear that hears badly." This suggests a loss of hearing rather than congenital deafness. The Egyptian notations about acquired deafness are also worth noting because congenital conditions were considered more puzzling than acquired ones,[3] as was the case in many cultures of the time.

[2] There is a growing body of work that looks comparatively and/or at other cultures (see [19, 21–23]). This interdisciplinary approach is important because it emphasizes the need for understanding human populations on both biological and cultural terms.

[3] Throughout this book deafness and loss of hearing are frequent topics. Therefore, it is important to note that people belonging to the Deaf Community do not consider deafness to be a disability and embrace the values upheld by the cultural model of Deafness [25].

Cited among the recorded observations we also find many eye problems even as far back as the *Ebers Papyrus,* again reinforcing that the Egyptians were keen observers. But, here too, there is no sense that they associated the retina with the optic nerve, for instance (see Fig. 2.5). Their documents do affirm they could, and did, observe the loss of visual perception and obvious biological degeneration or injury. Notations like a white opacity in the cornea of the eye show they could likewise identify physical symptoms of ailments such as leucoma, cataracts, and glaucoma. This is also evident from the cataloged treatments for eye diseases, although in many cases, the treatments themselves appear questionable (like using urine as an eyewash) [4].

The documentation makes it clear that ancient physicians were good observers but did not have enough information to consider healing and diagnostic options to the degree possible today. Obviously, they lacked the kinds of imaging tools we now have that allow us to see through opaque bodies and view brain processes. They also lacked any quantitative sense of electrophysiology, biochemistry, and the kinds of data we are beginning to accumulate on genetics, mutations, and the impact of environmental factors. Still, congenital abnormalities and "freaks" were catalogued, giving the impression that they saw this as a way to study them [26].

Along with the questionable treatments and diagnostic observations, we can also identify an array of other correct and incorrect assumptions. One frequently mentioned in neuroscientific histories is the Egyptian assumption that it was the heart, not the brain, that was the seat of perception, intellect, cognition, memory, and emotions. Indeed, like many Biblical and early Greek thinkers, we find the heart elevated over the brain in China, India, Egypt and throughout the world. Therefore, the heart was regarded as the spirit and soul of an individual. Phrases like "in my heart I know it's right" still reflect this way of thinking. Plato's tripartite theory of the brain is an exception here, although this in no way means all his ideas were error free. As discussed in Chap. 5, he had little regard for knowledge gained through our senses, although we know that our sensory system gives our brains knowledge about the world external to ourselves.

The importance of sensory experience in an individual's life was captured in an emotional response to blindness penned by an Egyptian painter Poi (c. 1304–1237 BCE) in a letter to his son. Poi's despondency over his vision loss shows that the despair that accompanies losing one's ability to see was as real then as now.

> Do not leave me. I live in despair … I live in darkness. My god Amon has left me. Bring me honey for my eyes, and fat … and genuine eye-paint, as soon as possible. Am I not your father? I wish to see, but my eyes are deserting me (p. 65, in [4]).

Artifacts and works of art additionally document maladies outside of the preserved medical literature. These, too, offer information about illness, health, emotions, and cultural views on well-being. When abnormalities and ailments are recorded, it is often unclear if the intention is to present symbolic ideas or if these serve as another example of how frequently cultural ideas are interwoven with views on biological change. Perhaps both.

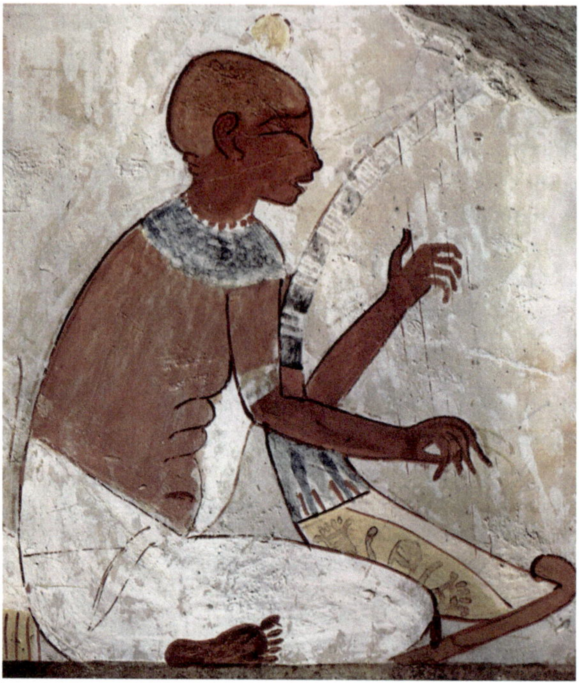

Fig. 1.1 A mural of a blind musician playing a harp, from the tomb of the ancient Egyptian scribe called Nakht. c. 1422–1411 BCE. Courtesy of the Yorck Project

The blind Egyptian harpist (Fig. 1.1), a common motif in Egyptian art, also reminds us that the meaning behind an artifact is often unclear or symbolic. These musicians are generally represented singularly as a wealthy, well-dressed, bald, fat man. The harpist's baldness associated him with priests who shaved their heads to achieve ritual purity. The difficulty in dating the songs precisely matches the difficulty in interpreting the songs and the paintings. When Lichtheim studied the harpists' songs in the mid-twentieth century, she concluded they reveal the history of Egyptian attitudes toward death and the afterlife [27]. Since none of the harpist songs mention blindness, the inference is that the motif is purely symbolic, one intended to urge people to enjoy life on earth and stress the good to come in the afterlife, as the following refrain preserved in two New Kingdom documents conveys.

> He is happy, this good prince!
> Death is a kindly fate.
> A generation passes,
> Another stays,
> Since the time of the ancestors (in [27]).

Debates as to what artistic works like the blind harpist say are not confined to their symbolic aspects. They also demonstrate that the brain interprets many factors when engaging with art, including the syntax of the presentation. In this case, the painter's syntax includes pigments, color, line, form, and so forth to convey and capture this work's message, which may include representations of higher-order symbolic ideas or abstract concepts that are culturally known, as it appears the blind harpists do. In linguistics, such as the poetic words comprising the harpist's song, we see another kind of syntax. Here words and morphemes are combined to form the language the brain interprets and by extension the abstract and symbolic ideas conveyed through it. Musical syntax would include pitch, notes, rhythm, melody, harmony, texture, timbre and so forth. Perhaps this song was read or sung with inflections that added a musical nuance to its linguistic, visual, and symbolic rendering?

1.4 The Ancient Greeks and Etymology

Ruminations on how the mind, the body, and nature functioned as a whole were also passionately discussed in Ancient Greece. From Pythagoras (c. 570 BCE–c. 500 BCE) and Thales (fl. 550 BCE) onward, thinkers and physicians were developing elaborate theories. One acknowledgement of the Greek contributions is found in current terminology that connects neuroscience and art to later fields of inquiry in the West. For example, the Greek term *psyche* (spirit or mind) later served as the root of psychology and psychiatry, two areas defined as stand-alone fields in the nineteenth century. Other terms associated with neuroscience and art that have Greek roots include neurology and aesthetics. Both are discussed shortly. *Therapeutikos*, the Greek root for therapy, means to cure or treat medically [28]. Therapeutics is the focus of Chap. 9.

The use of Greek etymology emphasizes how directly Western assumptions about the mind, the brain, and culture trace back to many Greeks ideas. Yet, it is important to note that despite some theoretical connectedness, the Greek worldview was intrinsically unlike our own, or even that of the Renaissance. How their cosmobiological assumptions differed is largely overlooked when discussing their impact on neuroscience and art [18, 29], and thus important to note.

A cosmobiological view is one in which everything is seen as a connected biological system. As the influential etymologist Henry Watson Fowler points out, both the words physics and physiology are derived from the Greek work *physiologia*:

> [*Physiks* and physiology] have now been narrowed and differentiated, physics retaining only the properties of matter and energy in inorganic nature, and physiology only the normal functions and phenomena of living beings (p. 427, [30]).

Physiologia, of course, is the root of physiology. When Jean Fernel (1497–1558), a sixteenth century French physician, introduced the term "physiology" to describe the study of the body's function, he said, "Anatomy is to physiology as geography is to history; it describes the theatre of events" (in [31]). When the term is used today to characterize how our bodies work, the term does not encapsulate the degree to which we live in a more bifurcated or specialized world than the Greeks, although theoretical conjectures surrounding it may. For the Greeks, however, physics (inorganic matter ***and*** nature) and physiology (nature ***and*** the body) were seen not just as working in tandem; they were seen as inseparable [18, 29, 32]. The vagrancies of mind, disease, and the inexplicable that appeared to violate the cosmic harmony were more of a mystery. How this seamless perspective was evident in the ever-changing world at-large was vigorously debated and influenced thinking about natural philosophy (medicine, science, and philosophy).

In other words, the Greeks, like the Egyptians, took a serious interest in how all things work, and work in tandem, even the quite mysterious congenital conditions, and even though their conjectures about human physiology were based largely on animal studies. Physicians and philosophers contributed to these efforts to understand life, where it came from and how it was lived. For example, Hippocrates studied chicken embryos. Aristotle classified animal species and considered whether they reproduced by means of eggs, spontaneous generation (i.e., in smaller species and insects) or viviparity [26]. The physician Galen, too, offered hypotheses. He asked whether the embryo came from the white or the yolk of the hen's egg. These kinds of investigations fostered speculation [26], particularly in regard to brain formation and function.

Another term derived from the Greeks is anatomy, the branch of biology concerned with the study of the structure of organisms and their parts. The word is derived from the Greek word for dissection. Early anatomical investigations echoed the cultural rubric that assumed the proper goal was understanding how to achieve harmony between the parts and the whole of organisms. Human dissection was controversial and forbidden by law so comparative analyses were made through animal dissections and vivisections [4, 6, 33]. Because dissection was frowned upon for religious, moral, and cultural reasons, physicians and anatomists instead relied upon knowledge gained from minor surgeries. Exceptions to the legal restrictions did occur as well, including the work of Herophilus (335–280 or 255 BCE) and Erasistratus (c. 304–c. 250 BCE). While at the Alexandrian Museum, they put together a tripartite brain theory that was more anatomically informed than Plato's and included beliefs, emotions, and desires. This later model was somewhat modified by Galen of Pergamon (130–200) in the second century [34], who is discussed shortly.

Arete, another Greek term, reflects how Greek cosmobiological cultural mores referenced functional parts (humans) within the whole (society and/or the cosmos). Although *arete* is generally translated into English as virtue, it is more ambiguous. Arete, like many Greek terms, was used in both the microcosmic and macrocosmic sense to convey how working with (or within) the proper functional mode harmonized the whole. The idea was that all of us and everything has a certain function that we are intended to perform for our common humanity to achieve excellence. It could be a skill or efficiency at a particular job. The arete of a doctor is different from that of a warrior. For a warrior, one's function includes dying in battle; just as a physician's function is healing.

Individual bodies were seen in similar terms. So, in health and well-being, limit and order were considered good. The well-being of the world (or its harmony) and that of every creature in it depended on a proper mingling of the elements of which it is comprised. With the proper mingling, all is in a state of *harmonia* [32]. We repeatedly see the urge for *harmonia* in descriptions of nature and the body's functions. One well known contributor here is Aristotle (384–322 BCE), Plato's student. His writings explain that every part of the body is formed for a purpose and that function could therefore be deduced from its structure.[4] *Harmonia's* medical formulation—humor theory—is credited to Hippocrates of Kos (460–377, BCE) and also connected to the Pythagoreans.

The Pythagorean tradition is important neuroculturally given that Pythagoras' (c. 570 BCE–490 BCE) most striking and enduring discoveries were in the field of music [32]. Although in many ways a mystical sect, the Pythagorean's also distinguished themselves from other Greek philosophers through their emphasis on form and order rather than the nature of matter. In addition, some say that Pythagorean thought influenced Plato in developing his ideas about Ideal Forms and the immortal soul. Pythagoras is particularly relevant neuroculturally because he identified that the pitch of a musical note is in inverse proportion to the length of the string that produces it, and that intervals between harmonious sound frequencies form simple numerical ratios [37]. When Hermann Helmholtz (1821–1894) published *On the Sensations of Tone as a Physiological Basis for the Theory of Music* (1863), as discussed in the next chapter, he referenced Pythagoras' contributions [38].

In addition, the Pythagorean idea of the music of the spheres or harmony of the spheres is still associated with his name, although the concept has various interpretations. In Chap. 12, when synesthesia is discussed, Pythagoras' influence is relevant again. At this point, his contributions are noteworthy in terms of medicine and humor theory because they influenced Greek thought overall. Indeed, his ideas about form and order flowed down through the ages, and were correlated with both music and the body through the idea of harmony:

> Pythagorean, ideas were still regarded as canonical throughout the Middle Ages and beyond… [in] the study of medicine. Limit and order are good, and the well-being of the world and of every creature in it depends on a right mingling (*krasis*) of the elements of which it is composed. It is then in a state of *harmonia*, the word which (in Greek as in English) primarily applied to music being extended to cover the whole field of nature. To the microcosm this doctrine was applied in the theory that bodily health was dependent on the rightly proportioned mixture of the physical opposites: hot and cold, wet and dry. If they are in a state of *harmonia* in the body, then, as the doctor in Plato's *Symposium* puts it, the most mutually hostile elements in it are reconciled and taught to live in amity: "and by the most hostile I mean the most sharply opposed, as hot to cold, bitter to sweet, dry to wet." This dogma of the importance of maintaining—or restoring in the case of sickness—the right quantitative relationships between opposite qualities became the cornerstone of Greek medicine, which started in a Pythagorean atmosphere with the work of Alcmaeon of Croton (pp. 38–39, [32]).

Work attributed to Hippocrates of Kos is also associated with humor theory, although his work is now seen as the compilation of research notes by a school of physicians

---

[4] Aristotle's three great books on biology were *History of Animals*, *Parts of Animals*, and *The Generation of Animals*. One fascinating part of his contributions is his diagram of the "ladder of nature" (*scala naturae*), demonstrating the breadth of his interests in the whole animal kingdom, plants, and inanimate materials [35, 36].

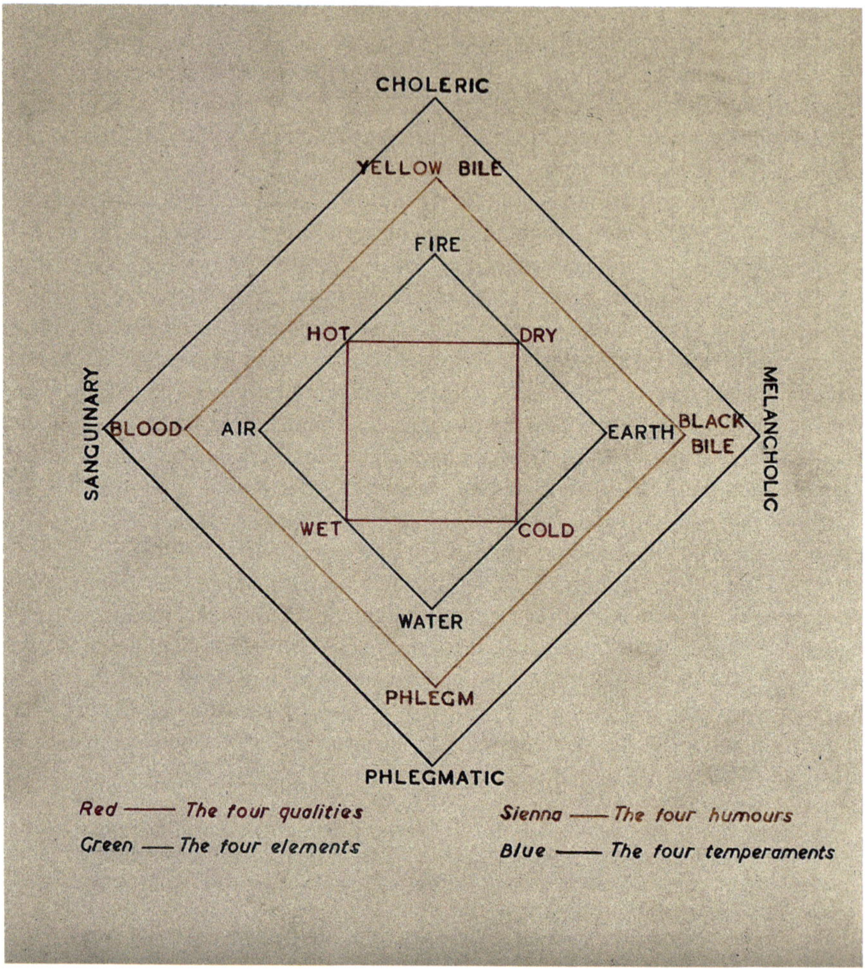

Fig. 1.2 The four qualities, elements, humors, and temperaments. Drawing, 19—. Courtesy of the Wellcome Collection

with which he was affiliated. They lived contemporaneously with Plato and because Hippocrates is often given credit for this work [6, 39], I will at times follow the convention that attributes this school's work to him.

This school's best-known contribution is the theory of "humors" (humor theory, or Humorism) (Fig. 1.2), which remained the prevailing view for millennium. It was a holistic prescription intended to balance human activity, achieve perfect health, and produce a balanced mind. Although these doctors only employed animal dissection in their studies, they recognized that the brain is involved in sensation, is the center of intelligence, and is the controlling center for the body [39]. These physicians kept careful notes as they studied patients and believed diseases have physical causes. Therefore, their legacy includes using case studies to bring new ideas to the table, and rejecting the view that affliction was a curse or a prophetic power [4, 39–41].

The basic idea of humor theory is that the body is composed of four humors: blood, black bile, yellow bile, and phlegm. Physicians of the time aimed to keep the humors in balance and to restore the balance if it was upset. Any disturbance in their ratios, as the theory goes, causes ill health. In practice we see the concept at work when it was assumed that a person with fever and a flushed face required bloodletting, while someone of a pale appearance would no doubt need blood-rich liver to restore a proper balance.

1.5 The Functional, the Neuroatypical, and the Senses

Looking back at Greek physiological characterizations and those of others in this period, it is clear that even before many conditions and abnormalities were defined by a distinct set of clinical and pathological features, there were strategies and norms for treatments. Within this, congenital conditions fared worse than those acquired, and confidence in eventually curing conditions was less evident than it is today. For example, while some measure of certainty currently exists that we are on our way to devising ways to prevent or delay conditions like Alzheimer's disease, this was not the view in ancient Greek society. Then, if you could not fulfill your function, you essentially did not have a place [42].

More specifically, biases against those categorized as not fully functioning individuals included the neuroatypical and those with acquired losses of sensory functions. We see this well in terms of the aged. In other words, Plato, Aristotle, and their peers spoke about *senile dementia* long before Alois Alzheimer described Alzheimer's disease in 1906. For the ancient Greek and Roman philosophers and physicians, dementia was largely a rather vague notion, a part of their view that mental decline inevitably occurred in old age [42]. For example, Aristotle argued, those with dementia had limited functional value:

> [T]here is not much left of the acumen of the mind which helped them in their youth, nor of the faculties which served the intellect, and which some call judgment, imagination, power of reasoning and memory. They see them gradually blunted by deterioration and see that they can hardly fulfill their function (p. 174, in [42]).

As alluded to earlier, how the brain constructs data received by the senses was largely theoretical and/or inaccurate in ancient models because the theories behind the models were based on what they could observe through gross anatomy and assumptions derived from rudimentary evaluations of human behaviors. For the Greeks, one peculiarity of life was when the body was materially altered by external objects. Another was how to handle sensation. Whereas Aristotle saw sensation as a judgment, and therefore nearer to reason than Plato, both distinguish sensation from thought. Briefly, Plato assumed sensory information was secondary to Ideal Forms because he held that what the senses perceive is unreliable. All did, however, draw conclusions about how the senses operate. For example, some, like Plato, associated hearing or sound with motion. According to Plato:

> In considering the third kind of sense, hearing, we must speak of the causes in which it originates. We may in general assume sound to be a blow which passes through the ears, and is transmitted by means of the air, the brain, and the blood, to the soul, and that hearing is the vibration of this blow, which begins in the head and ends in the region of the liver. The sound which moves swiftly is acute, and the sound which moves slowly is grave, and that which is regular is equable and smooth, and the reverse is harsh. A great body of sound is loud, and a small body of sound the reverse (p. 465, [43]).

Aristotle essentially concluded we have five distinct senses—touch, sight, hearing, smell, and taste. In his view, if a sensory faculty is lost, knowledge is lost as well. His theory of perception also emphasizes sensory integration, a process that is itself not sensory. In other words, a cup of tea is hot, might be sweetened, and has some color. Various specialized sensory organs respond respectively to each of these attributes: the temperature, taste, color, and consistency. But the experience is not of separate attributes [44, 45].

Aristotle's characterization of five distinct senses has echoed down through time, as discussed throughout this book, particularly his valuations of how the senses should work in a well-functioning being. A deviation from his rubric is synesthesia, a topic discussed in Chap. 12. As explained in that chapter, many life-long synesthetes see synesthesia more as a trait than an abnormal condition, and they describe their experiences in largely positive terms.

Aristotle claimed that the sense organs were located in the head (smell in the middle, sight above, and hearing to the side, while the source of sensation is the heart [46]. He held that to educate a well-functioning individual words were needed because "each word is a thought-symbol." His peers joined him in elevating language over input from sensory experience and in seeing words as the optimal communication tool. We see this when Aristotle explains hearing is more important than sight in the development of the intellect because the ability to use language is what makes one intelligent [47]:

> [I]t is hearing that contributes most to the growth of intelligence. For rational discourse is a cause of instruction in virtue of its being audible, which it is, not directly, but indirectly; since it is composed of words, and each word is a thought-symbol. Accordingly, of persons destitute from birth of either sense, the blind are more intelligent than the deaf and dumb [43].

One outcome culturally, for Aristotle was not alone here, was that the congenitally deaf were characterized as "senseless and incapable of reason," and "no better than the animals of the forest and unteachable" (p. 9, [48]); they were deemed a burden to society, dumb, and not amenable to education.[5] Because the deaf lacked lan-

[5] Some say this interpretation of Aristotle is the result of a misinterpretation of Aristotle's view in the original Greek. Even if the fault of a translator, this view cast the mold for how the deaf were characterized for later centuries (see [49]).

guage, they were also deemed inferior. In terms of medical and cultural history, Aristotelian views of the deaf were accepted by Galen of Pergamon (130–200 AD) and others of the scientific community for centuries despite evidence to the contrary, as discussed shortly. For now, suffice it to say, this history speaks to how assumptions become etched onto the cultural mind.

Galen is frequently characterized as the founder of experimental physiology and sensory psychology due to the extraordinary body of work he left behind. He, too, sprinkled misperceptions in with his insights. Many, such as his views on the deaf, expanded or built upon ideas introduced by Aristotle. The reason to emphasize this is that retrospectively there is a tendency to characterize Aristotle as a philosopher and Galen as a clinician, but this is somewhat misleading given how closely intertwined philosophy and medicine were in their time.

Working within the humoral paradigm, balancing the parts and the whole, Galen is remembered for the 500–600 ventricular localization treatises he published that had an impact on medicine until the sixteenth century. For 1300 years Galen's work was accepted as dogma by the State and later the Church. In fact, some medical students were still studying Galen's writings in the nineteenth century, including some of the practices that he advocated for in his writings such as bloodletting, which was still being used at that time [36].

In terms of the brain, unlike Aristotle, who had located the soul (or the brain) in the heart, Galen located imagination, intellect, and memory within the brain. Borrowing from Plato and others he enumerated three forms of the soul (rational, spirited, and appetitive), which he postulated corresponded to three types of *pneuma* or spirit, each of which originated in a different organ. The animal spirit corresponded to the rational soul and originated in the brain; the vital spirit to the spirited soul, originating in the heart, and the natural or physical spirit originated in the liver and corresponding to the appetitive soul [50]. Galen's many anatomical reports were based mainly on the dissection of Barbary apes, although his dissections and vivisections of animals included mice, birds, snakes, pigs, goats, oxen, horses, elephant, and so forth.

1.6 Greek Art and Aesthetics

As noted, the humoral theory of disease and its assumptions about balance paralleled the elevation of verbal communication and the supremacy of language. All of this was enfolded into how the culture approached health, art, disease, and education [12]. For example, the ancient Greeks and later the Romans did not have a conception of art comparable to our own. They neither gave it a special place, nor did they see art as an autonomous sphere that had an aesthetic quality separate from its intellectual, moral, religious, and practical functions. In addition, they did not approach art on its own terms, so to speak, or in terms that used an objective/subjective characterization [6, 12, 51]. They also did not see art in terms of beauty. The ancient Greeks did have a well-defined conception of beauty, but it was not an aesthetic quality or even used as a standard for grouping the fine arts together [51, 52].

Even the notion of Fine Art is an eighteenth century addition to art historical thinking. As Kristeller notes:

> The basic notion that the five "major arts" constitute an area all by themselves, clearly separated by crafts, the sciences and other human activities has been taken for granted by most writers on aesthetics from Kant to the present day. It is freely employed even by those who profess not to believe in "aesthetic"; and it is accepted as a matter of course by the general public of amateurs who assign to "Art" with a capital A that ever-narrowing area of modern life which is not occupied by science, religion, or practical pursuits (p. 498, [55]).

More specifically, Greek terms for art were equivalents to all kinds of human activities, and I largely adopt this view here. It includes what we would call crafts and inventive scientific problem-solving. So, when Hippocrates contrasts Art with Life, he is thinking of art as medicine. For Aristotle "aesthetics"is a kind of activity based on knowledge. Plotinus equated art and beauty in terms of a spiritual essence. The transition from the Greek time to the Renaissance largely adapted the Plotinus conclusion and placed art in a more spiritual realm. A second transition from the Renaissance to the eighteenth century ends with aesthetics and art history becoming different branches of study, much as philosophy and neuroscience did.

To oversimplify, the nature of beauty per se did not became a central intellectual question related to art until the emergence of the discipline known as aesthetics in the mid-eighteenth century, when this area of study was conceived by Alexander Baumgarten (1714–1762). In 1735 he called for a new science of perceptual knowledge, an *aestheticae*, which he introduced in *Aesthetica* [52, 53]. He used the word in terms like its original Greek meaning, or as an approach to perception and knowledge, which links it to synesthesia for they share the same root. Baumgarten's proposal was re-formulated by Immanuel Kant (1724–1804) in terms of "taste" or "good taste." The field of experimental aesthetics was founded in the nineteenth century by the German philosopher and physicist Gustav Theodor Fechner (1801–1887), who again links the concept to synesthesia through his studies in that area. More recently, Semir Zeki has reconfigured the aesthetics concept within a field he has defined as neuroesthetics. His work aims to characterize neural correlates of affective states, such as the experience of love, desire, and beauty that are generated by sensory inputs [54–56].

One example of what we now call Greek art that illustrates details about the culture is a gouache painting on a Greek aryballos that dates from about 480–460 BCE (Fig. 1.3). The image shows a surgeon letting blood from a man's arm, with cripples, wounded men, and a dwarf around them. A similar aryballos, now in the Louvre, delineated a fifth century Greek clinic that is believed to correspond to those found in the Hippocratic physicians' writings and other ancient sources. Many comparable works exist. For example, there is a depiction of the doctor Japyx healing Aeneas' wounded leg on an Ancient Roman fresco from the "House of Sirico" in Pompeii, Italy, from the mid-first century.

Art also offered those without verbal language an alternative way of integrating with the culture and communicating, as recorded by Pliny the Elder when speaking

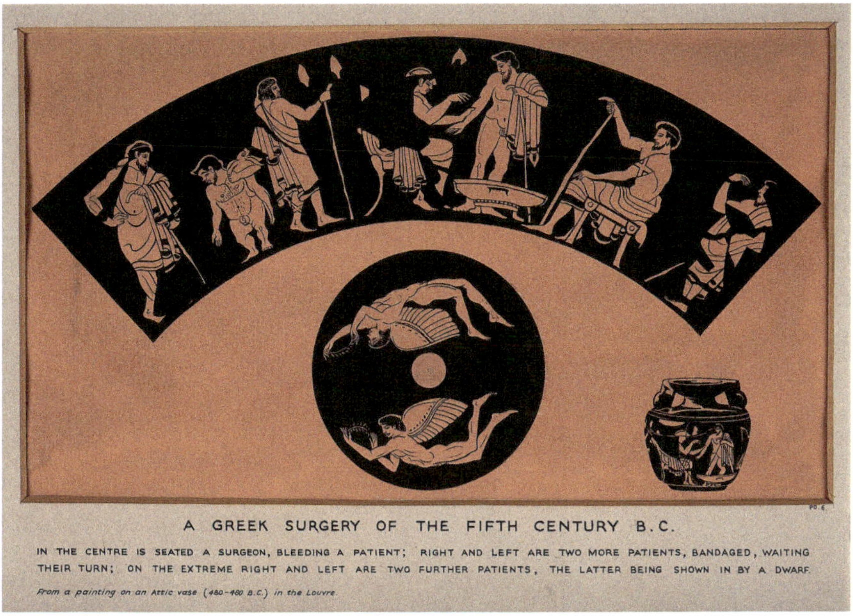

Fig. 1.3 A physician bleeding a patient, other patients are waiting to see him. Gouache painting. After a painting on a Greek aryballos, ca. 480–460 BCE. Courtesy of the Wellcome Collection

about the most eminent painters of Rome in his *Natural History*. In this section, Pliny includes a youth now characterized as the first mention of a deaf painter historically because of how frequently deafness and muteness were equated then. The painter was Quintus Pedius, grandson of Consul Quintus Pedius who was named in Caesar's will, and co-heir with Augustus. Unfortunately, this young painter died at the age of thirteen.

> Q. Pedius ... had a grandson, who being dumb from his birth, the orator Messala, to whose family his grandmother belonged, recommended he should be brought up as a painter, a proposal which was also approved of by the late Emperor Augustus. He died, however, in his youth, after having made great progress in the art (p. 231, [57]).

1.7 Art and Anatomy

Around the thirteen and fourteenth centuries interest in anatomy increased, including the appeal of illustrated depictions of anatomy. Both the artistic and medical anatomical Renaissance were most evident in Italy, as human dissection made an entry, initially with forensic postmortem examinations for legal purposes [58]. The first legal public dissection since ancient times was conducted by Mondino de Liuzzi (c. 1270–1326) in Bologna in the fourteenth century (Fig. 1.4) [58–60], and not long after the Florentine Academy of Art made an anatomy course compulsory

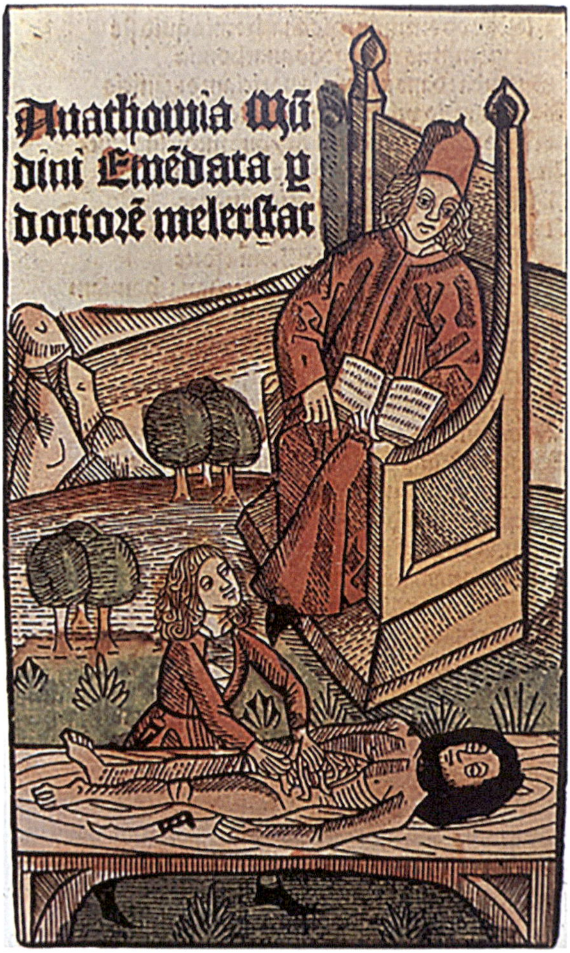

Fig. 1.4 Frontispiece of Mondino de Liuzzi's *Anathomia corporis humani* (1316) [52]. The original edition of his book carried no illustrations but was later visually updated by followers of de Liuzzi, including the anatomist Berengario da Carpi (1466–1530) and the German anatomist Johannes Dryander in 1541 [59]

for its artists. Presented within an academic framework, rather than as an adjunct to clinical practice, these public dissection events did little to unwedge the inaccuracies in Galenic theories and texts. Instead, if something was found that did not match Galen's description, it was dismissed as an anatomical deformity in the dissected human body, instead of a flaw in Galen's canon [58].

In the late Middle Ages, the Italian physician, Guido da Vigevano (1280–1349), a physician and inventor, following the work of Mondino de Liuzzi, his mentor, pioneered the art of illustration in the anatomical sciences, including drawings of neuroanatomical structures. His work was not based on dissection, but rather on written accounts from Hippocrates and Galen [58], with all their inaccuracies retained. We also find a Persian anatomist who created the world's first colored human anatomy atlas in the fourteenth century CE [58].

In the fifteenth century, the great Florentine sculptor, painter, and printmaker Antonio del Pollaiuolo (1429–1498) may have used cadavers as a tool for studying the human body, and it is said that his brother Piero was familiar with the dissection table as well [61]. Antonio's most characteristic works are of largely naked male figures in complicated poses of violent action, drawing from classical examples and often centered on a heroic Hercules. Giorgio Vasari (1511–1574), in his classic *Lives of Artists* [62], states that he was the "first master to skin many human bodies in order to investigate the muscles and understand the nude in a more modern way" (in [62, 63]). Notably, Laurenza disagrees with Vasari's claim that Antonio del Pollaiuolo used dissection as a tool for study, although he confirms that this artist did show more advanced anatomical studies than those recorded in the medical milieu of the fifteenth century. According to Laurenza, Pollaiuolo's figures were derived from a careful observer of living bodies and ancient statuary. In any case, Pollaiuolo's highly influential engraving of the *Battle of Naked Men* (Fig. 1.5) displays the kind of figures both are referencing. Here we see nude warriors with nearly flayed musculature presented in various fierce action poses and from different angles [63]. All of these features show that an interest in anatomy was beginning to emerge at this point in art history.

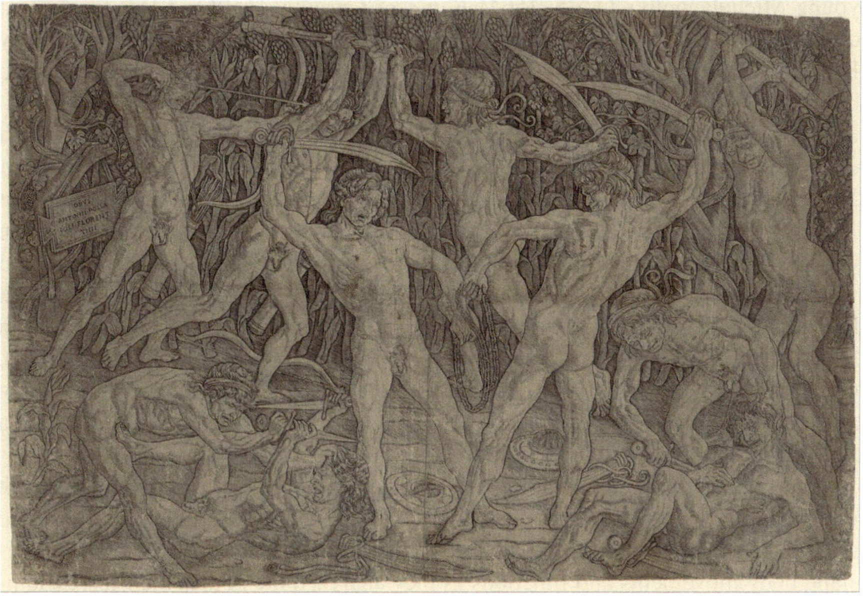

Fig. 1.5 *Battle of the Nude Men*. Artist: Antonio Pollaiuolo (Italian, Florence ca. 1432–1498 Rome). Date: ca. 1470–90. Engraving. Sheet: 15 1/8 × 23 3/16 in. (38.4 × 58.9 cm). Courtesy of the New York Metropolitan Museum Art, Joseph Pulitzer Bequest, 1917

The sixteenth century is often characterized as the "century of anatomy" because anatomy and anatomical dissections were practiced with greater freedom, with Italy still serving as the epicenter for both artists and scientists exploring how the body worked and functioned [64]. Vasari's comments on Antonio del Pollaiuolo are in line with his broader commentary on anatomy at that time. Indeed, he emphasized the importance of proper anatomical study for artists throughout his *Lives of Artists*, writing: "again having seen human bodies dissected one knows how the bones lie, and the muscles and sinews, and all order of conditions of anatomy…." [62]. Vasari's comments complement the literature showing that many artists are known to have attended live dissections when they were held by lecturers in universities, for they wanted to accurately represent anatomical structures such as muscles and tendons [6].

These artists additionally wanted to understand how the body was put together overall. When doing dissections, like others, they needed to deal with the rapid decomposition of the brain and body because at this time techniques for preservation had not yet been developed. Still, degeneration of functions and exposing patterns of disease were not their concerns artistically. Rather, they want to learn: they wanted to better conceptualize how the active body moved, the musculature of facial expressiveness, and so forth.

Many have claimed that Antonio Pollaiuolo's *Battle of the Nude* inspired the young Michelangelo's (1475–1564) *Battle of Cascina,* (1506), a lost fresco that depicts the 1364 Florentine victory over their rival Pisa. From a surviving copy by Bastiano [Aristotile] da Sangallo (1481–1551), mentioned by Vasari, we can see how much Michelangelo's representation and arrangement of the naked, male bodies resembles Pollaiuolo's earlier work. At this point Michelangelo would have had exposure to dissection according to his student Condivi who said that some of Michelangelo's dissection studies occurred in the convent of Santo Spirito in about 1494: "Michelangelo, to oblige the prior of Santo Spirito…made a wooden Crucifix…He was very intimate with the prior…who provided him with a room and with corpses for the study of anatomy" (p. 15, in [64]). Condivi also reported that between 1547 and 1553, in Rome, Michelangelo "began to discuss [dissection] with Messer R[e]aldo Colombo, a very superior anatomist and surgeon" (p. 16 in [64]). One fascinating difference between Pollaiuolo's *Battle of the Nude Men* engraving and the Michelangelo homage to it is that it appears that Michelangelo included a self-portrait on the figure in the middle.

What stands out here is that these two artists and others in this cohort were not referring to texts in their studies of gait, emotion, and so forth. Rather they were relying on information gathered through experiential and sensory systems to understand body movements and functions as they analyzed the bodies themselves, aiming to conceptualize how the human figure "worked" from different points of view in three-dimensional space. These artists had no concept to the effect that the brain does not process sensory cues independently or that when one commits to an action, various areas of the brain make assumptions as the action unfolds. For example, today we talk about how receptors analyze movement in three-dimensional space and about the sense of movement or kinesthesia, the result of cooperation among several sensors. Kinesthesia, which many characterize as one sense Aristotle's hypothesis of five distinct senses omitted, requires the brain to coherently reconstruct movement in the body and in the environment [65].

Suffice it to say, these artists, in studying how to represent movement were more interested conceptually in coherence among the sensors than what we garner from perceptual and motor disturbance. Thus, they were less interested in how the brain devises solutions to deal with discrepancies between sensory information and one's internal perception of it. They were also more interested in replicating the geometry of movements as the nervous system solved problems than in formulating solutions to deviations that interfered with problem-solving. Artists also recognized that a lifeless body did not mirror a living one. For example, Michelangelo made drawings of nudes and strictly anatomical drawings of muscles and bones. Another artist familiar with the body and dissection was Raphael (1483–1520). In one instance, using pen and ink, Raphael sketched a study of a lifeless body held up by cords to perceive both the static and dynamic equilibrium of the figure [64]. This untitled work is in the collection of The New York Metropolitan Museum of Art,

Looking at Michelangelo's studies for the Libyan Sibyl (Fig. 1.6), we see a learning sketch, in this case of a young male assistant posing in the artist's studio for the

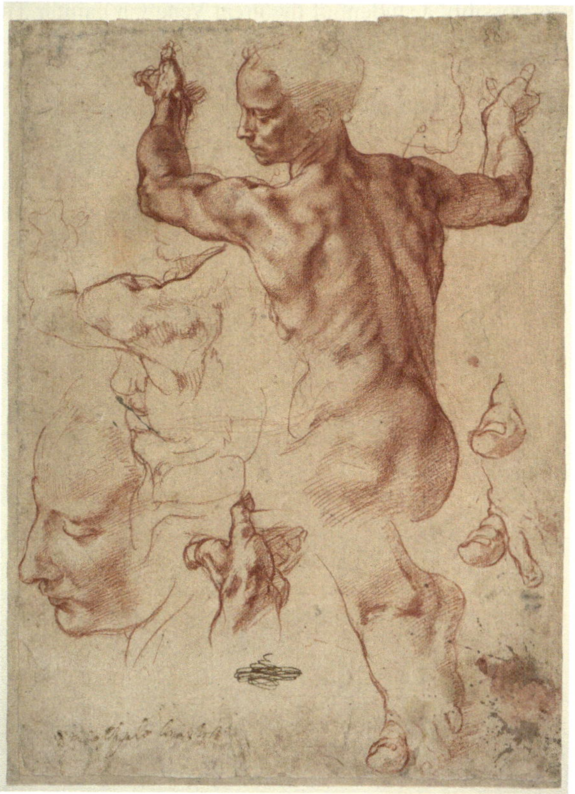

Fig. 1.6 Studies for the Libyan Sibyl (recto); Studies for the Libyan Sibyl and a small Sketch for a Seated Figure (verso). Artist: Michelangelo Buonarroti (Italian, Caprese 1475–1564 Rome). Date: ca. 1510–11. Medium: Red chalk, with small accents of white chalk on the left shoulder of the figure in the main study (recto); soft black chalk, or less probably charcoal (verso). Dimensions: Sheet: 11 3/8 × 8 7/16 in. (28.9 × 21.4 cm). Courtesy of the New York Metropolitan Museum of Art, Purchase, Joseph Pulitzer Bequest, 1924

monumental enthroned female figure painted in fresco on the north-east end of the Sistine Ceiling. On the ceiling, she is clothed except for her powerful shoulders and arms, and wears an elaborately braided coiffure. Evidently, her complex pose in the fresco required study in numerous drawings and was not merely an intuitive feat. We do not know precisely how his brain combined his dissection work with his nude sketches in producing the clothed figure that plays on the arrested motion of her stepping down from the throne, while holding an enormous open book of prophecy which she is about to close [64].

Artists also provided sketches and detailed drawings of discoveries for anatomists performing dissections. These were often unacknowledged. Suffice it to say, the study and teaching of anatomy through the ages would not have been possible without the artistic depictions of the placement of body parts, which played a crucial role in assisting those working with the human body. The unnamed include Andreas Vesalius' (1514–1564) collaboration with the Flemish painter Jan Stefan van Calcar (1499–1546/1550) on the revolutionary *De humani corporis fabrica* (*On the fabric of the human body*) (1543), discussed shortly. Vesalius' work similarly underscores that the entry of various forms of printing representations aided the anatomical revolution. Vesalius used the new state-of-the-art printing technologies to showcase this early work now remembered more for the magnificent drawings used to illustrate the text than the text itself (Fig. 1.7). Unfortunately, Calcar, who was possibly a student of the great Venetian painter Titian, is not acknowledged in the volume [6].

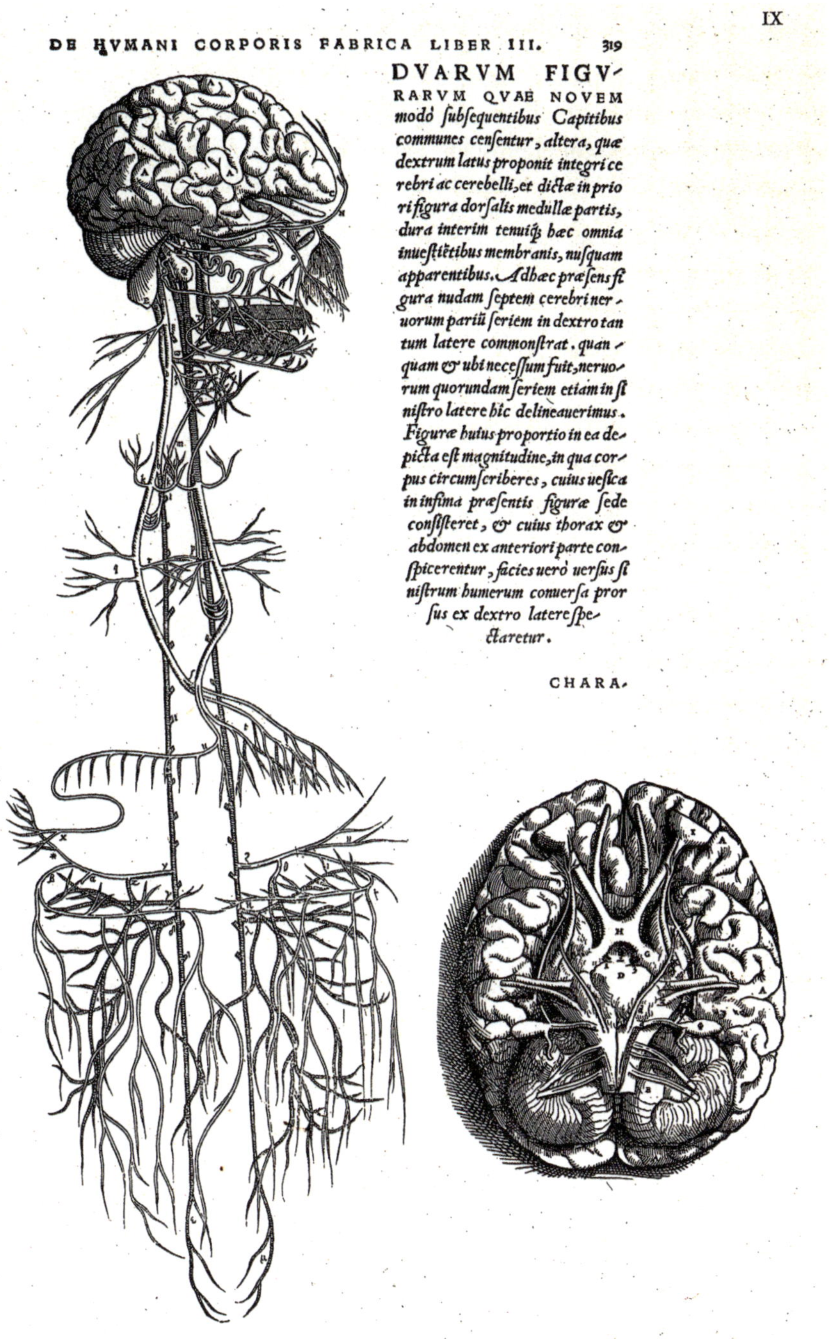
DE HVMANI CORPORIS FABRICA LIBER III. 319

IX

DVARVM FIGVRARVM QVAE NOVEM modò ſubſequentibus Capitibus communes cenſentur, altera, quæ dextrum latus proponit integri cerebri ac cerebelli, et dictæ in priori figura dorſalis medullæ partis, dura interim tenuiq; hæc omnia inueſtientibus membranis, nuſquam apparentibus. Adhæc præſens figura nudam ſeptem cerebri neruorum pariũ ſeriem in dextro tantum latere commonſtrat. quanquam & ubi neceſſum fuit, neruorum quorundam ſeriem etiam in ſiniſtro latere hic delineauerimus. Figuræ huius proportio in ea depicta eſt magnitudine, in qua corpus circumſcriberes, cuius ueſica in infima præſentis figuræ ſede conſiſteret, & cuius thorax & abdomen ex anteriori parte conſpicerentur, facies uerò uerſus ſiniſtrum humerum conuerſa prorſus ex dextro latere ſpectaretur.

CHARA-

Fig. 1.7 Andreas Vesalius (1514–1564). *De humani corporis fabrica* (1543), The brain, in right profile with the glossopharyngeal and vagus nerve and, to the right, a view of the base of the brain. Photolithograph, 1940, after a woodcut, 1543. Wellcome Collection. Public Domain Mark. Courtesy of the Wellcome Collection

1.8 Leonardo da Vinci: Anatomical Studies

Among the fifteenth century artists who explored the brain, the body, and the senses, Leonardo da Vinci (1452–1519) was the most comprehensive. Leonardo's research included both anatomical brain studies and work on expression that was aided by his interactions with deaf individuals. Combining all of these perspectives demonstrates that anatomical studies and applications may work in tandem even without being explicitly connected theoretically. According to Vasari:

> [Leonardo] applied himself with great diligence to human anatomy, helped in this by (and in turn helping) that excellent philosopher Marcantonio della Torre, who was then lecturing at Pavia and who wrote on this subject, and who was one of the first (as I have heard) to begin to illustrate matters of medicine by the teachings of Galen and to throw true light on anatomy, which up to then had been obscured by the great shadows of ignorance. In this he was wonderfully served by the intelligence, work, and hand of Leonardo, who composed a book drawn in red chalk and finished in pen, of bodies he had dissected himself. He depicted with great diligence all the bones and to what they are joined, and then in order all the nerves, covering them with the muscles—the first that are attached to the skeleton, the second that hold it firm and the third that move it in places [62].

The full extent of what Leonardo accomplished was largely unappreciated and hardly known during his lifetime, despite Vasari's recognition of it. Unpublished, this work barely circulated beyond his immediate circle [66]. In typical Leonardo fashion, his anatomical studies began in concert with his work as a painter and soon broke free into independent studies. Also typical of Leonardo was his urge to understand structural anatomy and physiology as well as conception, growth, emotional expression, the nature of the senses, and so on. The initial anatomical studies were on animals, a realistic choice since Leonardo only had limited access to human materials.

Leonardo's first phase of anatomical study ended by the early 1490s although it is unclear why [6]. Had he left only his sketches to us, this early work would be of interest, but not on the level of the studies that he began at the age of 55. His earliest surviving neuroanatomical studies consisted of a series of drawings of the skull that date from about 1489. Clayton and Philo hypothesize that his renderings of the brain may have been partially derived from those of oxen or a cow, particularly since the *rete mirabile* is included [66].

Notably, these drawings contain the first accurate depictions of the anterior and middle meningeal arteries and the anterior, middle, and posterior cranial fossae. He included the frontal vein, which was traditionally used by surgeons and barbers in bloodletting to treat head pain and mental conditions, and the optic, auditory, and other cranial nerves emerging from the bone [6, 67]. Another drawing shows the proportions of the skull, including the hole of the cranium. His inclusion of the frontal sinus, shown from the side, with the vault bisected, represents another original discovery. Also of note are Leonardo's drawings of the ventricles, for which he presents a much more modern appearance than the "cartoons" and diagrams of his predecessors. The technique that he developed—one that used solidifying waxes to determine structure—would not be applied to biology again until the work of Frederik Ruysch (1638–1731) [68]. Leonardo also made glass models of the aorta and casts of the ventricles of the heart to study their function [67]. Leonardo explained his modeling process as follows:

> Make two vent-holes in the horns of the greater ventricles and insert melted wax with a syringe, making a hole in the ventricle of memory [the fourth ventricle, at lower right in the side views of the brain]; and through this hole fill the three ventricles of the brain. Then when the wax has set, dissect off the brain and you will see the shape of the ventricles exactly. But first put fine tubes into the vent-holes so that the air which is in these ventricles is blown out and makes room for the wax which enters into the ventricles (p 144, in [66]).

The first Leonardo dissection and anatomical works that offer a more advanced inquiry began with his postmortem of an old man in the winter of 1507–8:

> This old man, a few hours before his death, told me that he was over a hundred years old, and that he felt nothing wrong with his body other than weakness. And thus, while sitting on a bed in the hospital of Santa Maria Nuova in Florence without any movement or other sign of any mishap, he passed from this life. And I dissected him to see the cause of so sweet a death (p. 17, in [66]).

Leonardo's images provide a powerful example of how visual art syntax offers an alternative form of communication. Clearly the brain studies demonstrate that visual data is not comparable to a "narrative" and they also show that, a "good picture" is invariably worth pages of writing (p. ix, [69]). Because the unpublished brain studies allowed him to produce an overall picture of the brain far beyond what science was "seeing" at that point in time, as I have noted elsewhere, three points are noteworthy:

> First, it is clear from his drawings that he gleaned more than his contemporaries from his brain studies, for he had seen the internal form more fully through the feat of injecting hot wax into the brain and thus providing a cast of the ventricles. Second, the wax experiments were the first known medical use of a solidifying medium to define the shape and size of an internal body structure. Finally, despite the clarity of his images, Leonardo was hesitant to break with the scientific specialists of his time in a theoretical sense. Although he gave us magnificent depictions of the ventricular system of the brain, he incorrectly held to the view that perception and cognition resided in the cavities of the brain, not in the brain substance itself (p. 59, [6]).

Leonardo's use of wax also shows that the tools of one profession can aid in conceptualizing information in another field of study. That said, it is noteworthy that Leonardo's use of this technique was not revolutionary on its own terms. Wax modeling was used by artists at least since ancient times. It was recorded in Pliny the Elder's *Natural History*, in the first century BCE [70], where he explained that wax casting was known to Greek artists. Then, it was used to create mimetic likeness and thus was used to help them create more naturalistic works. There was a renewed interest in wax modeling during the fourteenth century when Italian artists adopted it as a means to sculpt realistic figures.

Despite the clarity of Leonardo's magnificent drawings of the ventricular system of the brain based on his wax studies, as noted, he was like Vesalius, who is discussed shortly, in incorrectly holding onto the theoretical views of the time including the idea that perception and cognition resided in the cavities of the brain, not in the brain substance itself. Also, as noted, at this time there were no technologies for preserving brain tissue. With wax, Leonardo was able to counter this and do a comprehensive analysis of an actual brain without needing to preserve the tissue. Later, in the seventeenth century, scientists like the British chemist Robert Boyle (1627–1691), the Danish physician Frederik Ruysch (1665–1717) and the Scottish anatomist William Hunter (1718–1783) began to perfect chemical methods for embalming the brain's soft tissue [6].

Wax also continued as a viable tool. Although quite different methodologically from Leonardo's work, three-dimensional wax models, including some of the brain, were used by later anatomists as a tool of medical education and for research in the seventeenth and eighteenth centuries. Among the trained artists who excelled in producing these exquisite three-dimensional models (*écorchés*) were Gaetano Giulio Zumbo (1656–1701), an Italian sculptor of the Baroque era; the surgeon and anatomist Guillaume Desnoues (1650–1735); Giovanni Manzolini (1700–1755) and his wife Anna Morandi Manzolini (1714–1774), who had met as art students; and Marie Marguerite Bihéron (1719–1795) [6, 64, 65].

To close this section, it is now largely assumed that Leonardo had planned to publish an anatomical textbook in collaboration with Marcantonio della Torre (1481–1512), who was a Professor of Anatomy at the University of Padua. Unfortunately, della Torre's life was cut short by plague in 1511 and thus no treatise resulted. This magnificent and insightful work was subsequently lost to the world until the physician and anatomist William Hunter (1718–1783) examined the drawings that had made their way into George III's library. Hunter recommended that Albrecht von Haller (1708–1777), a Swiss anatomist, physiologist, naturalist, and poet include them in his eight-volume work on the history of anatomy [71]. Haller did not and the work remained unpublished until 1898 and 1916 editions were issued [66]. Had Leonardo produced a book when he lived there is no doubt it would have revolutionized anatomy and physiology. Instead, Andreas Vesalius is largely credited with the anatomical transformation that took place during the Renaissance.

1.9 Leonardo da Vinci: Expressive Studies and Deaf Painters

Leonardo also worked with and learned from the deaf. The de Predis family had hosted him in 1482 when he visited Milan for work under the service of Ludovico Sforza. At that time two of the de Predis brothers, Giovanni Ambrogio (c.1455–c.1508) and Evangelista (c.1460–1549), worked with him on the London version of his *Virgin of the Rocks*. In the de Predis home Leonardo met their brother, Cristoforo de Predis (~1440–1486), an Italian miniaturist and illuminator who was both deaf and mute from birth. Cristoforo's best known artistic works are the illustrations for the *Borromeo Book of Hours* and artistic illustrations in the book *De Sphaera d'Este*, in which astronomical motifs were illustrated in a manner similar to the contemporaneous astrological maps [72].

Writing about what he learned from Cristoforo and the deaf in *Treatise on Painting* (*Trattato della pittura*) [73], Leonardo talks about the value of keen observation of expression and character. Here he offers comparisons of laughing and weeping, noting that the only difference between the two emotions in terms of the "motion of the [facial] features" is "the ruffling of the brows, which is added in weeping, but more elevated and extended in laughing" [74].

Leonardo entries show a striking awareness of this deaf painter's artistic ability and how Cristoforo communicated with everyone through mimicry and signs [73]. Leonardo also makes it clear that he studied Cristoforo's gestures and mimicry—and those of the deaf in general [73]. Additionally, he advises artists to observe and learn from deaf-mutes:

> Don't hold it against me that I propose a teacher who does not speak, because he will teach you better with deeds than all the other masters with words. The good painter has two main things to paint, namely man and the concept of his mind. The first is easy, the second difficult, because one has to represent with gestures and movements of the limbs, and this must be learned from the dumb, who do them better than any other sort of men (see [73]).

It is unknown how Cristoforo learned to paint, although Wittgens suggested his teacher might have been Luchino Belbellow of Pavia [73, 75]; nor do we know how effectively Leonardo actually communicated with the deaf painter. Cristoforo's brothers found their brother quite skilled artistically, and the siblings all learned how to "talk" with one another using a form of sign language they devised. Deaf families and communities often conceived their own form of sign language before there were standard signing conventions [73].

As noted, at this time the deaf were considered morons, and thus not capable of taking care of themselves or allowed to own land, so Cristoforo's success is noteworthy on its own terms. Also noteworthy is that the painter needed a duke who appreciated his artistic skills to grant him a waiver so that he could live without a guardian overseeing him [73], much like the youth Quintus Pedius, grandson of Consul Quintus Pedius, needed special dispensation to learn to paint.

Cristoforo is not a singular example. Other deaf painters from this era also forged professional identities despite the restraints on the deaf population at that time. Juan Fernández de Navarrete (1526–1579) (*El Mudo* or The Mute) worked in the Spanish Mannerist school and studied in Italy, where he was influenced by Sebastiano del Piombo, Tintoretto, and Titian [76]. He was not congenitally deaf, but he suddenly became deaf in his early childhood, at the age of 3 and a half, so possibly prelingually deaf. He used drawing to compensate for his inability to hear and speak. It is said that his paintings exuded energy and preciseness, with very free usage of contrasts of light and darkness [72]. Ercole Sarti (1593–1639) was an Italian painter of the Baroque period, known by his nickname *Il muto de Ficarola* (The Mute from Ficarola), because of his mutism and congenital deafness [72].

1.10 Andreas Vesalius and Medical Dissection

Around the thirteenth century regulations forbidding human dissection began to ease, and within this Galen's supremacy in physiological and anatomical observations was questioned. As noted, Andreas Vesalius' (1514–1564) revolutionary *De humani corporis fabrica* (*On the fabric of the human body*) (1543) was key in revising Galen's work. Born 5 years before Leonardo da Vinci's death, Vesalius' book is considered a milestone in the history of scientific thought and biological illustration, although not all of the details within it were transformative. Nonetheless, it did provide a foundation for the modern disciplines of human and comparative anatomy and physiology, while also serving as an adjunct to biological research [77].

Working from human cadavers, Vesalius corrected anatomical errors found in the earlier studies that were based on animal dissections, but his theories largely kept to the status quo. For example, on the one hand, Vesalius' work on the brain allowed him to refute Galen's view that the vivifying force of the *pneuma physicon*, or "animal spirit," flowed down through the ventricles, and then through hollow nerves to nourish all parts of the body. Vesalius ruled this out on anatomical grounds after finding there is no physical outlet through the skull. Still, on the other hand, like Galen, Vesalius adopted a hydraulic view of the brain, seeing the brain in terms of the functional units being liquid-containing ventricles rather than grey and white matter. Therefore, he still searched for flow routes that would, for example, funnel "brain phlegm" into the nostrils.

What is particularly noteworthy here is that Vesalius, perceiving the brain in terms of functional units, was better at correcting anatomical errors than in identifying incorrect theoretical assumptions. His work shows how theories were pruned around the edges. A similar example is evident in terms of Aristotle's rubric of five distinct senses. Both European and Islamic medieval scholars added a set of "inner" or "internal" senses to the five "outer" or "external" senses (i.e., touch, taste, smell, hearing, and sight). These inner senses were widely believed to be housed in the ventricles of the brain. Suffice it to say here, these kinds of explanations remind us that physiologically views of both the brain and the body were quite limited, and while theories were re-worked, many continued to retain long-standing errors. Indeed, the longer an error was in place, the harder it was to dislodge it.

1.11 Theory and Applications

Theorical longevity is particularly evident in theories of hearing and views of the deaf. Here we see that even as Galen moved the paradigm toward more empirical study, and even showed an interest in sound in terms of acoustics, he still, like Aristotle, concluded that the congenital deaf were no better than animals based on the idea that hearing and speech were linked in the brain and if one capability was impaired the other would be impaired as well. From this he, too, concluded the deaf lacked intelligence and their muteness equaled dumbness. Galen decided that deafness was a disease brought about by an impaired function. More problematic in the scheme of things were the appended hypotheses. For example, he hypothesized that since the deaf were generally non-speaking people, they did not have souls, or maybe had damaged souls [72]. Fortunately, some like the deaf painters discussed above, were able to find a path that allowed them to break through some of the restraints and prosper.

There were broader influences on theories as well. To summarize, with the ascendency of the Church after the decline of the Roman Empire, western culture eventually moved into a top-down God-designed paradigm and the workings of the human body were seen as a part of God's plan. Many suppositions about our senses were ported over from the Greek view, even as some physiological conclusions were reframed through better observational tools. The Church's ascendency, which left the Greek cosmobiological paradigm behind, retained Humor Theory (Humorism). Indeed, Humorism was the preferred medical view for centuries. With Galvani's 1791 proposal that the nervous system was a generator of electricity, the orientation shifted from understanding illness holistically and in terms of an individual's humoral imbalance to the concept of organ-based diseases [78]. Humorism, however, was not conceptually disproven until the advent of germ theory in the 1850s when it became clear that microbes cause many diseases. Another way of envisaging this is to recall that in the Western world, the medical tradition, until at least the eighteenth century, explained disease largely in terms of an imbalance in the body's qualities, humors, or temperaments. Even still, within this, it was known that external factors such as environment, diet, behavior, and injury affected patients in different ways, and in accordance to their individual susceptibility, factors that we continue to acknowledge today. We also see that all theories did not fare equally. Again, this is particularly evident with the congenital deaf cohort.

By the Middle Ages, deafness was still among the most maligned and misunderstood ways of being. St. Augustine (354–430) declared faith comes by hearing—and hearing the word of God. This was interpreted to mean that those people without hearing could not learn, were incapable of reason, and uneducable. Since speech was linked with the soul, this created social and educational dilemmas for the deaf, their families, and the community. It is important to note that loss of hearing or acquired deafness, fared better than the congenital condition. Indeed, innovations were developed to try to help ease this type of acquired loss. For example, ear trumpets were known even in Galen's time [4] even as the congenital deaf—again, considered mute, dumb, and unable to learn—were either used as cheap labor or sent to monasteries where the monks cared for them [30]. This left the deaf in a more difficult position than the blind who were at least considered to have intelligence. As discussed earlier, some like the deaf painter Cristoforo de Predis (1440–1486) were able to break through. The larger point here is that although revisions were made to theoretical models concerning sound acoustics, and the senses generally, there was a lack of understanding regarding how the brain constructs knowledge and a lack of emphasis on explaining or examining cases that contradicted theoretical assumptions. This topic is pursued further in the next chapter. Still, a few final words are in order here. This was true generally, although deaf experience provides a particularly vivid example.

To expand this, first, the inability of the congenital deaf to speak and hear is unlike the kind of acquired aphasias described in the nineteenth century, and discussed in the next chapter. Acquired aphasias are language disorders caused by damage to specific areas of the brain that control language expression and comprehension. The problems of the congenital deaf were auditory, often these individuals were otherwise perfectly healthy.

Second, the ingrained theories created a vicious circle. In other words, because the literature deemed the deaf as dumb and uneducable, they were denied learning and integration. These theoretical biases co-existed with evidence that attempts were made to comprehend hearing physiologically and behaviorally. Deafness baffled Hippocrates, although somewhat akin to Aristotle, he assumed a common cerebral origin for speech and hearing [76]. This led him to surmise that there was something supernatural about deafness. Plato, who like Aristotle saw the deaf as dumb and as a burden to society, nonetheless observed in *Cratylus* that sign language was a tool that aided them in communicating with others:

> Answer this question for me: if we had neither voice nor tongue but we wished to indicate things to one another would we not act as dumb people do nowadays and try to make signs with our hands and head and the rest of our body? [79]

Third, families did step in and educate the luckier deaf individuals. Some physicians, theologians, and educators, particularly once the Renaissance emerged, did as well. Nonetheless, the view that those born deaf were nature's errors survived for centuries despite the evidence that it is possible to teach the deaf to communicate, or to speak with their hands, as Plato noted.

Finally, in the sixteenth century, paralleling the emerging interest in anatomy we begin to see an increase in efforts to educate the deaf despite the long-standing theoretical views that they were uneducable morons. Girolamo Cardano (1501–1576) was the first physician to recognize the ability of the deaf to reason. He recognized that the deaf "hear" through reading and "speak" through writing [80]. He was influenced by the Dutch humanist Rudolphus Agricola's efforts in teaching the deaf, as recorded in *De inventione dialectica* (1470) [81, 82]. Additional teachers included a sixteenth century anatomist and physician Benedictine monk Pedro Ponce de León who, like Cardano, used other sensory inputs to counter the auditory problem. For example, he used a method similar to what is now called the finger alphabet, which can be a tactile or visual exchange. Also, Juan Pablo Bonet (1579–1623) insisted that everyone living in a house with a deaf person use the manual alphabet and argued that the lack of early language training is an impediment to later speech development [83]. Hieronymus Mercurialis (1530–1606), a philologist and physician, wrote on speech and voice disorders, mutism, articulation, and fluency in *On the diseases of children* [84, 85]. He hypothesized that congenital deafness was a major cause in the underdevelopment of speech, astutely noting that children who were unable to hear anything did not receive the stimuli necessary for the brain to activate audition [72].

The subject of how education frequently counters implicit assumptions about brain functions and our sensory experience is one I return to throughout this book. It is also further examined in the next chapter, which turns to the seventeenth century and moves this chronology forward in time.

References

1. Sacks O. Musicophilia: tales of music and the brain. New York, NY: Alfred A. Knopf; 2010.
2. Critchley M, Henson RA, editors. Music and the brain: studies in the neurology of music. London: Heinemann Medical Books; 1977.
3. Henson RA. Neurological aspects of musical experience. In: Critchley M, Henson RA, editors. Music and the brain. London: Heinemann Medical Books; 1977. p. 3–21.
4. Finger S. Origins of neuroscience: a history of explorations into brain function. New York, NY: Oxford University Press; 1994.
5. Clarke E, Jacyna LS. Nineteenth-century origins of neuroscientific concepts. Berkeley, CA: University of California Press; 1987.
6. Ione A. Art and the brain: plasticity, embodiment, and the unclosed circle. Leiden: Brill; 2016.
7. Brazier MAB. A history of neurophysiology in the 17th and 18th centuries: from concept to experiment. New York, NY: Raven Press; 1984.
8. Brazier MAB. A history of neurophysiology in the 19th century. New York, NY: Raven Press; 1988.
9. Finger S, Tyler K, Boller F, editors. History of neurology: handbook of clinical neurology. St. Louis, MO: Elsevier Science; 2010.
10. Ione A. Multiple discovery. In: Runco MA, Pritzker SR, editors. Encyclopedia of creativity. 2: Academic Press; 1999. p. 261–71.
11. Ione A. Innovation in art and science. Trends Cogn Sci. 2001;5(4):140.
12. Ione A. Innovation and visualization: trajectories, strategies, and myths. Amsterdam: Rodopi; 2005.
13. Ione A. Neurology of the arts: painting, music, literature. Leonardo. 2005;38(1):69–70.
14. Ione A. Neurology, synaesthesia, and painting. In: Rose FC, editor. International review of neurobiology. 74: Academic Press; 2006. p. 69–78.
15. Ione A. Introduction: visual images and visualization in the neurosciences. J Hist Neurosci. 2008;17(3):257–9.
16. Ione A. Visual images and neurological illustration (chapter 19). In: Finger S, Boller F, Tyler K, editors. History of neurology, volume 95: Handbook of clinical neurology. 95. Elsevier; 2010. p. 271–87.
17. Ione A. Art and the brain: what does the evidence tell us? Leonardo. 2012;45:204.
18. Ione A. Nature exposed to our method of questioning. Berkeley, CA: The Diatrope Press; 2002.
19. Hacker M. History of pharmacology: from antiquity to the twentieth century. In: Hacker M, Messer W, Bachmann K, editors. Pharmacology. San Diego, CA: Academic Press; 2009. p. 1–7.
20. Conard NJ, Malina M, Münzel SC. New flutes document the earliest musical tradition in southwestern Germany. Nature. 2009;460(7256):737–40.
21. Chu N-S. Neurology and traditional Chinese medicine. In: Finger S, Boller F, Tyler K, editors. Handbook of clinical neurology, vol. 95. 2009/11/07. Amsterdam/New York, NY: Elsevier; 2010. p. 755–67.
22. Bigby J, American College of Physicians—American Society of Internal Medicine. Cross-cultural medicine. Philadelphia, PA: American College of Physicians; 2003.
23. Ong JA, Ahmed I. Cross-Cultural Medicine. In: Loue S, Sajatovic M, editors. Encyclopedia of immigrant health. New York, NY: Springer; 2012. p. 504–10.
24. York GK 3rd, Steinberg DA. Neurology in ancient Egypt. In: Finger S, Boller F, Tyler KL, editors. Handbook of clinical neurology, vol. 95. London: Elsevier; 2010. p. 29–36.

25. Zárate S. Who is our audience? Captioning and subtitling for d/Deaf and hard of hearing audiences. London: UCL Press; 2021. p. 21–38.
26. Horder T. History of developmental biology. Encyclopedia of life sciences; 2010.
27. Lichtheim M. Ancient Egyptian literature; a book of readings. Berkeley, CA: University of California Press; 1973.
28. Online Etymology Dictionary. Douglas Harper; 2014. Therapeutic (adj.). https://www.etymonline.com. Accessed 21 Apr 2023.
29. Greene WC. Moira: fate, good, and evil in Greek thought. Gloucester, MA: P. Smith; 1968.
30. Fowler HW. A dictionary of modern English usage. Oxford/London: Clarendon Press; 1926.
31. Tubbs RS. Anatomy is to physiology as geography is to history; it describes the theatre of events. Clin Anat (New York, NY). 2015;28(2):151.
32. Guthrie WKC. The Greek philosophers from Thales to Aristotle. London: Methuen; 1967.
33. Ghosh SK. Human cadaveric dissection: a historical account from ancient Greece to the modern era. Anat Cell Biol. 2015;48(3):153.
34. Smith CUM. The triune brain in antiquity: Plato, Aristotle, Erasistratus. J Hist Neurosci. 2010;19(1):1–14.
35. Aristotle. Aristotle: De partibus animalium (on the parts of animals). Clarendon Press; ~350 BC/1972.
36. West JB. Galen and the beginnings of Western physiology. Am J Phys Lung Cell Mol Phys. 2014;307(2):L121–L8.
37. Weiss P, Taruskin R. Music in the Western world: a history in documents. Belmont, CA: Thomson/Schirmer; 2008.
38. Helmholtz HV. On the sensations of tone as a physiological basis for the theory of music. Green: Longmans; 1885.
39. Nuland SB. Doctors: the biography of medicine. New York, NY: Alfred A. Knopf; 1988.
40. Garrison FH, McHenry LC. Garrison's history of neurology. Springfield, IL: Thomas; 1969.
41. Crivellato E, Ribatti D. Soul, mind, brain: Greek philosophy and the birth of neuroscience. Brain Res Bull. 2007;71(4):327–36.
42. Berchtold NC, Cotman CW. Evolution in the conceptualization of dementia and Alzheimer's disease: Greco-Roman period to the 1960s. Neurobiol Aging. 1998;19(3):173–89.
43. Plato. Timaeus. Cambridge: MIT Classics. 360 BCE
44. Aristotle. de Anima (On the Soul). In: Stevenson DC, editor. Web atomics: the internet classics archive; 350/1994–2000.
45. Aristotle. On sense and the sensible. In: Stevenson DC, editor. Web atomics: the internet classics archive; 350/1994–2000.
46. Sears E. Sensory perception and its metaphors in the time of Richard of Fournival. In: Bynum WF, Porter R, editors. Medicine and the five senses. Cambridge/New York, NY: Cambridge University Press; 1993. p. 17–39.
47. Winzer MA. The history of special education: from isolation to integration. Washington, DC: Gallaudet University Press; 1993.
48. Howe CJ. The deaf mutes of Canada; a history of their education, with an account of the deaf mute institutions of the dominion, and a description of all known finger and sign alphabets. Toronto, ON: CJ Howe Publisher; 1888.
49. Hull RH. Aural rehabilitation: serving children and adults. 4th ed. Australia and San Diego: Singular Thomson Learning; 2001.
50. Guerrini A. Animal bodies and human minds: the anatomy of the brain and the analogy of nature. In: Manning G, editor. Testimonies: states of mind and states of the body in the early modern period. Cham: Springer; 2020. p. 103–23.
51. Kristeller PO. The modern system of the arts: a study in the history of aesthetics part I. J Hist Ideas. 1951;12(4):496–527.
52. Konstan D. Beauty: the fortunes of an ancient Greek idea. Oxford: Oxford University Press; 2014.
53. Baumgarten AG. Theoretische Asthetik: Die Grundlegenden Abschnitte Aus Der "Aesthetica," (1750–58). Hamburg: Felix Meiner; 1983/1750–58.

54. Zeki S. Art and the brain. Daedalus. 1998;127(2):71.
55. Zeki S. Inner vision: an exploration of art and the brain. Oxford: Oxford University Press; 1999.
56. Zeki S. Art and the brain. J Consciousness Stud. 1999;6(6–7):160.
57. Pliny (the Elder). The Natural History of Pliny: H. G. Bohn; c 5AD/1857.
58. Chase M, Shalabi S. Theater of the body: a renaissance of human anatomy: University of St. Augustine for Health Sciences Library; 2023. https://library.usa.edu/theatre-body-exhibit. Accessed 12 March 2023.
59. Di Matteo B, Tarabella V, Filardo G, Mosca M, Lo Presti M, Viganò A, et al. Art in science: Mondino de' Liuzzi: the restorer of anatomy. Clin Orthop Relat Res. 2017;475(7):1791–5.
60. de Liuzzi M. Anathomia corporis humani. Padua: Petrus Maufer; 1316/1474.
61. Cruttwell M. Antonio Pollaiuolo. Duckworth and Company; 1907.
62. Vasari G. Lives of the most eminent painters, sculptures and architects. London: Philip Lee Warner; 1550/1915.
63. Metropolitan Museum of Art. Battle of the nude men by Antonio Pollaiuolo. New York: Metropolitan Museum of Art; 2023. https://www.metmuseum.org. Accessed 24 July 2023
64. Laurenza D. Art and anatomy in renaissance Italy: images from a scientific revolution. New York, NY: Metropolitan Museum of Art; 2012.
65. Berthoz A. The brain's sense of movement. Cambridge, MA: Harvard University Press; 2000.
66. Clayton M, Philo R. Leonardo da Vinci: anatomist. London: Royal Collection Publications; 2012.
67. Pevsner J. Leonardo da Vinci's contributions to neuroscience. Trends Neurosci. 2002;25:217–20.
68. Ginn SR, Lorusso L. Brain, mind, and body: interactions with art in renaissance Italy. J Hist Neurosci. 2008;17(3):295–313.
69. Cannon HG. A method of illustration for zoological papers. London: Association of British Zoologists; 1936.
70. Pliny. In: Riley HT, editor. The natural history of Pliny. London: H. G. Bohn; 1855.
71. Haller AV. Elementa Physiologiae Corporis Humani. Lausannae: Sumptibus Marci-Michael. Bousquet et Sociorum; 1757.
72. Radic-Sestic M, Dimic N, Sesum M. The beginnings of education of the deaf persons: renaissance Europe (XIV-XVI century). Specijalna edukacija i rehabilitacija. 2012;11(1):147–65.
73. Rossetti R, Folchi A. Cristoforo de Predis (o Preda). Milan: Storia di Milano; 2004. https://www.storiadimilano.it/arte/de_predis.htm. Accessed 12 May 2023
74. Leonardo PC. Leonardo da Vinci on painting: a lost book (Libro a) reassembled from the codex Vaticanus Urbinas 1270 and from the codex Leicester by Carlo Pedretti. Berkeley, CA: University of California Press; 1964.
75. Wittgens F. Cristoforo de Predis. Leo S. Olschki; 1935.
76. Peet HP. Memoir on the origin and early history of the art of instructing the deaf and dumb. Am Ann Deaf Dumb. 1851;3(3):129–60.
77. Singer C. Vesalius on the human brain. London: Oxford University Press; 1952.
78. Reeves C, Taylor D. A history of the optic nerve and its diseases. Eye. 2004;18(11):1096–109.
79. Plato. The collected dialogues of Plato: including the letters (Bollingen series LXXI). New York, NY: Pantheon Books; 2005.
80. Gannon JR. Deaf heritage: a narrative history of deaf America. Silver Spring, MD: National Association of the Deaf; 1981.
81. Agricola R. De Inventione Dialectica libri tres (1479): Haeredes Simonis Vincentii; 1479.
82. Rudolf Agricola AR. De inventione dialectica libri tres. Tübingen: Niemeyer; 1479/1992.
83. Moores DF. Educating the deaf: psychology, principles, and practices. 5th ed. Boston, MA: Houghton Mifflin; 2001.
84. Wollock J. Hieronymus Mercurialis, treatises on the diseases of children, Venice, 1583 (excerpted). J Commun Disord. 1977;10(1–2):127–40.
85. Wollock J. Communication disorder in renaissance Italy: an unreported case analysis by Hieronymus Mercurialis (1530-1606). J Commun Disord. 1990;23(1):1–30.

2 Early Experimentation, Theories, and Applications

In the mid- and late seventeenth century, European minds embraced the urge to look outward in order to gain a greater understanding of the Creator through the world He created. Some have proposed that this was due to the Church's inability to provide adequate means to address the bubonic plague pandemics that occurred in Western Eurasia and North Africa from 1346 to 1353 [1]. At this time natural philosophers—again, the term then used for both experimental scientists and discursive philosophers—began to probe brain mechanics and to develop techniques to learn more about the brain's anatomy, organization, and functions. Doing so led them to revise some theoretical interpretations as they evaluated empirical results. As knowledge of the human body complexified, one lively discussion about sense perception included whether matter was in itself active or passively mechanistic (inactive), quite a change from the cosmobiological assumptions of the Greeks discussed earlier.

2.1 Experimental Neurology, Philosophy of Mind, and the Seventeenth Century

A cleavage between experimental neurology and philosophy of mind began to emerge in the seventeenth century with the contributions of Thomas Willis (1621–1675), the "Father of Neurology," and René Descartes (1596–1650) the "Father of Modern Philosophy." Willis, a physician, was undoubtedly more aware of biological aspects and of how bodies and brains were changed in tandem due to his experience handling disease and injuries. His research paved the way for the localization theories adopted by the (now disdained) phrenologists and later experimental work in the nineteenth century.

René Descartes, by contrast, had a quite different impact on psychological (and neuropsychological) theories, perhaps because he was not educated as a physician.

A. Ione, *Neuroscience and Art*, Neurocultural Health and Wellbeing,
https://doi.org/10.1007/978-3-031-62336-3_2

Descartes' *cogito* argument: "*I think, therefore I am*" (cogito ergo sum) [2], probably his best-known contribution, focuses on obtaining knowledge through the mind alone. In general, his analyses ranged widely, and became foundational to theories that frame human sensibilities in subjective and objective terms. Many describe Descartes' physiology as "derivative" and built on theories of his predecessors [3, 4]. Hatfield, who largely agrees, emphasizes that his truly radical innovation came into his work through his reliance on mechanical categories alone in explaining how bodily functions are performed [4]. This mechanistic element separates him from the Greeks cosmobiological worldview, Plato's idealistic dualism, and from the kind of biologically-based inquires of Thomas Willis. Although, like Plato, Descartes saw the mind as nonphysical; for Descartes, the pineal gland in the brain allowed the mind to interact with the body.[1]

Thomas Willis earned the "Father of Neurology" sobriquet due to his laboratory work that overthrew concepts in place throughout the Middle Ages. Aided by a new medical chemistry and innovative instruments developed in his era, Willis and his colleagues provided detailed studies of the brain, nerves, and behavior through experiments conducted to develop neurologic treatments. This work also comprehensively described the structure of the brain: the brainstem, the cerebellum, the ventricles, and the cerebral hemispheres. Willis' legacy is memorialized in "the circle of Willis," the arterial anastomosis at the base of the brain, so beautifully depicted by the artist/architect Christopher Wren (Fig. 2.1) who is now better known for his architectural achievements [5].

Willis and Wren were a part of an Oxford group known as the *Virtuosi*. One invention used in their studies of the brain was a tool Wren devised that allowed him to trace the topography of the brain. Even so, his illustrations do not offer a photographic sense of any brain. Rather he portrayed a synthesis of what Willis and his collaborators had learned about the structure of the brain over a period of time. Robert Boyle, another member of the team, deduced that immersion in pure alcohol would halt the process of decay. Using Boyle's procedure, Willis could work with a brain for as long as he wished and study it slowly. In other words, Vesalius destroyed the brain as he studied it because he was using a primitive technique of cutting through the organ, hopefully before it rotted. Willis and his group, by contrast, were able to preserve the brain and look microscopically due to Boyle's preservation innovation. This allowed Willis to include a description of the arteries at the base of the brain and present his classification of the cranial nerves into nine pairs, demonstrating what a more refined handling of the brain was capable of revealing [5].

Included in Willis' contributions was his coinage of the word *neurologie*, or neurology, to define his studies of the nervous system. Although neurology is yet another term that derives its name from the Greeks, unlike the fields attached to *psyche*, the rationale behind Willis' choice of the word neurology is less clear because Thomas Willis did not adopt a Greek usage [6]:

---

[1] For a more detailed discussion of how Plato is both dualistic and cosmobiological, see *Nature Exposed to Our Method of Questioning* [1]. For more information comparing Plato and Descartes, see *Art and the Brain: Plasticity, Embodiment, and the Unclosed Circle* [5].

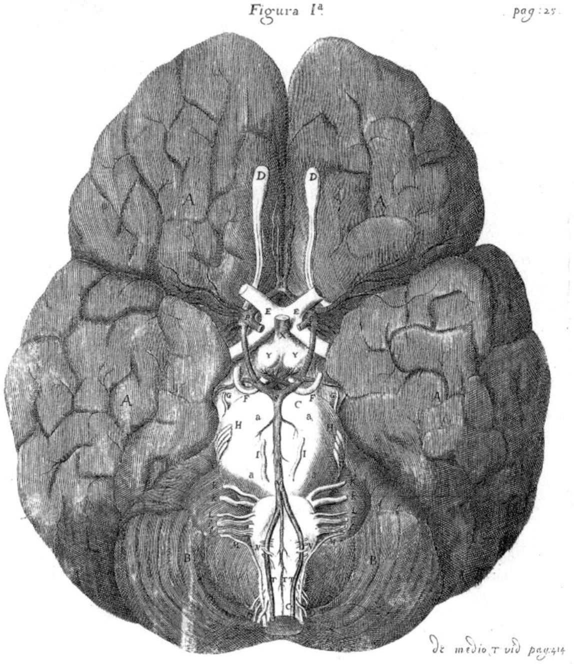

Fig. 2.1 Circle of Willis by Christopher Wren for Thomas Willis' *Cerebri Anatome*. Courtesy of the Wellcome Collection

> Although the term "nervous system" now refers collectively to the brain, spinal cord, and peripheral nerves, with the distinction of central and peripheral added for clarity, the classical Greek word [neuron]… referred to a plethora of objects—sinew, tendon, gut, and cord in the singular, sinews and nerves in the plural… and therefore originally it did not specifically have to do with the nervous systems (p. 374, [7]).

This ambiguity may explain why most were slow to adopt the term. The first drawing of a neuron as the nerve cell and its processes was published in 1865, in a posthumous work by Otto Friedrich Karl Deiters (1834–63). In the center, Dieters depicts the cell body with its nucleus. The illustration includes multiple dendrites and the single axon as well [7, 8]. The first neurologist to use the term was an American, Burt Green Wilder (1841–1925), in his Cartwright Lectures of 1884 [7].

In terms of the mind, the soul, the brain, and the body, Willis' work was also foundational to the eventual emergence of psychology and psychiatry due to his elucidation of mental illness and intellectual disability [6, 9, 10]. Indeed, he was one of the first clinicians to attribute hysterical, melancholic, and hypochondriac psychological states to differences in neural function. After Willis, clinicians also began to think of depression, epilepsy, and mania as "brain diseases" [5]. In addition, Willis probed neuropsychiatric diseases that occurred when both "sensitive and rational souls" were altered. His effort to explain mental illness led him to make some of the first systematic descriptions of neuropsychiatric diseases (e.g., cephalalgia, sleep disorders, mania, delusion, and melancholy). Here, too, he ushered in a new era. Willis wrote on hearing as well, hypothesizing that the external ear gathers together the particles constituting sounds and directs them inward to a thin membrane by way of twisting passages. According to Willis, sound shakes this membrane like a drum, and this delivers the impression to the sonorous particles in the cavity beyond it [11, 12].

Recent claims that Willis was the first pediatric neurologist [13] are noteworthy as well because his interest in determining causes of neurological impairment in children, including neonatal cases, speaks to the limited knowledge in those areas at this time. It also speaks to congenital and developmental conditions, the subject of the next chapter. Moreover, as I demonstrate throughout this book, physicians and educators who worked directly with populations born with congenital and developmental handicaps were able to educate them and compassionately change their lives even when prevailing theories characterized these cohorts as hopeless. Inquiries like Willis' into pediatric neurological impairments show that some experimentalists were probing for deeper understating.

As noted, Descartes' legacy is primarily tied to the philosophy of mind trajectory, an approach that was largely subsumed into cognitive science at the end of the twentieth century. His approach was both ontological and epistemological, with a strong bent toward mathematics and mechanics . His writings cover the nature of knowledge, the nature of mind, and the nature of being. We glimpse Descartes' range and some of the ways he viewed sensory experience in his writing on sound and music. This work shows his mechanistic approach well through his interest in physics and acoustics. He proposed a musical theory intended to identify the affections or properties of sound, as far as they may be known mathematically. As a part of this, Descartes wondered which sounds are capable of pleasing and arousing the passions. Toward this end he considered differences in duration (rhythm and meter) and differences in pitch [14]. His theory of vision held that light rays impressed subtle particles into the eyes and an image was transmitted to the pineal gland, his nexus between the mind and body [2].

Comparing these two "Fathers," Willis and Descartes, three points stand out. First, Descartes was more interested in knowledge for its own sake and universal principles than deviations from them. Willis, by contrast, connected knowledge to how the body functions, and this left more room for exploring deviations. Second, both men were operating within a "religious," pre-Darwinian sensibility. Therefore, even as their views laid the foundation for natural philosophy to morph into science and philosophy, both saw the brain, the body, the mind, and the soul as designed by God, and as evidence of his mastery. Third, their coupling in time demonstrates that in their era there was an increased emphasis on observation and experimentation. This observational behavior connected them with Vesalius, but their ways of thinking were distinctly unlike Vesalius' scholastic foundations.

John Bulwer (1606–1656), who also worked at this time, was among those who recognized the value of education and applications. Even before there was quantitative data identifying how sensory systems are repurposed, Bulwer conceptualized that this was possible, probably through his own work with individuals, including deaf individuals [15]. From the text of *Philocophus* (1648), it is clear that Bulwer personally knew deaf people and did not see them as errors of nature [16]. Rather he admired their capacities, including their capacity to use Sign language. His proposed model of deaf education was intended to allay cultural and legal discrimination [15]. He was interested in developing a learning academy [16], a project that was realized by Abbé Charles Michel de l'Épée (1712–1789), the "Father of the Deaf," who founded what would eventually become the first state-supported school for deaf children in Paris. Both conceptualized that the problem that needed to be solved was helping this cohort communicate. From Bulwer's text, it appears that he had also created the earliest system of Sign language, although credit is commonly given to Dr. John Wallis who demonstrated his method to the Royal Society in 1662 [17].

Bulwer's *Philocophus* [16] also offers evidence of deaf integration, citing an account of Sir Kenelm Digby of a meeting between the Prince of Wales and a deaf Spanish nobleman, Don Luis Velasco. Additionally, he draws heavily on accounts about deaf people living in Britain at that time. Some of his observations on the deaf community came from examples of their reactions to the vibrations produced by musical instruments by bone conduction through the teeth. We see one example on the frontispiece of *Philocophus* (Figs. 2.2 and 2.3) done by William Marshall, a most prolific engraver working in London. It shows, a deaf man "hearing" music by bone conduction through his teeth, biting the end of a viola de gamba. This is a method we know Ludwig van Beethoven (1770–1827) used in the next century, and one still used today, as explained in Chap. 4.

Bulwer's *Chirologia* (1644) and *Pathomyotamia* (1649) are also of interest. The latter was released in the same time frame as Descartes *Les Passions de l'âme)* (1649) [18], and it offers proposals to rename the muscles according to the facial expressions, emotions, or the behaviors in which they acted [19]. This was the first substantial English language work on the muscular basis of emotional expressions and foreshadowed by 200 years similar ideas of the French anatomist and electrophysiologist Duchenne de Boulogne's in *Mecanisme de la Physiognomie Humaine* (1862). Bulwer also predates illustrated works on expression by artist/neurologist Charles Bell and Charles Darwin (see Chap. 11). Darwin's work was largely patterned on Bell's expressions and done in collaboration with photographers such as the Swedish/British photographer Oscar Rejlander (1813–1875). Geen and Tassinary suggest Bulwer's impact was limited because of the book's outdated theoretical foundation [17].

William Holder (1616–1698), an English phonetician, music theorist, composer, mathematician, and church rector, was another who understood that what held the deaf back was not their inability to hear sound, their failure to speak verbal language, or an overall lack or intelligence. Rather they needed to develop alternative means of communication. In his *Elements of Speech* (1669) [20], he, too, refutes the hypotheses put forth by Aristotle and Galen that the deaf were unteachable. Like Bulwer, Holder, used bone conduction in his therapies.

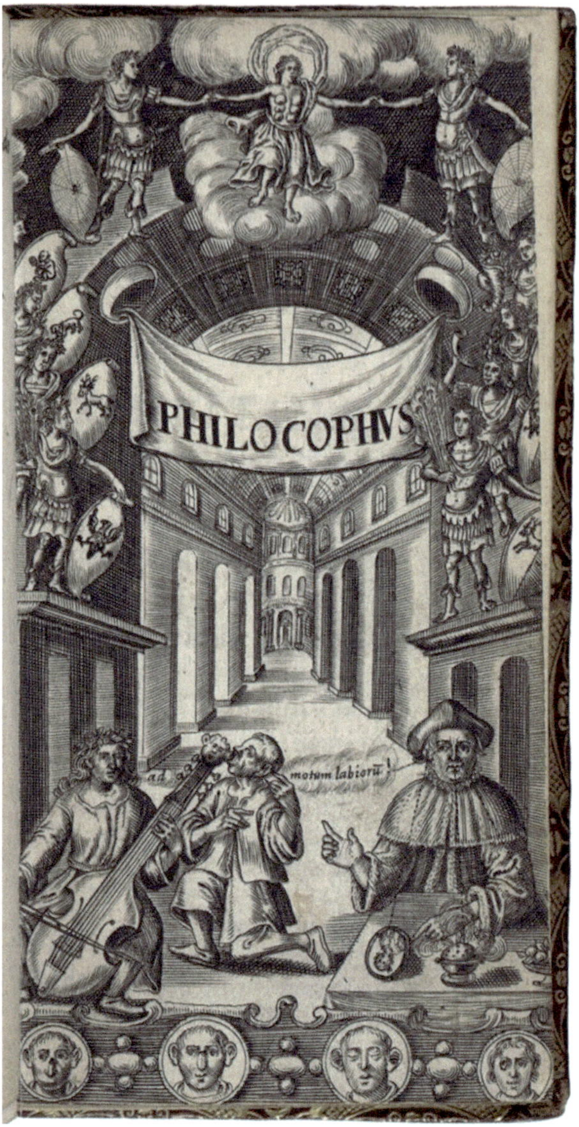

Fig. 2.2 *Philocophus: or, The deafe and dumbe mans friend* (frontispiece). This frontispiece depicts reactions to the vibrations produced by musical instruments and includes a deaf man listening to music by bone conduction through the teeth (on the lower left), biting the end of a viola de gamba. Figure 2.3 shows a detail of the deaf man listening on the lower left. Courtesy of the Folger Shakespeare Library

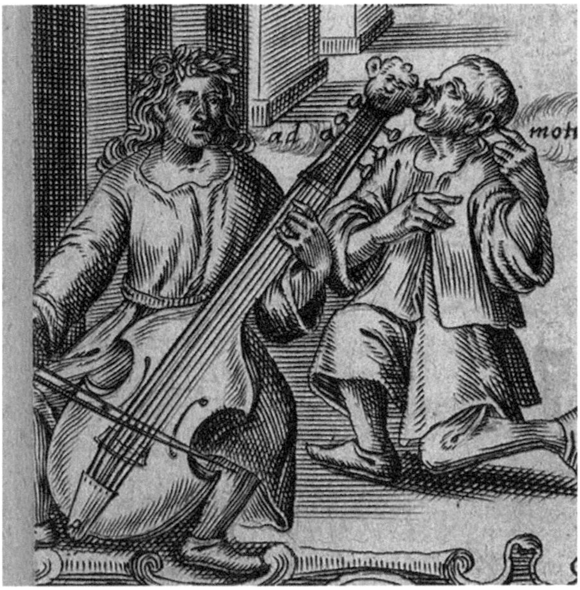

Fig. 2.3 *Philocophus: or, The deafe and dumbe mans friend* (frontispiece detail). This frontispiece depicts reactions to the vibrations produced by musical instruments and this detail shows the deaf man listening to music by bone conduction through the teeth, biting the end of a viola de gamba. Figure 2.2 shows the full image. Courtesy Folger Shakespeare Library

2.2 Language: Aphasia and Sign Language

At the beginning of the last chapter, I noted that Critchley and Henson found that applications among those who used music had flourished in the absence of neurological case studies. By the seventeenth and eighteenth centuries we find a similar bifurcation emerging regarding the deaf. Theoretically, the deaf were still considered unteachable and as errors of nature [15], yet, as the above examples showed, some physicians, theologians, and educators were beginning to develop strategies to help them learn. Thus, at this point, the hearing paradigm is moving in more than one direction, with theoretical ideas about the acquired loss of hearing and congenital deafness on one path and deaf education on another.

Galvani's eighteenth century studies on the effects of electricity on muscular contractions in a frog added a natural animating feature to the neuroscientific toolbox and built on the employment of medical electricity that was increasingly deployed therapeutically in the eighteenth century. This showed that animal tissues are endowed with intrinsic electricity. In other words, it was now clear that electrophysiology is involved in fundamental physiological processes such as nerve conduction and muscle contraction. In effect, this was one impetus for moving neuroscientific observation into more instrumentally based investigations. So, while much of the general neurological discussion centered around whether cerebral localization was the proper vantage point for investigating brain function, parallel work moved to develop technological instruments to aid research.

As noted, helping the congenital deaf was largely tangential to the work investigating brain function and developing technologies to aid this research. Thus, the deaf offer a vivid point of entry into how chasms between theory and application can exist in tandem. In this case, hands-on applications were implemented for this cohort that changed their ability to socially integrate. Building on Abbé Charles Michel de l'Épée's efforts, residential schools and the development of standardized signing conventions, the Deaf Community developed a shared language. Sign, this language allowed the deaf to hear through reading and to speak through writing. Sign, although not an art form like painting or music, does demonstrate that nonverbal languages, like many arts, provide a syntax and offer a means of communication, or an alternative means of communication if necessary.

In the last chapter I pointed out that behaviorally congenital deafness manifests in some ways like acquired aphasias given the problems the deaf have with speech, although it is not aphasia. Aphasias are defined as language disorders. The British neurologist John Hughlings Jackson (1835–1911) captured the nature of what the aphasic communication problem is early in the twentieth century, when stating that one with aphasia was unable to "propositionize" [21], or communicate the abstract meanings of words and to join them together in a cohesive manner.

> We do not either speak or think in words or signs only, but in words or signs referring to one another in a particular manner.... Without a proper interrelation of its parts, a verbal utterance would be a mere succession of names, a word-heap, embodying no proposition.... The unit of speech is a proposition. Loss of speech (aphasia) is, therefore, the loss of power to propositionize...not only loss of power to propositionize aloud (to talk), but to propositionize either internally or externally.... The speechless patient has lost speech, not only in the popular sense that he cannot speak aloud, but in the fullest sense. We speak not only to tell other people what we think, but to tell ourselves what we think. Speech is a part of thought (in [21]).

Being unable to "propositionize" was also the problem the deaf had until Sign gave them a language they could use to tell other people what they thought. Towards the end of the twentieth century, William C. Stokoe (1919–2000) and Ursula Bellugi (1931–2022) began to perceive that Sign language, in this case American Sign Language (ASL), was not simply a broken or simplified version of English. Rather, once adopted by the Deaf Community, it had become a complex and thriving natural language, one with an independent syntax and grammar as functional and powerful as any found in the oral languages of the world [22–24]. The importance of sign to art and the neurosciences is that it underscores that verbal language is neither the only form of communication, nor a superior form. Rather, as the examples in this book show, art, too, provides alternative forms of communication that many use to propositionize—to communicate. Each art is a language with its own syntax.

Later, Bellugi, a linguist and cognitive neuroscientist, and her collaborations showed, in *What the Hands Reveal About the Brain* (1987) [25], that left hemisphere damage to Broca's area (where speech is produced), causes Sign language aphasias, parallel to spoken language aphasias, but right hemisphere damage does not, despite the visual–spatial nature of Sign language. She and her colleagues also demonstrated the plasticity of the nervous system through research showing that ASL is processed by the left hemisphere of the brain, as are spoken languages.[2] This key discovery revealed that what is critical to the control of language by the left hemisphere is linguistic structure, not auditory processing, or vocal articulation—and that Sign had a linguistic structure. Over time they also revealed similar within-hemisphere neural organization for signed and spoken languages, but some Sign language functions engaged the right hemisphere as well (particularly the use of spatial classifier constructions).

With other contemporaneous studies, this research definitively refuted many biases discussed above about deaf capabilities, and intelligence. It further confirmed that deaf people used language, and, like hearing speakers, they exhibit distinct language disturbances when left hemisphere cortical regions are damaged. Indeed, signers who have never spoken indicate similar compromised behavior with their signed speech when there is damage to either the Broca or Wernicke areas. With Broca damage, their signing is halted. With receptive aphasia, or Wernicke's aphasia, individuals can no longer understand the signs someone is making to them [26].

Equally important is that when we say that an individual with Broca's aphasia can no longer speak language normally, but can understand what is said, we are referencing the motor component of speech, not the essence of speech itself.[3] In a more general sense, Broca's area is considered a language area that maps the form of any language to "meaning." It is primarily of the left cerebral cortex, and we now have information showing that Broca area aphasias are evident in the nonverbal deaf as well. Although they do not speak verbally, they communicate through Sign language and deaf signing speech is evident in the dominant left hemisphere, as with hearing people. This shows that the brain in effect interprets signing language like spoken language [27].[4]

There are three larger points here. First, art media, like Sign, have their own syntax and serve as communication tools too, as discussed throughout this volume. Second, once some recognized that the deaf had a communication problem it was possible to build programs based on practices they used with their own community and families to create a standardized language to aid this cohort to learn and integrate socially. Third, this was accomplished by using a language that came naturally to them, for Sign language is a language that speaks with the hands. It is fascinating to think that Plato had noted as far back as ancient Greece how easily this cohort communicates with their hands and nonetheless the assumption that they were unable to communicate largely prevailed.

---

[2] She and her colleagues also went on to discover that musical rhythm processing was relatively spared in individuals with Williams syndrome, supporting the idea that musical ability constitutes an independent intelligence.

[3] Broca's area is also involved in language comprehension, in motor-related activities associated with hand movements, and in sensorimotor learning and integration, a topic beyond the scope of this interlude.

[4] *Prosody*, the "musical" properties of a language, is said to be a function of the right or nondominant hemisphere, a point not explored here.

2.3 Vision and Brain Function

Like theories about the deaf, theories about vision have a long history. As noted, those born blind were generally able to hear and thus learned to communicate verbally, so on the whole they fared better than the deaf and occupied a different kind of cultural space in terms of the mind, body, and brain connectiveness. Essentially, this is because seeing was not hypothesized as intrinsically connected to another faculty the way it was assumed that hearing was connected to speech. People did understand the importance of touch to the blind conceptually, but it was not theoretically seen as fused with the inability to see the way hearing was fused with the inability to speak.

Early visual debates centered around *extramission* and *intromission*. Those who believed in extramission theorized that emissions from the eyes allow us to see an object. Intromission theorists said emissions traveled from the object to the eyes. The nuances were often blurry; thinkers often disagreed even in their own writings. For example, sometimes Aristotle argued in favor of intromission (*On Sense and Sensible Objects*, 438a), but at other times he supported the notion of extramission, such as when he refers to "the ray leaving the eye" (*On the Soul*, 435a). One influential figure was the medieval Islamic scholar, Ibn al-Haytham or Alhazen (965–1040), who threw his support to intromission. Within this, ideas about vision were conceptually like ideas about hearing and sound. With vision, the propagation source was also debated. With sound, even as late as the seventeenth century many philosophers and scientists believed that sound propagated via invisible particles originating at the source of the sound and moving through space to affect the ear of the observer.

Bernardo Strozzi's *Tobias Curing His Father's Blindness* (1630–35), Fig. 2.4, much like the blind Egyptian harpists discussed earlier, reminds us that humans often communicate symbolically with art using its syntax of color, line, form, and so forth to bring viewers into the artwork. These symbolic renderings also remind us retrospectively that seeing and constructing a visual experience are not precisely the same thing. The theme of this Strozzi painting was drawn from the apocryphal Book of Tobit, which tells how the archangel Raphael directed Tobias to cure his father's blindness with the gall of the fish he had caught. Instructed by St. Raphael behind him, Tobias rubs the gall of a fish into his father Tobit's eyes, curing his blindness. On his left (on the right in the painting) are his wife and the dog that accompanied Raphael and Tobias on their journey to Media. The story is told in the Deuterocanonical Book of Tobit:

> Then the dog, which had been with them in the way, ran before, and coming as if he had brought the news, shewed his joy by his fawning and wagging his tail. And his father that was blind, rising up, began to run stumbling with his feet: and giving a servant his hand, went to meet his son. And receiving him kissed him, as did also his wife, and they began to weep for joy. And when they had adored God, and given him thanks, they sat down together. Then Tobias taking of the gall of the fish, anointed his father's eyes. And he stayed about half an hour: and a white skin began to come out of his eyes, like the skin of an egg. And Tobias took hold of it, and drew it from his eyes, and immediately he recovered his sight. And they glorified God, both he and his wife and all that knew them (Chap. 11:9–16).

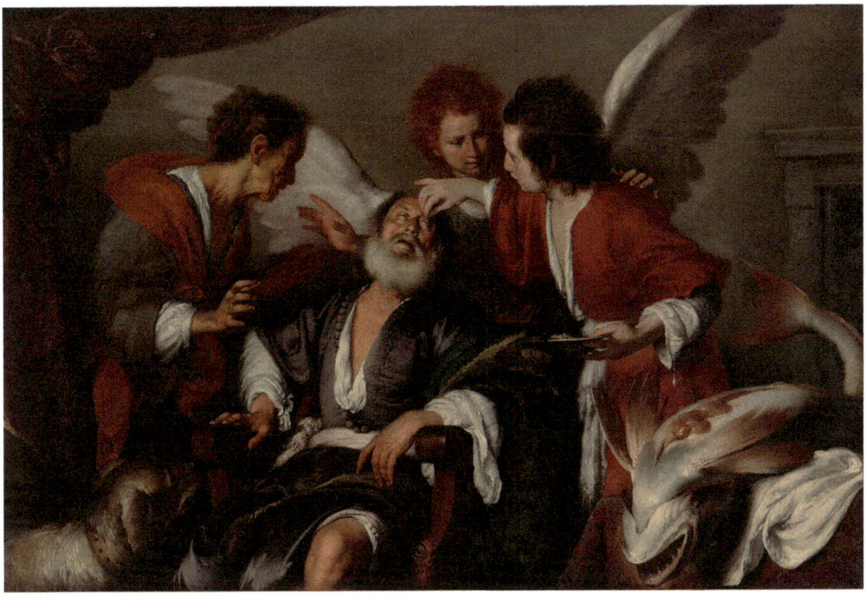

Fig. 2.4 Tobias Curing His Father's Blindness. Artist: Bernardo Strozzi. Oil on canvas. 57.5x88″ (146.1x223.5 cm). Date: 1630–35. Courtesy of the New York Metropolitan Museum of Art. Purchase, Mary Wetmore Shively Bequest, in memory of her husband, Henry L. Shively, M.D., 1957

By assuming a cure for blindness is possible and immediate, this work does not differentiate between sight and vision, or being able to see and having the ability to automatically comprehend what is seen. Nor did ophthalmologists of that era although it was during the Renaissance that scientists transferred the organ of vision from the lens to the optic nerve, which was generally believed to be on the axis of the eye. Johannes Kepler (1571–1630) was a key figure in updating understanding of visual construction or how the eye sees. He used the optics of the camera obscura, which has a lens like the eye, and found that the visual image is not "caught" by the lens, so to speak, but was painted onto the retina, the eye's truly sensitive element. He also correctly recognized that the retinal image, which he called a *pictura*, must be inverted and reversed. But it is with the discovery of the picture on the reddish-white concave surface of the retina that his optics ends. How the reversed and inverted image is corrected was beyond the scope of his science [28].

The acuity of central vision (at the optic disc) was explained at this time by the concentration of visual spirit where the optic nerve met the retina. The concept of nerve force replaced visual spirit in the seventeenth century, and it later became associated with electricity traveling along nerve fibers. At this time studies of anatomy and the influence of mechanistic thinking also influenced the overall discourse on the nature of vision and of the nervous system [29]. For example, when the blind spot was discovered in the seventeenth century, it led to much physiological and philosophical discussion concerning the imperceptibility or "filling-in" phenomena that is responsible for the completion of missing information [11, 29]. At this time, we can also see questions emerging in the philosophical sphere, asking how congenital blindness differed from the acquired condition. Or, more specifically, could one cure congenital blindness? So, rather than putting blindness into the "incurable" box as was the case with congenital deafness, people began to ask questions about how seeing happens and is experienced.

One was posed by the Irish scientist and politician William Molyneux (1656–1698), whose wife was blind. He wrote a letter to John Locke on 7 July 1688 putting forward a problem that continues to engage those who study sensory experiences. His question, known as Molyneux's problem, was this: if a person born blind might immediately identify a shape previously familiar to them only by touch if they were made to see. John Locke interposed the question within the second edition of his *An Essay Concerning Human Understanding*:

> Suppose a Man born blind, and now adult, and taught by his touch to distinguish between a Cube, and a Sphere of the same metal, and nighly of the same bigness, so as to tell, when he felt one and t'other, which is the Cube, which the Sphere. Suppose then the Cube and Sphere placed on a Table, and the Blind Man to be made to see. Quære, Whether by his sight, before he touched them, he could now distinguish, and tell, which is the Globe, which the Cube (p. 100, [30]).

Curiously, while deafness was not perceived as "curable" it was still coupled with blindness conceptually. We see this when Locke, who was a part of Thomas Willis' *Virtuosi* research group early in his career, is contemplating Molyneux's question. He asks to what degree does one who is blind or deaf knows the experience of those who are not similarly impaired:

> I know not how men, who have the same idea under different names, or different ideas under the same name, can in that case talk with one another; any more than a man who, not being blind or deaf, has distinct ideas of the color of scarlet and the sound of a trumpet, could discourse concerning scarlet color with the blind mind I mentioned in another place, who fancied that the idea of scarlet was like the sound of a trumpet (p. 136, [30]).

Clinical cases of that time investigated the nature of blindness as well. Unlike Tobias who sought to cure his father's blindness with the gall of the fish he had caught, William Cheselden is credited with performing the first known case of full recovery from blindness in 1728. He operated on a blind 13-year-old boy who gained his sight after removal of the lenses rendered opaque by cataract from birth [31]. Despite his youth, the boy encountered profound difficulties with the simplest visual perceptions. He could not judge distances, space, or size. Drawings and paintings confused him because he had no concept of what a two-dimensional representation is. He found it easiest to deal with the world when he could connect visual experiences with what he knew tactilely [31, 32].

Richard Gregory and Jean Wallace began to study a similar case in 1959, S.B. (Sidney Bradford), after Wallace saw in a local paper that a man blind from birth and then age 52 was about to receive corneal grafts that might allow him to see. S.B. told them he believed he became blind at the age of ten months —the age given in the records of the blind school—and his only visual memories were of three colors—red, white, and black. He remembered no other visual phenomena [33].

Oliver Sacks reported a third case in 1984, a 50-year-old man named Virgil, who had been blind since early childhood. He could recognize letters, for he knew them by touch, but could not read or even see words except by touch. He also needed touch to distinguish a cat from a dog, no matter how hard he tried. Objects of variable appearance were difficult as well. For example, he often gave colors the wrong names. Sacks compared him to an infant grappling with his primal construction of the world, and subsequently having difficulty doing the kinds of operations that are seamless for most of us. Virgil was also torn in his relationship with the world he now encountered, not knowing if he should look at it or feel it as was his custom [32], much as was the case with Cheselden's patient.

All of these cases remind us that our knowledge of how seeing, like hearing, is a construction of the brain is relatively recent and still growing. One key point here is that Virgil and S.B. both reported that they were disappointed with the results after being "cured." They were shocked by the imperfections of the visual world. They also found the venues that opened to them after their surgeries afforded less than they had imagined would be the case. What was particularly heart-rendering is that their operations created auxiliary and unexpected problems after a lifetime of living without sight.

Virgil, who had always been healthy and active, died at 54, 2 years after his operation. He had come to see sight as a curse. Virgil also experienced a range of medical problems—retinal, cortical, psychological—before a lobar pneumonia put him in the hospital. This led to a waxing and waning of his vision, and an eventual return to blindness, which he saw as a gift; it was a release from the glaring, confusing world of sight [32].

Gregory, too, writes about the importance of other senses when reporting on S.B.'s life.

> Vision was generally thought to be separate from the other senses and is still mainly studied in isolation; yet Bradford showed that exploratory touch—as well no doubt as taste, sound and other sensory experiences—gives richness and meaning to retinal images. For optical images are but ghosts, materialized into objects by perceptual experience of the non-visual properties of things (p. 836, [33]).

2.4 Physiology, Sensation, Perception

Subsequent chapters build on these historical ideas about the senses and add to them when introducing historical vignettes related to new topics such as dyslexia, autism, Down syndrome, brain injuries, shell shock, and so forth. Before moving on to the topical discussions, it is also important to add some thoughts on how investigators in the nineteenth and twentieth centuries shifted their research orientation to instrumentally based investigations. This was an approach that favored probing commonalities within the neurotypical populations, even as it operated in tandem with specific pathological research. As discussed in the last chapter, by the nineteenth century, psychology, psychiatry, and neurology were all staking out their own investigative territory as ways of studying the brain, the mind, and behavior expanded. All of these fields existed in tandem with fields like physiology and all were providing their own versions (or visions) on how the mind, the body, and nature fit together. Some of these investigations touched on art practice, art history, and aesthetics, now distinct fields as well, particularly when sensory perception was a part of the equation.

Hermann Helmholtz (1821–1894) offers a prototypical example of how this worked, and of how things began to change at this point, because his own research reached across disciplines over the course of his career. He is often paired with Johannes Müller (1801–1855) in terms of auditory sensations, visual sensations, and perception because Müller helped to launch and shape Helmholtz's career [34, 35]. Both contributed to our understanding of sensation and perception. Both had more of a neurotypical focus and a physiological emphasis.

In his *Handbuch der Physiologie des Menschen für Vorlesungen* (*Manuel of human physiology for lectures*), Müller, for example, devoted three chapters to hearing. He viewed the sensory nerves as highly specialized structures that are designed to deal with specific inputs appropriate to each sensory system. He had the nerves playing an active role in sensation, although he also maintained that the "sensorium" must recognize the changes that take place in the sensory nerves [35, 36]. Finger and Wade point out that Müller was concerned with both the sensorium and the mind:

> That which, through the medium of our senses is actually perceived by the sensorium is indeed merely a property or change of condition of our nerves; but the imagination and reason are ready to interpret the modifications in the state of the nerves produced by the external influences as properties of the external bodies (p. 245, in [35]; see [36]).

Helmholtz conducted studies on a wide range of topics—the speed of neural transmission, the natures of sound and color, the nature of our perceptions of them, and so forth. In the three volumes of *Handbuch der physiologischen Optik* Helmholtz shows the breadth of his research in physics, physiology, and the psychology of vision. Volume 1 covers the anatomy and optics of the eye, with consideration of image formation and optical aberrations. In Volume 2 he turns to the sensations of vision, dealing principally with color and contrast phenomena. Volume 3 presents the theory of visual perception and addresses eye movements, visual direction, and binocular vision [37, 38]. Figure 2.5, from Volume 1, (it is Fig. 17 in the first Helmholtz volume) shows the muscles of the eyeballs viewed from above. Note that the inferior rectus and inferior oblique are not visible in the figure: the superior rectus is removed from the right eyeball in order to show the optic nerve "N."

In terms of acoustics, the science of sounds, Helmholtz argued for clarifying how sound was perceived, rather than merely investigating vibrating bodies that happened to be audible but were not different in any respect from other vibrations. *On the Sensations of Tone as a Physiological Basis for the Theory of Music* (1863) [39],

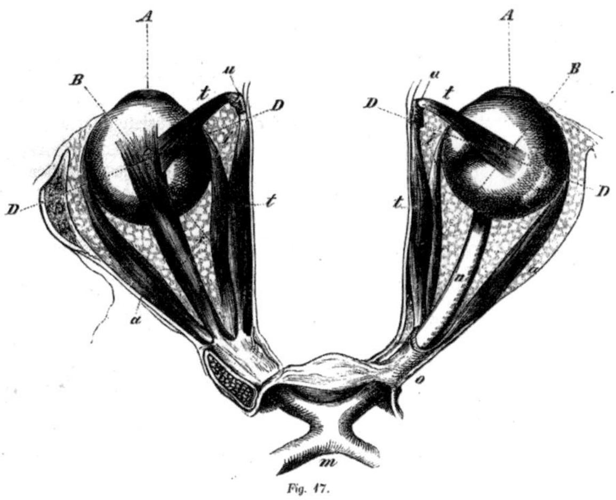

Fig. 2.5 Hermann von Helmholtz. Figure 17 in *Handbuch der physiologischen Optik, Volume 1* [37]. Note that the inferior rectus and inferior oblique are not visible in the figure: the superior rectus is removed from the right eyeball in order to show the optic nerve "N"

he even promised to revolutionize music theory by combining physical and physiological acoustics with musical science and aesthetics. In the introduction, he stated that his aim was to:

> …connect the boundaries of two sciences, which, although drawn toward each other by many natural affinities, have hitherto remained practically distinct—I mean the boundaries of physical and physiological acoustics on the one side, and of musical science and esthetics on the other (p. 353, [39, 40]).

In this book Helmholtz also referenced Pythagoras' contributions, discussed in the last chapter:

> The enigma which, about 2500 years ago, Pythagoras proposed to science, which investigates the reasons of things, "Why is consonance determined by the ratios of small whole numbers?" has been solved by the discovery that the ear resolves all complex sounds into pendular oscillations, according to the laws of sympathetic vibration, and that it regards as harmonious only such excitements of the nerves as continues without disturbance (p. 229, [39]).

Helmholtz's far-reaching and pioneering work included some thoughts on the Renaissance composer Giovanni Pierluigi da Palestrina (1525–1594), the construction of the ear, how soundwaves are propagated, and how an individual conceives tones, concluding that the concept of the ear as a nonnegligible medium was essential to the phenomenon of combination tones. One reason his approach is so relevant is that he forced physicists who dealt with sounds to pay attention to the body as a physical object, with physiological conditions of sound and hearing [40].

One of Helmholtz's popular lectures was titled "The eye as an optical instrument," a conception of the eye now accepted as a fundamental feature of visual science [28]. In analyzing the optics of the eye, Helmholtz constructed some novel optical instruments for studying the eye. Optometers, ophthalmometers, and ophthalmoscopes are presented historically with emphasis on how these instruments and camera analogies helped scientists to understand the functions of the eye, especially the enigma of accommodation. These kinds of mechanistic analogies bring Descartes to mind. On the biological side, with the invention of the ophthalmoscope in 1850 it became possible to peer into the living eye. Before the introduction of the ophthalmoscope into clinical practice, case studies concerning blindness documented the sudden sight loss following suppression of other bodily secretions, such as perspiration or pus. These were cited in popular ophthalmological textbooks during the first half of the nineteenth century [29]. Within 20 years of the ophthalmoscope's introduction, diseases once believed to cause sight loss by sympathetic irritation were being reclassified either as neurological diseases or as conditions that exerted their primary effect on the nervous system.

Helmholtz was also among those who helped develop the path to learning how the brain constructs our experience of vision, so his work is particularly compelling in terms of how his insights have overridden the notion that we see intuitively. As explained earlier, it was long held that we simply saw our world automatically, and thus efforts to return seeing to the blind had a semblance of logic behind them. Moreover, this kind of mechanical visual model was reinforced by technologies like the camera obscura and camera lucida. The idea that the brain might reconstruct what we see and how we develop a vocabulary for modalities like color and shape was not a part of this framework, a topic discussed throughout this book.

We see questions surrounding this insight taking shape with the reactions to various case studies related to vision. One example is the reaction to a paper published by Louis Verrey, a Swiss neurologist, entitled *Hémiachromatopsie droite absolue* (absolute right hemiachromatopsia), describing a case of a semi-achromatopsia [41], or the loss of color perception in only one half of the visual field. This discussion of color-blindness is expanded in Chap. 4 when color-blind painters are presented. What is important in this historical section is that in 1888 Verrey recorded his examination of a patient's lesion [41, 42]. He was able to delineate the exact area of the visual cortex that had been damaged by her stroke, and reports of problems with color perception and the determination of the lesions that caused her problems [43]. Yet, at the time, there was great resistance to this work for it seemed to violate the idea that we simply "know" what we see with all its color, form, movement, and depth. The idea of a central color area in the brain seemed non-sensical. Other factors suggesting a problem with the study were at work as well. One was that the legendary neuroscientist Gordon Holmes (1876–1965) and his team produced complete topographical mapping of the visual field in the primary cortical vision center from the cases of two hundred war injuries. Holmes emphatically noted that none of these showed isolated deficiencies in color perception [42–44]).

Later, vision research in the twentieth century included topics such as orientation, motion, color, and disparities. In the late 1950s and early 1960s David Hubel and Torsten Wiesel definitively pierced the idea that perception is given to us fully formed [45, 46] in research that earned them the Nobel Prize in 1981 (which they shared with Roger Sperry). They showed that there were cells and columns of cells in the visual cortex that act as feature detectors of the visual field. Representations were thus not given to us fully formed but constructed in perception by the brain through the seamless integration of a huge number of components. With this, it was clear that vision was an analytic process involving the differing sensitives within the cerebral and retinal systems [43, 45, 46]. Semir Zeki's discovery of specific cells sensitive to wavelength and color [47, 48] validated both Verrey's research and the neurological condition achromatopsia Verrey described. In Chap. 4, when color-blind painters are examined, I will return to this topic and also discuss how knowing more about the relationship of the eye and the brain now allows us to distinguish retinal color-blindness from both acquired and congenital achromatopsia.

In terms of neuroscience and art, the terrain remains more dynamic and fluid because art takes so many forms. For example, Semir Zeki moved from his earlier neurobiological studies of the primate visual brain to neuroesthetics, a field he founded. With this work Zeki began to investigate the neural correlates of affective states, such as the experience of love, desire and beauty that are generated by sensory inputs [49–53]. This recent research appears aimed at verifying historical philosophical theories about art and beauty, particularly those postulated by Schopenhauer in the eighteenth century [5, 50, 53]. Curiously, his ideas seem largely independent of the art and psychology approaches that emerged in the nineteenth and twentieth centuries with the elevation of psychology, psychiatry, and neurology [54, 55].

Finally, it should be noted that the dynamic and fluid nineteenth and twentieth centuries approaches to art were interwoven with neuroscience in a number of ways with the elevation of psychology, psychiatry, and neurology [54, 55]. Art and psychology were frequently intertwined with Gestalt Psychology, an approach that featured the idea that the psychological "whole" has priority and that the "parts" are defined by the structure of the whole, an approach that found expression in Rudolf Arnheim's (1904–2007) influential writings [56–59]. What we now call art therapy, and there are many schools of it, also began to develop and thrive at the end of the nineteenth century. More discussion about art therapy is introduced in Chap. 9, where the focus is on therapeutics and applications. Methodologies devised by the various schools are now frequently deployed to aid those with neurodegenerative diseases, in particular Parkinson and Alzheimer's [38]. These topics are also discussed elsewhere. In the twentieth century we find interest in developing studies that examined art and science, or sometimes art, science, and technology. We also see that these intersect with traditional art history, studies of art practice, multisensory projects (Chap. 12), and the growing interest in how media now impacts our lives.

In closing, this chapter identified how implicit assumptions about brain functions and our sensory experience have taken hold historically, evolved, and intersected with perspectives on communication, education, and therapeutic avenues. The next chapter, on biological and genetic perspectives will focus on congenital and developmental case studies, while also emphasizing that even here we can identify environmental influences.

References

1. Ione A. Nature exposed to our method of questioning. Berkeley, CA: The Diatrope Press; 2002.
2. Descartes R. Selections from the Principles of Philosophy (Project Gutenberg). Munich: C. H. Beck; 1644/1901.
3. Descartes R, Hall TS. Treatise of man. Amherst: Prometheus Books; 2003.
4. Hatfield G. Descartes' physiology and its relation to his psychology. In: Cottingham J, editor. The Cambridge companion to Descartes. Cambridge companions to philosophy. Cambridge: Cambridge University Press; 1992. p. 335–70.
5. Ione A. Art and the brain: plasticity, embodiment, and the unclosed circle. Leiden: Brill; 2016.
6. Willis T. The anatomy of the brain and nerves (Cerebri Anatome). Montreal: McGill University Press; 1664/1964.
7. Mehta AR, Mehta PR, Anderson SP, MacKinnon BLH, Compston A. Etymology and the neuron(e). Brain. 2020;143(1):374–9.
8. Deiters O, Schultze M. Untersuchungen über Gehirn und Rückenmark des Menschen und der Säugethiere. Braunschweig: Vieweg & Sohn; 1865.
9. Wilson MD. The life and times of Thomas Willis and his impact on contemporary medicine. J Hist Neurosci. 2012;21(2):127–31.
10. Willis T. Two discourses concerning the Soul of Brutes, which is that of the vital and sensitive of man. Gainesville, FL: Scholars' Facsimiles & Reprints; 1683/1971.
11. Finger S. Origins of neuroscience: a history of explorations into brain function. Oxford: Oxford University Press; 1994.
12. Willis T. Two discourses concerning the Soul of Brutes, which is that of the vital and sensitive of man. The first is physiological, shewing the nature, parts, powers, and affections of the same. The other is pathological, which unfolds the diseases which affect it and its primary seat; to wit, the brain and nervous stock, and treats of their cures: with copper cuts. London Printed for Thomas Dring etc.; 1672.
13. Williams AN, Sunderland R. Thomas Willis: the first paediatric neurologist? Arch Dis Child. 2001;85(6):506–9.
14. Buzon F. Compendium of music. In: Nolan L, editor. The Cambridge Descartes Lexicon. Cambridge: Cambridge University Press; 2015. p. 143–5.
15. Bearden EB. Before normal, there was natural: John Bulwer, disability, and natural signing in early modern England and beyond. PMLA. 2017;132(1):33–50.
16. Bulwer J. Philocophus: or, the deafe and dumbe mans friend. London: Humphrey Moseley; 1648.
17. Geen TR, Tassinary LG. The mechanization of emotional expression in John Bulwer's "pathomyotomia" (1649). Am J Psychol. 2002;115(2):275–99.
18. Descartes R. Passions of the soul. Earlymoderntexts.com: Early Modern Texts; 1649/2010.
19. Bulwer J. Anthropometamorphosis: man transform'd, or the artificial changeling. London: J. Hardesty; 1650.
20. Holder W. Elements of speech, an essay of inquiry into the natural production of letters: with an appendix concerning persons deaf and dumb. London: John Martyn; 1669.
21. Sacks O. Mysteries of the deaf. New York Review of Books; March 27, 1986: 22.
22. Stokoe WC. Sign and culture: a reader for students of American sign language. Silver Spring: Linstok Press; 1980.
23. Stokoe WC. Language in hand: why sign came before speech. Washington, DC: Gallaudet University Press; 2002.
24. Stokoe WC, Armstrong DF, Karchmer MA, Van Cleve JV. The study of signed languages: essays in honor of William C. Stokoe. Washington, DC: Gallaudet University Press; 2002.
25. Poizner H. What the hands reveal about the brain. Cambridge: MIT Press; 1987.

26. Corina DP, Blau S. Chapter 36—Neurobiology of sign languages. In: Hickok G, Small SL, editors. Neurobiology of language. San Diego, CA: Academic Press; 2016. p. 431–43.
27. Trettenbrein PC, Papitto G, Friederici AD, Zaccarella E. Functional neuroanatomy of language without speech: an ALE meta-analysis of sign language. Hum Brain Mapp. 2021;42(3):699–712.
28. Wade NJ, Finger S. The eye as an optical instrument: from camera obscura to Helmholtz's perspective. Perception. 2001;30(10):1157–77.
29. Reeves C, Taylor D. A history of the optic nerve and its diseases. Eye. 2004;18(11):1096–109.
30. Locke J, Locke's essays. An essay concerning human understanding. And a treatise on the conduct of the understanding. With the author's last additions. T. Ellwood Zell; 1689/1854.
31. Cheselden WVII. An account of some observations made by a young gentleman, who was born blind, or lost his sight so early, that he had no remembrance of ever having seen, and was couch'd between 13 and 14 years of age. Philos Trans R Soc London. 1728;35(402):447–50.
32. Sacks O. To see and not see. The New Yorker; May 10, 1993. p. 59–73.
33. Gregory R. The blind leading the sighted. Nature. 2004;430(7002):836.
34. Finger S, Wade NJ. The neuroscience of Helmholtz and the theories of Johannes Müller. Part 1: nerve cell structure, vitalism, and the nerve impulse. J Hist Neurosci. 2002;11(2):136–55.
35. Finger S, Wade NJ. The neuroscience of Helmholtz and the theories of Johannes Müller. J Hist Neurosci. 2002;11(3):234–54.
36. Müller J. Handbuch der Physiologie des Menschen für Vorlesungen (2 vols). Bonn: J. Hölscher; 1833, 1843
37. Helmholtz H. Handbuch der physiologischen Optik. Voss; 1867.
38. Wade NJ. The vision of Helmholtz. J Hist Neurosci. 2021;30(4):405–24.
39. Helmholtz H. On the sensations of tone as a physiological basis for the theory of music. Longmans, Green; 1885.
40. Kursell J. A third note: Helmholtz, Palestrina, and the early history of musicology. Isis. 2015;106(2):353–66.
41. Verrey L. Hémiachromatopsie droite absolue, conservation partielle de la perception lumineuse et des formes, ancien kyste hémorrhagique de la partie inférieure du lobe occipital gauche, par le Dr Verrey. G. Steinheil; 1893.
42. Zeki S. Improbable areas in color vision. In: Chalupa LM, Werner JS, editors. The visual neurosciences, 2-vol set. Cambridge: The MIT Press; 2003.
43. Sacks O. The river of consciousness. New York, NY: Alfred A. Knopf; 2017.
44. Fishman RS. Gordon Holmes, the cortical retina, and the wounds of war. Doc Ophthalmol. 1997;93(1):9–28.
45. Hubel DH, Wiesel TN. Receptive fields of single neurones in the cat's striate cortex. J Physiol. 1959;148(3):574–91.
46. Hubel DH, Wiesel TN. Receptive fields, binocular interaction and functional architecture in the cat's visual cortex. J Physiol. 1962;160(1):106–54.
47. Zeki SM. Representation of central visual fields in prestriate cortex of monkey. Brain Res. 1969;14(2):271–91.
48. Zeki SM. Cortical projections from two prestriate areas in the monkey. Brain Res. 1971;34(1):19–35.
49. Ione A. Innovation in art and science: a reply to Semir Zeki. Trends Cogn Sci. 2001;5(4):140.
50. Ione A. Examining Semir Zekis' neural concept formation and art: Dante, Michelangelo, Wagner. J Conscious Stud. 2003;10(2):58–66.
51. Zeki S. Art and the brain. Daedalus. 1998;127(2):71.
52. Zeki S. Inner vision: an exploration of art and the brain. Oxford: Oxford University Press; 1999.
53. Zeki S. Neural concept formation & art: Dante, Michelangelo, Wagner. J Conscious Stud. 2002;9(3):53–76.

54. Gombrich EH. Art and illusion: a study in the psychology of pictorial representation. Princeton, NJ: Princeton University Press; 2000.
55. Kandel ER. The age of insight: the quest to understand the unconscious in art, mind, and brain: from Vienna 1900 to the present. New York, NY: Random House; 2012.
56. Arnheim R. Art and visual perception: a psychology of the creative eye. Berkeley, CA: University of California Press; 1954.
57. Arnheim R. Visual thinking. London: Faber; 1970.
58. Arnheim R. New essays on the psychology of art. Berkeley, CA: University of California Press; 1986.
59. Verstegen I. Arnheim, Gestalt, and art: a psychological theory. Wien, Austria/New York, NY: Springer; 2005.

Biological and Genetic Perspectives 3

A United States government report in 2019 estimated that about 5% of the country's children under the age of 18 have disabilities (over three million children), and the number is increasing. Cognitive difficulty was the most prominent and it was not uncommon for children to have multiple disabilities [1].

> Some children, for example, may have difficulty with sensory activities, such as hearing or seeing, while others may have difficulty with mobility, requiring assistive devices such as a wheelchair. Children may also have difficulty with cognitive tasks such as speaking or understanding speech, reading text, or concentrating. In some cases, childhood disability can have implications for later well-being, including the level of education one is able to obtain, the type of work one is able to undertake, and the social networks one is able to build (p. 1, [1]).

Five percent is a relatively small number and possibly incorrect as well. It does not account for the many cases where someone has a neuroatypical condition that is not discovered until later in life, as is often the case with conditions like dyslexia, autism, prosopagnosia (face blindness), and synesthesia. Fortunately, many adapt to their condition even before explicitly identifying it. In the case of synesthesia, many synesthetes say they are glad to put a name to the condition, even if it seems more like a trait to them. Indeed, most synesthetes report that they enjoy this condition and feel it adds to their lives, as discussed in Chap. 12.

The report cited above also says that only 15% of people with disabilities are born with their impairments and one third of those with disabilities are over the age of 65. A World Health Organization (WHO) report combining adults and children tells us that more than a billion people in the world today experience disability [2].

None of these statistics, however, encapsulate what it means to live with a disability [3], a neuroatypical trait, and when or if therapies are in order. This chapter begins to probe what is behind these kinds of statistics, examining select cases through the lens of art. Since the unique and profound structure of an individual's

A. Ione, *Neuroscience and Art*, Neurocultural Health and Wellbeing, https://doi.org/10.1007/978-3-031-62336-3_3

brain is not fully set in place at birth, distinguishing between congenital baselines and acquired conditions is useful in terms of a condition's impact as one's life develops. For this reason, this book puts biological and genetic perspectives in this chapter and social and environmental modifications in the next one. This separation is intended to emphasize that our baseline matters.

For example, conditions like dyslexia (or developmental dyslexia) involve difficulty reading due to problems identifying speech sounds and learning. The problem is how an individual relates to letters and words (decoding). There are strategies one can use to work with the condition, and it is not one that is due to problems with intelligence, hearing, vision, or brain atrophy. Moreover, it is quite unlike "acquired dyslexia" (or alexia), which develops due to a traumatic brain injury, stroke, or dementia. This topic is discussed in more detail in Chap. 10.

Art offers a particularly robust entry point for examining the complex variations and strategies used to help those with a neuroatypical condition because it has an overall versatility and offers a variety of syntactical languages. These serve as means of expression, avenues for innovation, and offer therapeutic options as well. In other words, disparate genres activate and stimulate latent areas of the body and brain in a multitude of ways. In terms of human health and well-being it is also helpful to think about this in terms of how our physiologies change over the course of our lives, for how an individual uses and experiences art is likely to vary and change over the course of one's life as well.

3.1 Autism

On June 15, 2023, Donald Triplett (1933–2023) died at the age of 89. As a child he was "Case 1" in the history of autism diagnoses [4]—and as an adult he became an influential case study on how people with autism can find fulfillment. He was known for his savant abilities, particularly the ability to name musical notes played on a piano and the ability to perform rapid mental multiplication [5]. His case was first published in 1943, when he was ten. Triplett's death, among other things, shows that Autism Spectrum Disorder (ASD) entered the developmental disability lexicon relatively recently.

ASD is now defined as a condition that can cause significant social, communication, and behavioral challenges. Before Triplett met Dr. Leo Kanner, who designated him as "Case 1," Triplett spent a year in an institution, as was common for kids like him in the 1930s. After seeing him listless and often motionless when they visited, his parents decided to take him home. Then they found Dr. Leo Kanner and, as they say, the rest is history.

When Kanner's 1943 paper "Autistic Disturbances of Affective Contact" described Triplett with ten others [4], the doctor said these children illustrated a condition that differed "markedly and uniquely from anything reported so far" in the annals of psychology." Sixty-three years later, in 2006, about one in 110 children was said to have the condition. This remarkable increase was topped in 2020 when it was estimated by the United States Centers for Disease Control and Prevention (CDC) that that one in 36 has the condition [6].

Autism is a lifelong neurogenetic condition. Symptoms generally appear during the first 3 years after birth, although there are exceptions to this global characterization. As noted, the CDC currently estimates that an average of 1 in 36 children is on the autism spectrum [7]. In China, where it was first recognized as a neurological disability in 2006, the prevalence rate was estimated at about 0.25–1% of the population in 2019 (or about 3.9–14.5 million out of 1.4 billion individuals) [8]. Recent researchers claim this is probably an underestimation [8, 9]. Worldwide, the number is estimated at 1 in 100 children [10].

Autism affects the brain in four major areas of functioning: language/communication, social skills, sensory systems, and behavior. As is often the case with neurological differences, the cause of autism is not known. Current research suggests it may arise from genetics, environmental insults, or both [11]. Generally, people with Autism Spectrum Disorder, to varying degrees, exhibit difficulty with social interactions, have restricted interests, present remarkable focus, and show an aptitude for patterns and details. Challenges that are present include a hypersensitivity to light, sound, tastes, and so forth. Often there is noticeable difficulty with conversational dialogue, difficulty with nonverbal communication, uncoordinated movements or clumsiness, and anxiety and depression.

The spectrum is a loosely defined cluster of neurodevelopmental disorders that includes autism and Asperger's. One of the oddest elements of the spectrum is its diversity. A 2001 taxonomy estimated that those with Asperger's syndrome have an average or even very high IQ. Seventy percent of those with other autistic disorders suffer from mild to severe mental retardation [12, 13]. Obviously, this broad range shows there is some definitional ambiguity. It is also now known that regardless of where one lands on the spectrum, how autistic individuals grow and learn depends on how well they develop their communication and social skills.

Although Kanner claimed in 1943 that he was reporting a new condition, he also confirmed there was a literature on it. Looking closely shows that the history of autism does extend back into myths and legends. A limited taxonomy includes one Brother at a medieval monastery described as naively innocent, lacking social intuition and common sense [13, 14]. Kanner's seminal article quotes a sixteenth century account offered by Martin Luther in a later book, *A History of the Care and Study of the Mentally Retarded* [15]. The tale was recorded in Luther's *Table Talk* (German: *Tischreden*), compiled by Johannes Mathesius, which brings together a collection of sayings based on notes taken by various students of Luther between 1531 and 1546.

> Martin Luther referred to the feebleminded as godless and reported this occurrence in one of his *Table Talks:* "Eight years ago, there was one at Dessau whom I, Martinus Luther saw and grappled with. He was twelve years old, had the use of his eyes and all senses, so that one might think that he was a normal child. But he did nothing but gorge himself as much as four peasants or threshers. He ate, defecated, and drooled and, if anyone tackled him, he screamed. If things didn't go well, he wept. So I said to the Prince of Anhalt: 'If I were the Prince I should take this child to the Moldau River which flows near Dessau and drown him.' But the Prince of Anhalt and the Prince of Saxony, who happened to be present, refused to follow my advice" [15].

A detailed 1747 court case that fits into this literature is that of Hugh Blair of Borgue, nicknamed "the daft lad of Borgue." In this instance, Hugh Blair's brother successfully petitioned to annul Hugh's marriage in order to gain his inheritance [16]. Documentation for the case is now thought to describe Hugh Blair in a manner that is consistent with a modern diagnosis of Autistic Spectrum Disorder. Another well-known case is Victor the "Wild Boy" of Aveyron, a feral child caught in 1798. Now considered autistic, Victor was educated by the French physician Jean-Marc-Gaspard Itard (1774–1838) in the nineteenth century [13, 17].

More recently, Oliver Sacks suggested Henry Cavendish (1731–1810) had Asperger characteristics in a particularly intriguing publication because Sacks has described himself as an individual with Asperger's [18, 19]. Cavendish, a prolific natural philosopher first published in 1766. Contemporaneous biographers describe Cavendish as eccentric and "peculiarly shy." Those who wrote about him so long ago would have had no diagnostic presuppositions. According to Sacks:

> Many of the characteristics that distinguished Cavendish are almost pathognomic of Asperger's syndrome: a striking literalness and directness of mind, extreme single-mindedness, a passion for calculation and quantitative exactitude, unconventional, stubbornly held ideas, and a disposition to use rigorously exact (rather than figurative) language—even in his rare nonscientific communication—coupled with a virtual incomprehension of social behaviors and human relationships (p. 1347, [20]).

Despite the abundance of vivid descriptions, no one saw these kinds of cases as connected until the nineteenth century. Then the pioneering English psychiatrist Henry Maudsley (1835–1918) suggested they might be linked to a childhood psychosis. Psychoses at the time were considered to be due to physical causes affecting the brain [13, 21].

Eugen Bleuler (1857–1939) first used the word *autismus* (English: autism) and he began to describe it as a condition in his paper "On the theory of schizophrenic negativism" [22]. Kanner's 1943 evaluation of a group of eleven children mentioned above disputes Bleuler's connection of schizophrenia and autism:

> "We must, then, assume that these children have come into the world with innate inability to form the usual, biologically provided affective contact with people, just as other children come into the world with innate physical or intellectual handicaps. …[W]e seem to have pure-culture examples of *inborn autistic disturbances of affective contact* (*his italics*, p. 250, [4]).

Despite Kanner's separation of the two conditions, a position that he does not consistently adhere to, the schizophrenic label continued to affect the field well into the 1970s, particularly in the period when psychoanalytic treatments were the recommended therapy. Based on the techniques available in the 1940s and 1950s, researchers were assuming that emotional damage must be the cause since no specific physical brain abnormalities or lesions had been revealed [13].

The Austrian pediatrician Hans Asperger (1906–1980) described Asperger's condition in a paper he wrote around 1944, over 30 years after Bleuler's work. While Asperger's is now considered a part of the autism spectrum, it was largely ignored at the time because he worked in Austria while it was under Nazi control. Notably, Asperger used Bleuler's terminology "autistic psychopaths" in a lecture in German about child psychology. Asperger died in 1980 and the term Asperger syndrome was coined a year later, by Lorna Wing, who also translated his original paper into English in 1991. She was also among the first to build upon Asperger's intuition that even certain gifted children might also be autistic, and she thus described the disorder as a continuum [13].

We can likely add Asperger's to the list of people who had Autism Disorder Spectrum. Some have suggested he began working in this area because he, himself, fit the Asperger profile [23, 24]. According to Uta Frith, Franz Grillparzer (1791–1872), an important nineteenth century Austrian dramatist that Asperger liked to quote may have also fit into this category [23]. Grillparzer was described as a quiet, reticent man, steeped in the humanist tradition, with an extensive knowledge of classics, history, art, and literature. It is said he took after both his parents. His father was a reserved man with a cold and brusque behavior, whereas his mother was a very kind woman who, like the dramatist, had a great interest in music. Grillparzer suffered from depression, had very strange and unsuccessful relations with females, was a hypochondriac, and was unable to maintain close relationships [24].

In other words, as noted, disagreements abound in regard to how to characterize this continuum in which we find both the severest cases of infantile autism (perhaps accompanied by retardation and various neurological problems) and the most gifted, high-functioning individuals like Temple Grandin, discussed below. According to Isabelle Rapin (1927–2017), a neurologist who specialized in autism, the two conditions (autism and Asperger's) may be separate at the biological level even if they are sometimes similar at the behavioral level [25]. Her thinking is that although Asperger's syndrome may be radically different from classical infantile autism, in a child of three all forms of autism may look the same.

Lorna Wing (1928–2014) and Isabelle Rapin are among those who were at the forefront in debunking the misconception that autism was a single disease, a psychiatric syndrome caused by so-called refrigerator (emotionally "cold") mothers [13, 26–30]. Wing, who had trained as a medical doctor specializing in psychiatry, narrowed her focus to childhood developmental disorders in 1959 after she recognized that her daughter Susie (1956–2005) was autistic. This change of emphasis led her to become involved in researching developmental disorders, particularly Autistic Spectrum Disorders [31]. At the time there were also groups forming where mothers began to compare note and organize autistic networks. This, too, was influential in changing stereotypes [13].

By contrast, Rapin, a pediatric neurologist, found herself intrigued by children with speech and hearing disorders, recognizing that what ailed these nonverbal children was not obvious:

> After evaluating hundreds of autistic children, I became convinced that the report by one-third of parents of autistic preschoolers, of a very early language and behavioral regression, is real and deserving of biologic investigation [32].

Now called the mother of autism, Rapin detected early in her investigations that behavioral audiology was unreliable at diagnosing the severity of hearing losses in very young children, and, in retrospect, concluded that autism would no doubt have been the correct diagnosis in some puzzling "emotionally disturbed" mute children [32]. With others, she has also done research showing that congenital deafness and congenital blindness caused profoundly deviant sensory experience and massive reorganization of brain connectivity [33–35].

Although parents are no longer blamed for an autistic child's difficulties, their lives are impacted, nonetheless. Poet Mathew Zapruder, the father of an autistic son, conveyed this in his recently published *A Story of a Poem* [36]. He had intended this collection of poetry to record what it means to live in a time of turmoil. In the end, his poetry helped him resolve the unexplored pain, unexpected joys, and all he was confronting in the wake of his son's autism diagnosis. After abandoning the drinking that gradually increased with the information, he decided to write a single prose poem, as slowly as possible, in his effort to decipher the question in his mind: "What is the relation between making poems and learning to be the father of this atypical child?" (p. 26, [36]). Introducing the project Zapruder explains in the third person how he, as a parent, felt upon learning his child was atypical.

> The love he felt for his son collided practically every moment with intense fear and exhaustion. So many things that came easily to other children—speech, social play a naturally growing understanding of the world—had to be painstakingly broken down into steps and taught. Sometimes there was no discernable progress. What seemed to be joyful for other parents was such a source of worry. All this permeated not just by intense love, but also admiration for his son's joyful spirit, his fierce affection for those close to him, his sensitivity to music, his astounding memory, his joy and facility in building … what did it all mean? Where was it going? Those questions could not be silenced, and there could be no answers (p. 23, [36]).

3.2 Savant Syndrome

Savant syndrome is a phrase frequently used to characterize unusual skills—most often in memory, music, art, and calculation—in the presence of disability in other areas. It is sometimes grouped with Autism Spectrum Disorder because the skills may be associated with the unusual eye for detail typical of autism. In some cases, a specific skill might exist, while in others there may be several skills that co-exist simultaneously.

As is the case with autism, the brain basis of savant syndrome is unknown. In the congenital form, the savant skill surfaces in childhood. This skill is often—but not always—accompanied by an underlying developmental disability. Acquired savant syndrome occurs in previously neurotypical individuals who suffer head injury, stroke, dementia, or other central nervous system events or disorders, with injury to the brain often in the left hemisphere but not always [37, 38]. Both congenital and acquired savant syndrome are rare, with their documented case totals being 287 and 32, respectively as of 2021 [38]. Treffert, a researcher in this area, surmises that we can explain the condition as an example of the 3 Rs: rewiring, recruitment, and release [37–40].

Most of the time, the savant's abilities are due to a recruitment of undamaged tissue in the right hemisphere and rewiring accounts for the release of dormant potentials, whether it is music, art, or other abilities. According to Happé and Wallace, there is a genetic predisposition that interacts with environmental exposure and extensive practice [41]. Neuroimaging studies as of 2012 had not suggested recruitment of atypical regions [41].

The cases described in this chapter speak to congenital savant syndrome. Chapter 10 describes an acquired case, Tommy McHugh, who suffered a bilateral stroke and acquired savant syndrome. Also included there are some notations on neurotypical persons whose savant skills emerged suddenly. These individuals—without underlying disability or brain injury—were also without prior interest or ability in the newly emerged skill areas [38, 42].

In terms of congenital savant syndrome, John Langdon Down who is credited with describing Down syndrome, discussed in the next section, also introduced the term *idiot savant* (French for "learned idiot") into the literature in 1887. He used it to juxtapose the evidence of severe mental handicap with prodigious mental ability [43, 44]. At that time the term idiot was used alongside intellectual disability and was a technical term for those of IQ less than 20–25. It was in common use in the nineteenth century [44].

In terms of art, the psychological and neurological literature includes many well-described artistic savants such as Nadia and Stephen Wiltshire as well as more general studies (see [41, 45–48]). These compelling findings document to varying degrees that born savants tend to exhibit extraordinary talent in one area along with profound linguistic and social impairment. Most striking is that their developmental sequence differs from that of normal children. With visual art, in normal children one sees a sequence of random scribbling to schematic and geometric figures to "tadpole" figures. The savants, by contrast, produced highly recognizable, detailed representational drawings from the start. According to Happé and Wallace, their exceptional focus, intense interest, and an ability to remember detail may involve loss of function in the left temporal lobe with enhanced function of the posterior neocortex [41].

Musical savants include the blind prodigy and composer Thomas Greene Bethune (1849–1908) and the blind musician Leslie Lemke (b. 1952) who was born prematurely in Milwaukee, Wisconsin. As a complication of his premature birth, Lemke developed retinal problems, then glaucoma, and his eyes had to be surgically removed in the first 6 months of life. He also had brain damage that caused cerebral palsy and severe mental retardation. His birth mother gave him up for adoption, and the county asked 52-year-old May Lemke, a nurse-governess, if she would take the infant into her home. She agreed [37, 49]. Leslie was 12 before he learned to stand, and 15 before he learned to walk. Mary would put his hands over hers on a piano she bought for him at about age eight or nine. Through this process he learned to play some simple tunes. After hearing Tchaikovsky's Piano Concerto No. 1 on television, he played it flawlessly from beginning to end. Soon he played other instruments as well. Normally, his spasticity was so severe that he could not hold eating utensils. His handicap disappeared completely when playing the piano [37, 39].

Darold Treffert and Lemcke first met in 1980. Later, in 1986, Treffert carried out a complete neurologic examination on Lemcke, including imaging and neuropsychological tests:

> A CT scan showed marked left-sided abnormality especially in the left frontal lobe. There was of course evidence of the eyes having been surgically removed; the eye sockets showed some deformity along with deformity of the sinus area, both primarily on the left side. The left frontal area showed an area of atrophy consistent with either some previous vascular damage or residual damage from a previous brain abscess. There were areas of atrophy in the anterior and posterior portions of the left parietal lobe, and, to a minor degree, damage as well in the left and right occipital lobes of the brain (p. 51, [37]).

Thomas Greene Bethune, popularly known as Blind Tom or Blind Tom Wiggins, was born a century earlier. Tom was cognitively disabled, blind, and untrained. The most prominent and highly paid worldwide black entertainer in the 1800s, he was also the first African American artist to perform at the White House when he played for President James Buchanan in 1860. At the end of his life, he had an incredible repertoire of over 7000 pieces, and eventually composed his own pieces, including some as long as twenty pages [49].

3.3 Temple Grandin

Temple Grandin (b. 1947), who is autistic and an advocate for neurodiversity, is sometimes characterized in terms of the savant syndrome. Grandin, however, disputes this. She characterizes herself a visual thinker in a culture that is largely geared to language and the verbal. This also aligns her with the visual arts and art was a tool used in her childhood to help her learn. Oliver Sacks memorialized Grandin as "An anthropologist on Mars" in his writings about her. The moniker was an outcome of a conversation they had about myths, drama, and Shakespeare's plays:

> She was bewildered, she said, by Romeo and Juliet ("I never knew what they were up to"), and with *Hamlet* she got lost with the back-and-forth of the play. Though she ascribed these problems to "sequencing difficulties," they seemed to arise from her failure to empathize with the characters, to follow the intricate play of motive and intention. She said that she could understand "simple, strong, universal" emotions but was stumped by more complex emotions and the games people play. "Much of the time," she said, "I feel like an anthropologist on Mars" (p. 112, [25]).

Grandin met Oliver Sacks at the urging of Uta Frith, an autism researcher. He had consulted Frith about the autistic artist Stephen Wiltshire and other savants [50]. According to Sacks, he was so taken aback by Grandin's brilliance and the degree to which she differed from the autistic children he met in hospitals and clinics that he arranged to spend some time with her. Moved by her obvious intelligence and yearning to communicate, he found her personality quite unlike the passivity of the autistic artist Stephen Wiltshire.

A few things stand out in Temple Grandin's biography that are worth underscoring in terms of how we address neurodiversity, art, therapies, and cases of neuroatypical individuals. First, born in the mid-twentieth century 14 years after Donald Triplett and 4 years after Kanner's seminal paper, it is unsurprising that Temple Grandin was not formally diagnosed with autism until adulthood [11, 51–54]. When she was two, the only formal diagnosis given to her was "brain damage" [52]. Second, she did not speak until she was almost four. Third, and perhaps most importantly, her success, in her view, was built in her early years by a supportive family, mentors, and an empathetic neurologist who steered her mother in an alternative direction [11, 51–54].

In other words, Temple Grandin's achievements could have easily never happened. Her father favored putting her into an institution, a standard "solution" at that time. Fortunately, her mother, who was strongly opposed to the idea, prevailed [54]. On a neurologist's advice, her mother located a recommended speech therapist. From the age of two and a half, Temple attended a speech therapy nursery school to supplement the personalized training. With the help of a nanny, her mother played educational games with her. Her mother additionally encouraged her interest in drawing. Recalling this period Grandin told Sacks:

> No less crucial was the emergence of some genuine playfulness and creativity—painting, drawing, making cardboard models and sculptures, as well as "unique and creative ways of being naughty." At eight, Temple was starting to achieve the "pretend play" that normal children achieve as toddlers but the lower-functioning autistic child never achieves at all (p. 166, [25]).

Later, Temple was placed in a boarding school capable of nurturing her special needs and abilities [11]. She eventually received her PhD in Animal Science and to date has published more than 60 scientific papers and over 20 books on both autism and animal studies.

3.4 Down Syndrome

A quite different congenital disorder is Down syndrome. It is the most common chromosomal condition diagnosed in the United States. Recent estimates are that Down syndrome occurs in about 1 in every 700–1000 babies in the United States. In 2015, the syndrome was present in 5.4 million individuals globally [55]. Almost a century after it was named, genetic analysis by Lejeune, Gautier, and Turpin [56] revealed that this syndrome was due to an extra copy of Chromosome 21 (trisomy 21), a random error in cell division, generally resulting from nondisjunction during maternal meiosis I (Fig. 3.1) [57]. This extra copy changes how the baby's body and brain develops, which can cause both mental and physical challenges. To date, no behavioral activity of the parents or environmental factor is known to cause the syndrome.

The nineteenth century physicians Jean-Étienne-Dominique Esquirol (in 1838) and Edouard Séguin (in 1846) first published descriptions of individuals with Down syndrome while working to improve the social and educational conditions of individuals living in mental institutions [58–60]. The first comprehensive description was provided in 1866 by John Langdon Down (1828–1896), an Englishman who assumed that parental tuberculosis caused this disorder [61]. His short paper "Observations on an ethnic classification of idiots" noted that these patients have a facial resemblance with individuals of the Mongolian people [62]. As a result, the term Mongolism was coined in 1908. It remained in medical usage until the 1950s when it was finally deprecated as misleading in 1961 by a group of 19 genetic experts in an article in *Lancet*. It is now largely abandoned [63] (Fig. 3.1).

Reginald Down, John Down's oldest son, had a son who was born with the syndrome in 1905, 9 years after the death of his grandfather and 39 years after John Down's classification paper. At that time, institutionalization was common for such cases but Reginald, also a physician, responded by raising his son at home. He also committed himself to improving the medical, social, and educational experience for this cohort. In addition, as a physician Reginald Down sought to broadly help by devising a comprehensive educational program comprised of physical exercise, sensory stimulation, and dramatic play. Today these are the kinds of programs professionals would likely classify as physical, occupational, speech, and play therapy [60].

Individuals with Down syndrome have more opportunities these days no doubt due to the efforts of people like Reginal Down. Indeed, as I was working on this chapter, James Martin, an Irish actor who was born with Down syndrome, was celebrated when *An Irish Goodbye*, a film in which he starred, was awarded the best

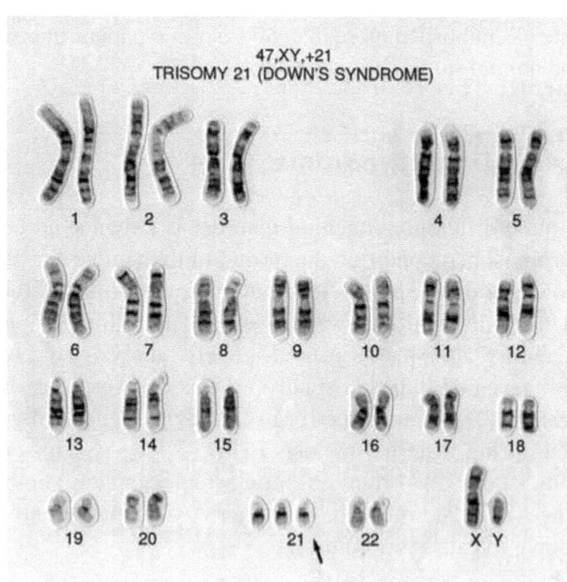

Fig. 3.1 Down syndrome human karyotype 47, XY, +21. Wessex Reg. Genetics Centre. Attribution 4.0 International (CC BY 4.0). Courtesy of the Wellcome Collection

live action short winner at the Academy Awards. The film told the story of two brothers who come together after their mother's death. One of the brothers has Down syndrome. Other award-winning actors with Down syndrome include Tommy Jessop and Zack Gottsagen [64–67].

3.5 Art History and Down Syndrome

Researchers have asked if artwork exists to validate a longer history for this condition because it is often the case that conditions are illustrated long before they are described (see [68–72]). For example, in 1968 a debate on Down syndrome took place in the medical journal *Lancet*. One author, Mirkinson, hypothesized that it is a modern disease, claiming it rarely was depicted in art [73]. A second author, Richards, basically agreed and estimated that there were not more than one hundred persons with Down syndrome in the whole of England when Shakespeare was born in 1564. He added there were probably one hundred in the hospital where he was working at the time, St. Lawrence's Hospital, Caterham, Surry [74]. Zellweger, a third author, took issue with his colleagues, pointing to a painting by Jacob Jordaens (1593–1678) that shows two children with Down syndrome. Art scholars have hypothesized these children were likely his own [75].

Additional, and compelling historical evidence adds to Zellweger's example. Documentation includes historical paintings, figurines, and pottery found in the Greco-Roman world, in many Central and South American pre-Columbian cultures, and in Khmer temples (see [72, 76]). There are also studies on artifacts that establish the origin and pathology of chromosome 21 as far back as 30–50 million years ago. This suggests trisomy 21 has essentially existed since time immemorial [72, 76].

One particularly articulate historical example is a Flemish nativity painting from the sixteenth century, dated approximately 1515, *The Adoration of the Christ Child* (Figs. 3.2 and 3.3), attributed to a follower of Jan Joest of Kalkar [70]. While a typical Nativity in many ways, it is unusual in showing the scene at night, with the divine infant illuminating the onlookers from below. According to art historians, the most important element of the Saint's story for painters was the illumination emanating from the Christ Child to which the sun could not be compared and the divine splendor that totally annihilated the material splendor of the candle shielded by Joseph's hands [77].

Close examination shows that one angelic figure and an onlooker appear to share the appearance of Down syndrome, as highlighted in Figs. 3.2 and 3.3. Focusing in on the triplet of angels in the foreground we see that the angel next to Mary, whose single visible wing is behind the head, has a flattened facial midface, epicanthal folds, upslanted palpebral fissures, small and upturned nasal tip and the downward curving of the mouth corners. The hands crossed over the breast have short fingers, in sharp contrast to the elongated tapering fingers of the others. The earthly admirer in the center behind the angels has similar facial features.

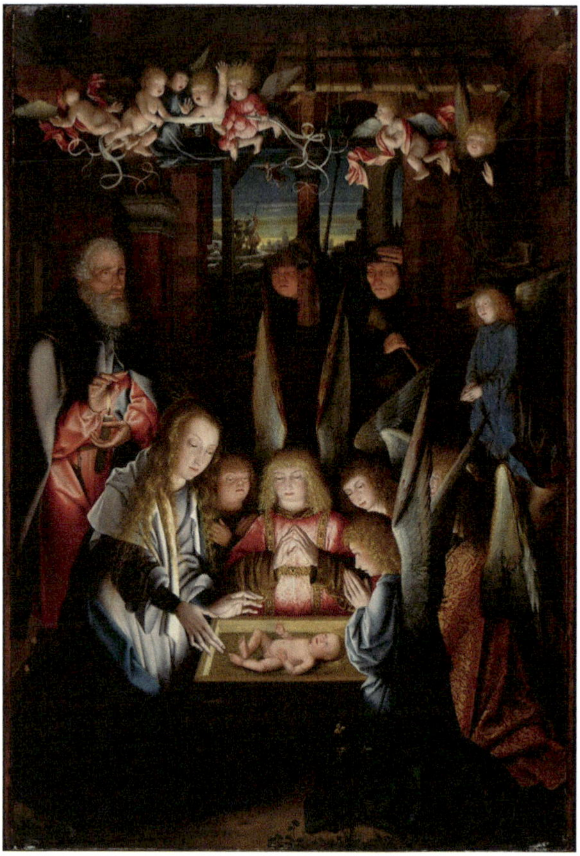

Fig. 3.2 *The Adoration of the Christ Child*. Follower of Jan Joest of Kalkar (Netherlandish, active ca. 1515). Oil on wood. Overall measurement: 41 × 28 1/4 in. (104.1 × 71.8 cm); painted surface 41 × 27 5/8 in. (104.1 × 70.2 cm). Courtesy of the New York Metropolitan Museum of Art, The Jack and Belle Linsky Collection, 1982

Based on the long history of using family and friends as models it is possible that those depicted were known to the artist. Of particular cultural interest here is that depicting a Down-faced angel goes against the attitudes of the time. In those days, disabilities were often presented in comical terms or as symbols of evil [70].

Before moving on to Judith Scott, a contemporary artist with Down syndrome, it should be noted that while the term mentally retarded is increasingly considered offensive, it was the dominant term until recently for these types of neuroatypical cases. In terms of cultural behaviors, many children with disabilities like Down syndrome were abandoned to death or servitude. Sometimes these children were given to monasteries in a practice known as oblation, or placed in foundling hospitals that had high levels of mortality [70].

Institutionalization remains an outcome even in recent times, as was the case with the artist Judith Scott, whose case study follows. Her experience further aids in conceptualizing that art is a means of communication that can further individual development. Her biography additionally highlights the difference between treating a person humanely and as an individual, as opposed to treating a person as an unsolvable chronic health problem.

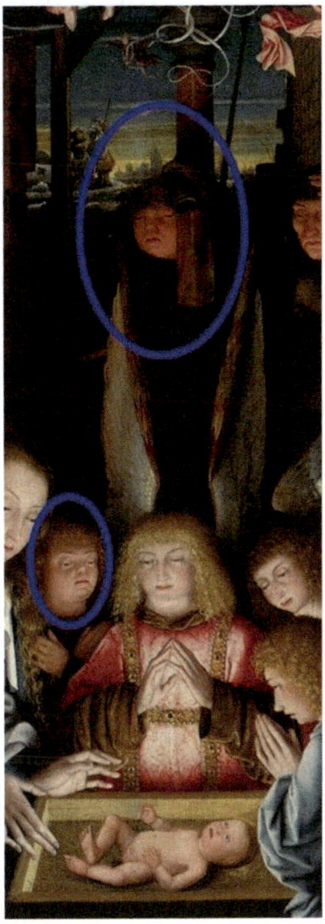

Fig. 3.3 Detail: *The Adoration of the Christ Child*. Follower of Jan Joest of Kalkar (Netherlandish, active ca. 1515). Oil on wood. Overall measurement: 41 × 28 1/4 in. (104.1 × 71.8 cm); painted surface 41 × 27 5/8 in. (104.1 × 70.2 cm). Courtesy of the New York Metropolitan Museum of Art, The Jack and Belle Linsky Collection, 1982

3.6 Judith Scott: Fiber Artist

Celebrated fiber artist Judith Scott (1943–2005) was born 4 years before Temple Grandin. Their lives were not at all similar, although both are characterized as neuroatypical. One compelling aspect of Scott's biography is that whereas Judith was born with the genetic disorder Down syndrome and became deaf prelingually, her fraternal twin Joyce was perfectly normal. Despite Judith's neurological challenges, after she turned forty, she became an artist, eventually earning international recognition for her fiber sculpture.

In a general sense, we know that fraternal twins on average have a 50% percent chance of inheriting the same version of any particular gene from each parent, so they are genetically similar to siblings [78]. In this case only one twin, Judith, was born with the genetic disorder, considered disabled and characterized as uneducable. We also know that physically, the brain of a child with Down syndrome develops differently from a normal one, attaining a form reduced in size and altered in configuration [68]. A complex condition, Down syndrome can result in multiple lifelong health problems, including varying degrees of intellectual disability and delays in speech, memory, and learning [79]. The anomaly of chromosome 21 can also cause intellectual disability, short stature, and characteristic facies [71]. While sometimes inherited, it is unlikely that either of Scott's parents were carriers of the anomalous chromosome since neither her twin nor her other siblings developed the profile. Therefore, Judy's Down syndrome mosaicism presumably resulted from nondisjunction (when chromosomes fail to pass to separate cells) during cell division in the embryo [71].

Despite their genetic differences, the Scott twins were inseparable until the age of seven (Fig. 3.4). Then one night, and without warning, Judith was whisked away from the bed they shared and taken to live in an institution for the mentally disabled. Joyce only found out when she awoke, confused by Judith's absence. As it turned out, the entire Scott family was affected by young Judith's removal.

The impact on their mother was quite noticeable. Mrs. Scott had what was described as a case of nerves a few months after Judith left for her first institution, for there were subsequently several. Initially she seemed to lose touch with reality, but this subsided a bit as the months passed. Ultimately, she underwent dramatic personality changes [80].

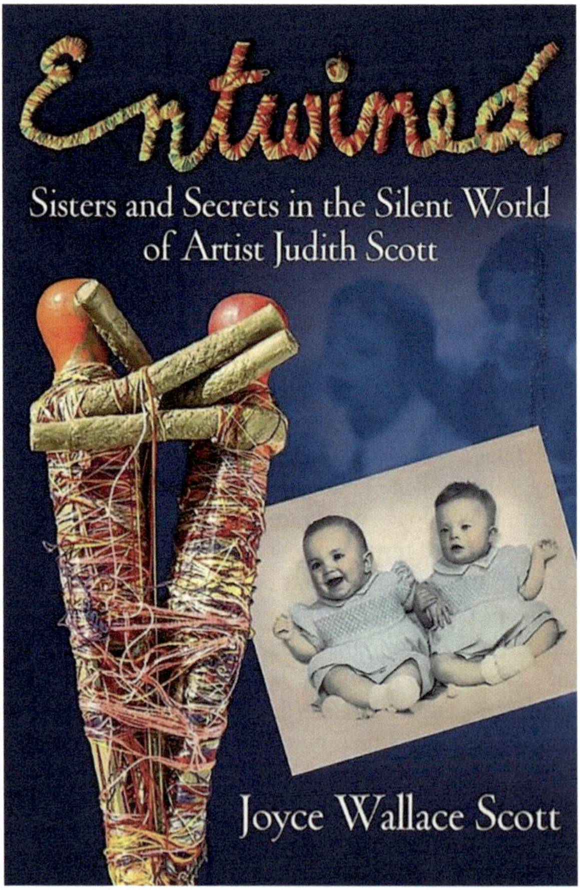

Fig. 3.4 Book cover: *Entwined: Sisters and Secrets in the Silent World of Artist Judith Scott* by Joyce Scott. Courtesy of Beacon Press [80]

Joyce felt she had lost a piece of herself, later recalling that they had shared everything as if they were one, through touch—and without Joyce's realization that Judith was deaf or different in any meaningful way [80]. As an adult, Joyce wondered if their tactile exchange had helped Judith navigate her world before she was put away, another instance in which the loss of one sense (hearing) perhaps strengthened others. Joyce also wondered whether touch had strengthened the bond between them: They not only were always touching each other, they also touched forms around them together. Thirty-six years later, Joyce gained guardianship of her twin and removed her from institutional facilities.

We do not know specifically how Judith felt upon the separation. Since she did not speak and could not hear one can only imagine her state of mind when she was abruptly moved to a bleak Ohio institution at the recommendation of doctors and clergy—and lost the tactile connection to her twin as well [80]. No doubt Judith shared Joyce's sense of traumatic loss to some degree.

The initial institutional record read: "Not appropriate for any educational program. Mongoloid idiot—Profoundly retarded—IQ 30" (p. 99, [80]). Later institutional records are sketchy. They do say she had behavioral problems and was subject to tantrums [80]. Reviewing them during the process of gaining guardianship, Joyce discovered that Judith was given electric shock and other behavior-modification therapies. She was also given experimental drugs.

Institutionalization was unkind in other ways as well. At one point when Joyce visited her sister in their thirties, she inquired about Judith's missing teeth. The staff revealed that all of Judith's teeth were removed by the institution to avoid paying for dental work in the future. She was also told that Judith had no medical or dental problems that would warrant their removal. It was also in their thirties that a nurse realized Judith was deaf.

Behavior problems are common with people who lack communication tools, and apparently this was the case with Judith. One particularly heartbreaking entry stands out in the sketchy records. Judith, who later became an internationally celebrated artist, was denied the chance to participate in a small art activity. On the one occasion when she wanted to draw with a colored pencil it was taken from her! She left the room in tears in effect confirming the behavioral problems notations throughout her medical profile. Considering her capacity to make friends and work with others after her release from the institutional system, it is hard to reconcile that her tantrums at institutions that afforded her so little expressive opportunity only enhanced their conclusion that she was uneducable [80].

Judith's life also had challenges added on for uncertain reasons. As noted, her hearing loss in childhood was not diagnosed until she was in her thirties. It was probably caused by a case of Scarlet Fever when she was about three. Whatever the cause of the missed diagnosis, the result was that it was sadly assumed that she was more developmentally disabled than she was. One outcome of the missed diagnosis was that she never learned Sign language. Another was that because her loss of hearing was prelingual, she had never learned to speak. Research shows that prelingual loss of hearing loss has a greater impact on cognitive, social, and language development than in cases where a child has more exposure to language [81]. Yet, according to Lynne Cooke, curator of special projects at the National Gallery of Art in Washington DC, Scott is one of the most remarkable instances of an individual overcoming psychic alienation (p. 209, [82]).

Historically, many individuals with Down syndrome were killed, abandoned, or ostracized from society. In the twentieth century, it was still common for these individuals to be institutionalized, so Judith Scott was no exception in this regard. This raises the question of how did her artistic accomplishments come about?

It was not all smooth sailing. Once Joyce became her guardian, she found that Judith's hearing loss made it more difficult to place her in a half-way house. Since Judith had never learned Sign language, she was ineligible for houses that took the deaf. Houses for more general disabilities rejected her twin because she could not hear. They only wanted people with developmental disabilities, not those with deafness.

Judith Scott began to find her own communication tools in 1987. Now in her forties and under Joyce's care, Judith entered the Creative Growth Studio in Oakland, California, a venue designed to help adults with developmental disabilities flourish through artistic practice. Judith did not flourish immediately. At first, she showed little evidence of artistic interest, ability, or creativity. During her first few months she drew colorful, repetitive circular and spiral shapes on paper, sometimes incorporating images cut from magazines, a type of manufactured item that appealed to Judith throughout her life. Judith's early lack of interest and lackluster results led Joyce to fear they might not let her sister retain the studio space at the Center.

It is worth noting here that the kinds of therapies first offered to Judith were based on successful art therapies used with Alzheimer's patients where drawing and painting are used to stimulate them. This approach had become commonly employed with Down syndrome patients because the two disorders appear to have a genetic linkage, which is supported by the triplication of the gene that codes for amyloid beta (A4) precursor protein (APP) in persons with Down syndrome [60]. A 2023 study looked at how well this therapy fared over a decade to see if it was effectively repurposed. According to the study: "translation of specific findings into effective therapeutic strategies has been disappointingly slow and has failed in many cases at the clinical level" [83]. Scott's case was one example of an individual who failed to show results.

After 2 years at the Center, Scott did find an approach that made sense to her—and this allowed her artistic journey to begin. Her inspiration was an artist workshop with fiber artist Sylvia Seventy. After Scott began to experiment on her own the uninspired art projects gave way to inventive sculptures. At this point Judith began to thrive and the isolation during some of her key developmental years no longer fully held her back.

On a personal level, it appears she found making her work rewarding. Her artistic process was fastidious, and she had a passionate involvement with each piece once she began working. Interpersonally, once she began creating her remarkable mixed media sculptures, she found a means to make artistic statements that other people felt. She could not articulate what they meant to her—or what she was "saying"—in any other form, but others could relate to their visual power and sense of mystery.

Like many artists with an ongoing practice, Scott developed a varied range of techniques over the years, ultimately producing over 200 breathtaking creations until her death in 2005 [84]. Fiber and fabric were her passion and media of choice, although she used other materials as well, often deftly woven over an armature of quotidian items she discovered around her such as keys, plastic tubing, bicycle wheels, and a shopping cart. Her practice was marked by her independence, self-direction, and for never repeating a form or color scheme. Figure 3.4 features one of her works. Some sculptures are relatively spare and monochromatic. Others are exuberant forms composed of colorfully printed fabric strips, knotted, and fraying with tactile appeal. A sophisticated feeling for juxtapositions of color and material—bold dualities, distinct tonal schemes, and complex layering of monochrome palettes—became a hallmark of her work [84].

Her final piece stands alone in terms of how we might decipher her array of silent communications. After many productive years she one day made a small black sculpture and handed her sister her beloved stack of magazines. The next day she died of heart failure [80].

Obviously, Judith Scott does not provide a standard Down syndrome medical profile because her undiagnosed hearing loss complicated her life adding another and a long undetected obstacle to her development. What we can say is that communication and language development are areas that are more difficult for all young children with Down syndrome. Judith, however, missed out on the kind of caregiver interaction that influences language development for those in this cohort because she could not hear [85]. So, while many with Down syndrome show mild deviations from the neurotypical in early childhood, their development is still delayed [86].

Another element that stands out when sifting through the totality of her life is the degree to which the mature Judith was aided by the humane support system backing her up: the Center, her family, and her community. John MacGregor, writing of her epiphany to work with fiber sums up her life well:

> No one could have predicted the nature or the intensity of her response to the new materials. Clearly incapable of conforming to expectations, of following instructions or imitating what the other students were doing, she simply invented something totally new [87].

Another element that stands out is that Judith's sculptures reflect little cultural input. They are highly individualistic and yet they reach out and touch others. Initially she acquired fame as an Outsider artist because Outsider artists are generally self-taught and operate outside of the mainstream art community. With the passage of time, however, the focus on her disabilities faded to become merely a footnote in the life story of a world-famous artist of great distinction [88].

> Today, Judy is no longer recognized primarily as a leading outsider artist. She has successfully crossed the nebulous divide that has separated art brut from the formal art world. Judy is now recognized as a brilliant and original contemporary artist. As her works have been acquired by museums around the world … she has paved the way for other artists to transcend the original conceptual limits placed on their art. In terms of her artistic reputation, her disabilities are now consigned to mere biographical footnotes (p. 209, [80]).

Finally, I have included Judith Scott in this chapter because her Down syndrome is clearly a genetic disorder. Yet, her case underscores that having options for doing something personally enriching and the ability to communicate in some fashion makes a difference in one's life. Finding and using tools that bring the mind and body together is developmental whether one is neurotypical or neuroatypical. In this respect, her case study shows social and environmental modification as well.

3.7 The Carlin Brothers

The Carlin Brothers expand the historical discussion of the deaf presented in Chaps. 1 and 2. Like Judith Scott, the Carlin brothers, John and Andrew, were similarly not defined by their disability, their congenital deafness. They also show how deaf education benefitted individuals in nineteenth century America, a topic mentioned in the last chapter.

Born congenitally deaf and mute in Philadelphia, Pennsylvania, both John and Andrew attended The Pennsylvania Institute for the Deaf and Dumb. Both, too, found success as portrait and miniature painters after graduation. The lessor known of the two, Andrew B. Carlin (1816–1871), is now largely remembered as an Outsider or Folk Artist. His painting of 3-year-old Thomas Wister Grookett shows he proudly identified himself as mute for on the back of this painting; the inscription reads: "Painted by AB Carlin, a Mute, Fox Chase, PA, 1844," The 1860 United States Census records Andrew Carlin and his wife Anne, as "Deaf," living with their five children in Philadelphia [89].

His use of the word mute firmly places him in the nineteenth century for the word mute is now considered offensive. Similarly, terms like deaf-and-dumb are now considered negative labels because they presume that a person could not develop cognitive abilities [90]. Yet many have shown this is not the case. Many communicate with Sign language, lip-reading, vocalizations, and so on. In addition, as discussed in Chaps. 1 and 2, the growth of institutions and Sign language enabled individuals like the Carlin brothers to forge better lives and to integrate with the community.

John Carlin (1813–1891) graduated in 1825 from The Pennsylvania Institute for the Deaf and Dumb and was one of the school's 16 original students. In Philadelphia, from 1833–1834, he studied portraiture and drawing (Figs. 3.5 and 3.6). From there he went to London in 1838, studying at the National Museum. Following this, John studied with Paul Delaroche (1797–1856) in Paris, returning to the United States in 1841 to practice as a miniaturist in New York City. He was untiring in his efforts to help people who were similarly deaf and unable to speak. One example of his commitment to the Deaf Community was his role in establishing the National Deaf-Mute College (now Gallaudet University) in Washington DC in 1864.

John was known as a poet and a painter of miniatures on ivory before turning to landscapes and genre scenes once photography made miniature painting less attractive commercially. Figure 3.5, *After a Long Cruise*, painted in 1857, displays the bawdy humor quite popular in the mid-nineteenth century. Like the term mute,

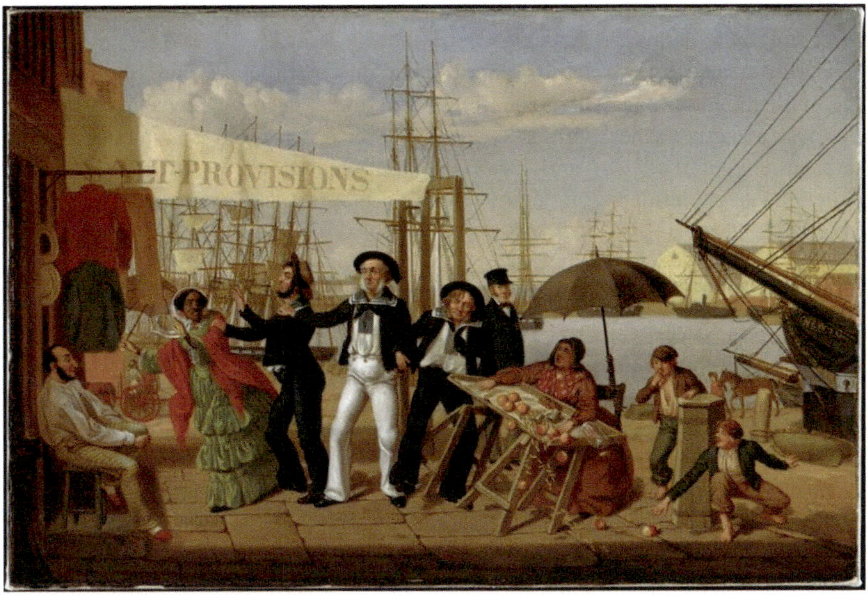

Fig. 3.5 *After a Long Cruise*. Artist: John Carlin (American, Philadelphia, Pennsylvania 1813–1891 New York), 1857. Medium: Oil on canvas. 20 × 30 in. (50.8 × 76.2 cm). Courtesy of the New York Metropolitan Museum of Art, Maria DeWitt Jesup Fund, 1949

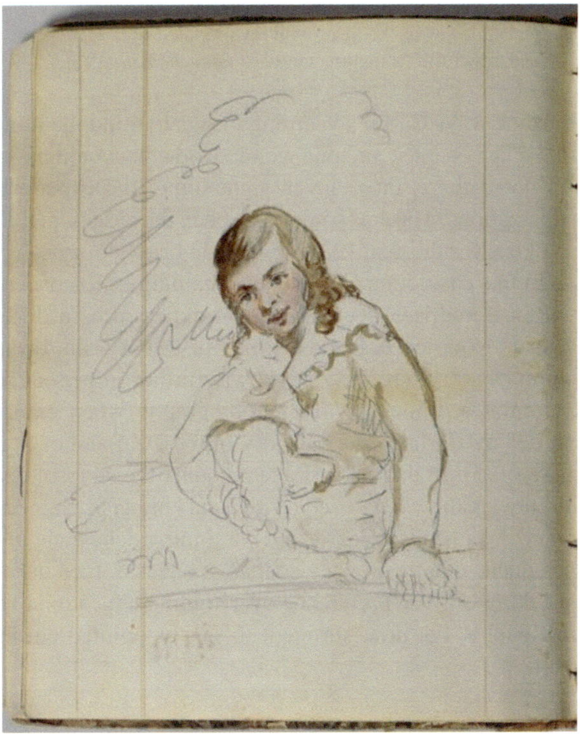

Fig. 3.6 Sketch from John Carlin's 1835–56 account book ledger. Ink on paper. Dimensions: 7 1/2 × 6 1/4 in. (19.1 × 15.9 cm). Courtesy of the New York Metropolitan Museum of Art, Dale T. Johnson Fund, 2009

some might consider this compositional theme demeaning now. His vibrant colors and attention to detail are particularly striking.

Carlin is also known for his literary pursuits, particularly his poetry, a rare accomplishment for one with no hearing at that time. *The Mute's Lament* (Appendix A) first appeared in the *Saturday Courier* of Philadelphia and was included in the first volume of the *Deaf and Dumb* annals prepared by the instructors of the American Asylum in 1847. It begins "I move a silent exile on this earth." Its closing lines read:

> That when [my soul] leaves this sphere replete with woes
> My ears shall be unsealed, and I shall hear;
> My tongue shall be unbound, and I shall speak,
> And happy with the angels sing forever! [91]

The Mute's Lament conveys that Carlin feels a sense of loss that music and speech were unknown and unknowable to him. It is clear he yearns to perceive what he has learned through reading about the texture of a kind of sensory experience others know, and he could not. The poetry is even more poignant when one considers how well he can communicate through his paintings and writings. When John alludes to melody, music, song, and sound-related experience we feel his sense that elemental aspects of life are missing from his experience, although it is also clear that he has an abstract sense of them:

> The touches-lyric of the lute divine,
> Obedient to the rise, the cadence soft,
> And the deep pause of maiden's pensive song,
> While swells the heart with love's elated life,
> Draw forth its mellow tones—*I hear them not* [91].

In some ways Carlin's yearning brings to mind the blind men given sight discussed in the last chapter, and one wonders how he would have responded if given hearing. As noted there, those given sight found the experiences did not live up to their expectations. This has sometimes been the case with the quite controversial cochlear implants for the deaf [92–94], a topic briefly discussed in Chap. 9.

In this case, John Carlin is also helping us to conceptualize what it is like to live within a multisensory culture that experiences a different range from one's own sensory experience. Upon learning of John Carlin, I was struck by how little I knew about Deaf Culture. Recently, I was further surprised to find Oliver Sacks expressed a similar sentiment in his autobiography when explaining how his interest was piqued by Harlan Lane's history of the Deaf and of Sign language, *When the Mind Hears* [95]. In this book, Lane recounts the story of deaf people; their unique, rich culture and how it is based on a visual language, Sign, as briefly discussed in Chaps. 1 and 2. Lane also records the continuing debate about whether deaf people should be educated in their own visual language or forced into "oralism," a term used when speaking is given preference over signing educationally. These topics are discussed in Chap. 9. For now, suffice it to say, as pointed out in Chaps. 1 and 2, the use of hand communication has a long history. Among the examples introduced so far are Plato's words on those who talk with their hands and the deaf painter Cristoforo de Predis, who communicated with Leonardo da Vinci and his family this way.

Finally, the stress on biological and genetic perspectives in this chapter is intended to underscore the complexity of brain development. This is also the case with the social and environmental modifications topics discussed in the next chapter. Before moving on, as both chapters point out, these are not separate topics in individual lives. Individuals who are going to thrive despite genetic and congenital conditions require social and developmental responses that meet their needs. Similarly, although an acquired condition is clearly biological and possibly related to genetics or environmental factors, it often creates a social rupture that distinguishes it from a condition one knows from birth and infancy. These topics are picked up again and expanded on in later chapters.

References

1. Young NAE. Childhood disability in the United States: 2019 [Report Number ACSBR-006]. American community survey briefs. Washington, DC: US Census Bureau; 2021. p. 14.
2. World Health Organization, World Bank. World report on disability. Geneva: World Health Organization; 2011.
3. Davis LJ. Enforcing normalcy: disability, deafness, and the body. Verso Books; 1995.
4. Kanner L. Autistic disturbances of affective contact. Nervous Child. 1943;2(3):217–50.
5. Traub A. Donald Triplett, 'Case 1' in the Study of Autism, Dies at 89. New York Times. 2023 June 20, 2023; Sect. A.
6. Center for Disease Control and Prevention (CDC). Autism spectrum disorder (ASD). Washington, DC: U.S. Department of Health and Human Services (HHS); 2023.
7. Center for Disease Control and Prevention (CDC). What is autism and how do I recognize a kid who might be diagnosed as having autism spectrum disorder? National Center on Birth Defects and Developmental Disabilities, Centers for Disease Control and Prevention Internet]. 2012. https://www.cdc.gov. Accessed 8 March 2023.
8. Sun X, Allison C, Wei L, Matthews FE, Auyeung B, Wu YY, et al. Autism prevalence in China is comparable to Western prevalence. Mol Autism. 2019;10(1):1–9.
9. Zhou H, Xu X, Yan W, Zou X, Wu L, Luo X, et al. Prevalence of autism spectrum disorder in China: a nationwide multi-center population-based study among children aged 6 to 12 years. Neurosci Bull. 2020;36(9):961–71.
10. World Health Organization. Autism: key facts. Geneva: World Health Organization; 2023.
11. Grandin T. The way I see it: a personal look at autism and Asperger's. Arlington, TX: Future Horizons Inc.; 2011.
12. Silberman S. The Geek syndrome. Wired Magaz. 2001;9:12.
13. Wing L. The history of ideas on autism: legends, myths and reality. Autism. 1997;1(1):13–23.
14. Frith U. Autism: explaining the enigma. Br J Dev Psychol. 1989;3:465–8.
15. Kanner L. A history of the care and study of the mentally retarded. C. C. Thomas; 1964.
16. Houston RA, Frith U. Autism in history: the case of Hugh Blair of Borgue. Oxford: Blackwell; 2000.
17. Itard JMG. Rapport fait à son excellence le Ministre de l'intérieur sur les nouveaux développemens et l'état actuel du sauvage de l'Aveyron. Paris: Imprimérie impériale; 1807.
18. Sacks O. The mind's eye: a man of letters. New York, NY: Alfred A. Knopf; 2010. p. 53–81.
19. Sacks O. A neurologist's notebook: face-blind: why are some of us terrible at recognizing faces? The New Yorker; 2010. p. 36–43.

20. Sacks O. Henry Cavendish: an early case of Asperger's syndrome? Neurology. 2001;57(7):1347.
21. Maudsley H. Insanity in early life. Physiology and pathology of the mind. New York, NY: D. Appleton and company; 1867. p. 259–386.
22. Bleuler E. Zur Theorie desschizophrenen Negativismus. In: Psychiatrisch-neurologische Wochenschrift. Halle a.S: Marhold; 1910. p. 171–7.
23. Frith U. Autism and Asperger syndrome. Cambridge: Cambridge University Press; 1991.
24. Lyons V, Fitzgerald M. Did Hans Asperger (1906–1980) have Asperger syndrome? J Autism Dev Disord. 2007;37(10):2020–1.
25. Sacks O. An anthropologist on Mars. The New Yorker; 1993. p. 106–25.
26. Rapin I, Autism and Language Disorders Nosology Project. Preschool children with inadequate communication: developmental language disorder, autism, low IQ. London: Mac Keith; 1996.
27. Tuchman R, Rapin I. Autism: a neurological disorder of early brain development. London: MacKeith Press; 2006.
28. Elgar S, Wing L. Teaching autistic children. London: College of Special Education; 1969.
29. Wing L. Autistic children: a guide for parents and professionals. New York, NY: Brunner/Mazel; 1985.
30. Wing L, Wing JK. Early childhood autism: clinical, educational, and social aspects. Oxford: Pergamon Press; 1976.
31. Silberman S. Neurotribes: the legacy of autism and the future of neurodiversity. New York, NY: Avery; 2015.
32. Rapin I. Isabelle Rapin: an autobiography. J Child Neurol. 2001;16(5):352–6.
33. Jure R, Pogonza R, Rapin I. Autism spectrum disorders (ASD) in blind children: very high prevalence, potentially better outlook. J Autism Dev Disord. 2016;46(3):749–59.
34. Xiao Z, Qiu T, Ke X, Xiao X, Xiao T, Liang F, et al. Autism spectrum disorder as early neurodevelopmental disorder: evidence from the brain imaging abnormalities in 2-3 years old toddlers. J Autism Dev Disord. 2014;44(7):1633–40.
35. Rapin I. Effects of early blindness and deafness on cognition. Res Publ Assoc Res Nerv Ment Dis. 1979;57:189–245.
36. Zapruder M. Story of a poem. Los Angeles, CA: Unnamed Press; 2023.
37. Treffert DA. Islands of genius: the bountiful mind of the autistic, acquired, and sudden savant. Jessica Kingsley Publishers; 2011.
38. Treffert DA. The sudden savant: a new form of extraordinary abilities. Wis Med J. 2021;120(1):69–73.
39. Treffert DA. Extraordinary people: understanding "idiot savants". New York, NY: Harper & Row; 1989.
40. Treffert DA. Accidental genius, TEDx talks. 2017.
41. Happé F, Wallace GL. Savant syndrome. In: Ramachandran VS, editor. Encyclopedia of human behavior. 2nd ed. San Diego, CA: Academic Press; 2012. p. 281–5.
42. Main-Brackenridge CS. The empty chair—Revisited: Tommy McHugh. 2014:16, 3.
43. Down JL. On some of the mental affections of childhood and youth. London: Churchill; 1887.
44. Treffert DA. The idiot savant: a review of the syndrome. Am J Psychiatry. 1988;145(5):563–72.
45. Hou C, Miller BL, Cummings JL, Goldberg M, Mychack P, Bottino V, Benson DF. Autistic savants. [correction of artistic]. Neuropsychiatry Neuropsychol Behav Neurol. 2000;13(1):29–38.
46. Wiltshire S. Floating cities: Venice, Amsterdam, Leningrad, and Moscow. New York, NY: Summit Books; 1991.
47. Sacks O. Prodigies. An anthropologist on Mars: seven paradoxical tales. New York, NY: Alfred A. Knopf; 1995. p. 188–243.
48. Selfe L. Nadia revisited: a longitudinal study of an autistic savant. Taylor & Francis; 2012.
49. Treffert DA. To creation and beyond: the remarkable life of Leslie Lemke. Wisconsin Academy of Arts and Science Magazine. 2014. https://www.wisconsinacademy.org/magazine. Accessed 17 Apr 2023.
50. Sacks O. On the move: a life. New York, NY: Alfred A. Knopf; 2015.

51. Grandin T. Thinking in pictures: and other reports from my life with autism. New York, NY: Vintage Books; 2006.
52. Grandin T, Scariano M. Emergence, labeled autistic. Novato, CA: Arena Press; 1986.
53. Rozsa M. Interview with Temple Grandin: autism, genetics and the steep price of being intelligent. Salon; 2020.
54. White R. Interview with Temple Grandin. Medscape. Psychiatry. 2005;10:1.
55. Vos T. Global, regional, and national incidence, prevalence, and years lived with disability for 310 diseases and injuries, 1990–2015: a systematic analysis for the global burden of disease study 2015. Lancet. 2016;388(10053):1545–602.
56. Lejeune J, Gautier M, Turpin R. Study of somatic chromosomes from 9 mongoloid children. C R Hebd Seances Acad Sci. 1959;248(11):1721–2.
57. Taeusch HW, Barrard RA, Gleason CA. Avery's diseases of the newborn. Philadelphia, PA: Elsevier Saunders; 2005.
58. Dominique EJ-E. Des maladies mentales considérées sous les rapports médical, hygiénique et médico-légal. Belgium: Tircher; 1838.
59. Séguin É. Traitement moral, hygiène et éducation des idiots et des autres enfants arrièrés ou rétardés dans leur développement, etc. Paris: J.B. Baillière; 1846.
60. Salehi A, Ashford JW, Mufson EJ. The link between Alzheimer's disease and Down syndrome. A historical perspective. Curr Alzheimer Res. 2016;13(1):2–6.
61. Van Robays J. John Langdon Down (1828–1896). Facts Views Vis Obgyn. 2016;8(2):131–6.
62. Down JL. Observations on an ethnic classification of idiots. Clinical lectures and reports by the medical and surgical staff of the London hospital. Lond Hosp Rep. 1866;3:259–62.
63. Allen G, Benda CE, Böök JA, Carter CO, Ford CE, Chu EH, et al. Mongolism. Lancet (Letters to the Editor). 1961;1(775):426.
64. O'Kane C. Oscars audience sings "happy birthday" to James Martin, actor with Down syndrome, as he accepts award. Disability Horizon. 2023. https://disabilityhorizons.com/. Accessed 14 March 2023.
65. Purcell E. James Martin: an award-winning actor with Down's syndrome. Disability Horizon. 2020.
66. Purcell E. Zack Gottsagen: an actor with Down's syndrome tipped for Oscar glory. Disability Horizon. 2019. https://disabilityhorizons.com. Accessed 14 March 2023.
67. Purcell E. Down's syndrome actor Tommy Jessop stars in award-winning crime drama innocence. Disability Horizon. 2020.
68. Becker L, Mito T, Takashima S, Onodera K. Growth and development of the brain in Down syndrome. Prog Clin Biol Res. 1991;373:133–52.
69. Berg JM, Korossy M. Down syndrome before Down: a retrospect. Am J Med Genet. 2001;102(2):205–11.
70. Levitas AS, Reid CS. An angel with Down syndrome in a sixteenth century Flemish nativity painting. Am J Med Genet A. 2003;116a(4):399–405.
71. Powell-Hamilton NN. Down syndrome (trisomy 21). The MSD manual. Rahway, NJ: Merck & Co.; 2022.
72. Stahl A, Tourame P. Trisomy 21 in visual art. Arch Pediatr. 2013;20(12):1342–51.
73. Mirkinson AE. Is Down's syndrome a modern disease? Lancet. 1968;2(7565):458.
74. Richards BW. Is Down's syndrome a modern disease? Lancet. 1968;2(7563):353–4.
75. Zellweger H. Is Down's syndrome a modern disease? Lancet. 1968;2(7565):458.
76. Nippert I, Kunze J. Genetics and malformations in art. Grosse Verlag; 1986.
77. Ainsworth MW, Christiansen K. From Van Eyck to Bruegel: early Netherlandish painting in the Metropolitan Museum of art. New York, NY: Metropolitan Museum of Art; 1998.
78. Linden DJ. Unique: the new science of human individuality. New York, NY: Basic Books; 2020.
79. Baburamani AA, Patkee PA, Arichi T, Rutherford MA. New approaches to studying early brain development in Down syndrome. Dev Med Child Neurol. 2019;61(8):867–79.
80. Scott J. Entwined: sisters and secrets in the silent world of artist Judith Scott. Boston, MA: Beacon Press; 2016.

81. Gleason JB, Ratner NB. The development of language. Boston, MA: Pearson Education; 2012.
82. Cooke L. Orthodoxies undermined. In: Percy A, editor. Great and mighty things: outsider art from the Jill and Sheldon Bonovitz collection. New Haven, CT: Yale University Press; 2013.
83. Lorenzon N, Musoles-Lleó J, Turrisi F, Gomis-González M, De La Torre R, Dierssen M. State-of-the-art therapy for Down syndrome. Dev Med Child Neurol. 2023;65:870.
84. Morris C, Higgs M. Judith Scott: bound and unbound. Brooklyn, NY: Brooklyn Museum; DelMonico Books-Prestel; 2014.
85. O'Toole C, Lee AS, Gibbon FE, van Bysterveldt AK, Hart NJ. Parent-mediated interventions for promoting communication and language development in young children with Down syndrome. Cochrane Database Syst Rev. 2018;10(10):Cd012089.
86. Grieco J, Pulsifer M, Seligsohn K, Skotko B, Schwartz A. Down syndrome: cognitive and behavioral functioning across the lifespan. Am J Med Genet C Semin Med Genet. 2015;169(2):135–49.
87. MacGregor JM. Metamorphosis: the fiber art of Judith Scott: the outsider artist and the experience of Down's syndrome. Oakland, CA: Creative Growth Art Center; 1999.
88. Marchini G. Judith Scott. Outsider Art Now. 2014. https://www.outsiderartnow.com/judith-scott. Accessed 6 Feb 2023.
89. Austin T. Miller American Antiques. Thomas Wister Grookett, Aged Three Years, Fox Chase, Pennsylvania, 1844. Catalog entry US Folk Art. 1844. https://usfolkart.com. Accessed 23 March 2023.
90. Gannon JR. Deaf heritage: a narrative history of deaf America. Silver Spring: National Association of the Deaf; 1981.
91. Carlin J. The Mute's Lament. In: The Instructors of the American Asylum, editor. American annals of the deaf and dumb. 1. Convention of American Instructors of the Deaf and Conference of Executives of American Schools for the Deaf: Hartford; 1848.
92. Blume S. The artificial ear: cochlear implants and the culture of deafness. New Brunswick, NJ: Rutgers University Press; 2009.
93. Mauldin L. Made to hear: cochlear implants and raising deaf children. Minneapolis, MN: University of Minnesota Press; 2016.
94. Novic S. A clearer message on cochlear implants. The New York Times. November 21, 2018.
95. Lane HL. When the mind hears: a history of the deaf. New York, NY: Random House; 1984.

Social and Environmental Modifications: Dynamics of Change

4

When comparing individuals, we find a spectrum of information about brain functioning within and across cohorts. For example, if we turn again to hearing we find that the World Health Organization estimates that by 2050 nearly 2.5 billion people will have some degree of hearing loss [1]. People with hearing loss cover a range from minimal to severe loss that might result in total deafness. Many born prelingually, profoundly deaf see their deafness as a part of who they are, how they were born. Hearing loss later in life invariably has a quite different texture. Individuals in the latter category have known sound and are generally comfortable communicating through spoken language. This distinction between our baseline physiology and acquired features extends broadly. With this distinction in mind, this chapter emphasizes acquired disorders and how one adapts to neuroatypical conditions later in life. Thus changing one's relationship to society and one's environment.

4.1 Hearing Loss: Ludwig van Beethoven

One historical individual with acquired hearing loss was Ludwig van Beethoven (1770–1827, Fig. 4.1). Despite over 250 years of discussion since his death in 1823, the cause of Beethoven's deafness remains unresolved. Part of our fascination with his creativity stems in part from our knowledge that he continued to produce awe-inspiring musical work even after going deaf. Yet, although his name is likely the first to come to mind when we think about a composer's hearing loss and deafness, he is only one of many composers with hearing loss who continued to compose. Other less well-known composers include William Boyce (1711–1779), Bedřich Smetana (1824–1884), Gabriel Fauré (1845–1924), and Ralph Vaughan Williams (1872–1958).

Before outlining Beethoven's case, it is imperative to emphasize that he placed himself in the social and environmental modifications category, frequently writing

A. Ione, *Neuroscience and Art*, Neurocultural Health and Wellbeing,
https://doi.org/10.1007/978-3-031-62336-3_4

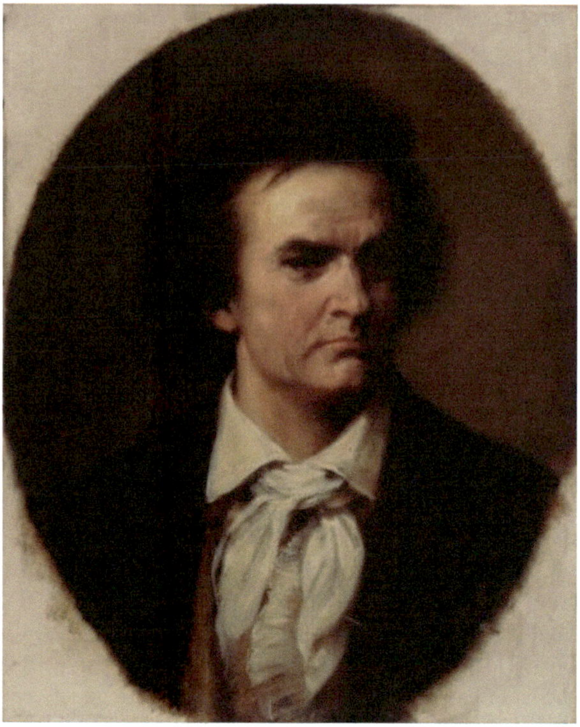

Fig. 4.1 Ludwig Van Beethoven. Artist: Henry Ulke, 1875. Oil on canvas, 26 1/4 × 21 1/4 in. (66.7 × 54.0 cm.), Courtesy of the Smithsonian American Art Museum, Gift of Titus Ulke, 1998.44

that he experienced his increasing deafness as a social problem. For example, in the Heilgenstadt Testament, penned in 1802 and discovered the day after he died in 1827, Beethoven wrote:

> Though born with an active, fiery temperament, and receptive of the diversions of society, I was soon compelled to withdraw, to live life alone ... it was impossible for me to say to people, "speak louder, shout, for I am deaf" (p. 118, [2]).

The Heilgenstadt Testament and many other contemporaneous documents reveal how much his inability to hear conversations and listen to music altered his social life and professional integration. Initially the loss of his hearing presented a difficulty in hearing sounds. Ultimately, cophosis, the most serious form of deafness developed. In cophosis it is impossible to perceive sounds and the loss of hearing capacity is total. Yet, even in cophosis, Beethoven continued composing.

Beethoven initially became deaf in his left ear, and the right was affected soon thereafter. By the time he was 31, one year after he wrote the letter quoted above, he had lost 60% of his hearing. Beginning in his forties, he used "communication notebooks" that provide us today with both medical and musical documentation. From these we know how his hearing loss was experienced biologically, how he experienced it as a composer, and how he experienced it as an individual [3].

Beethoven first noted symptoms of hearing loss around 1796. These he associated with tinnitus and a bout of typhoid fever during his youth [4, 5]. The loss of higher-frequency tones and tinnitus was joined with a depression that emerged in 1787 shortly after the death of his mother and then his sister [4]. In 1801 he contemplated suicide [6] and feared that his deafness, isolation, and illness would soon cripple his work [2]. In 1815, Louis Spohr, the leader of an Austrian orchestra and a good friend of the composer wrote that in order to communicate with the composer he now had to shout so loudly that his voice could be heard three rooms away. Beethoven stopped playing the piano publicly by 1816.

By 1817, now age 37 and still 10 years before his death, Beethoven had bilateral deafness [4]. When his Ninth Symphony was first performed in Vienna in 1824 his hearing impairment was so bad that he had to turn and look at the cheering crowd to see their enthusiastic applause. I have found no documentation to support the oft quoted statement that on his death bed the composer exclaimed: "Ich werde im Himmel hören!" ("I will hear in heaven!") but, given his oft expressed sense of loss, it is easy to think it could have happened.

Over the span of his life, Beethoven's extraordinary output included piano sonatas, string quartets, symphonies, operas, and so forth. Among his nine symphonies the *Fifth* is often cited as the most popular today. It was crafted between 1804 and 1808, so a few years after a letter written on June 29, 1800, to his personal physician Dr. F. G. Wegeler, quoted shortly. This letter, too, conveys the degree to which the composer's hearing loss was already weighing on his mind around the age of 30. Here, too, we see the social impact:

> [T]he humming in my ears continues day and night without ceasing. I may truly say that my life is a wretched one. For the last 2 years I have avoided all society, for it is impossible for me to say to people, "I am deaf." Were my profession any other, it would not so much matter, but in my profession it is a terrible thing; and my enemies, of whom there are not a few, what would they say to this? To give you an idea of this extraordinary deafness, I tell you that when at the theatre, I am obliged to lean forward close to the orchestra, in order to understand what is being said on the stage. When somewhat at a distance I cannot hear the high tones of instruments, voices. In speaking it is not surprising that there are people who have never noticed it, for as a rule I am absent-minded, and they account for it in that way. Often I can scarcely hear any one speaking to me; the tones yes, but not the actual words; yet as soon as any one shouts, it is unbearable.…I have often cursed my existence (p. 32, [7]).

Beethoven tried to adjust to his increasing loss of hearing for many years, and others attempted to come to his aid by developing therapeutic devices. Stories differ regarding how often he used these technologies, and the lack of documentation, here too, makes it difficult to decipher the facts. Two known developmental efforts are attributed to Johan Nepomuk Maelzel (1772–1838) and Conrad Graf (1782–1851).

Maelzel, a German inventor, engineer, and showman is best known for manufacturing a metronome and several music-playing automatons. He designed at least six hearing trumpets for Beethoven, each one attempting to resolve problems in earlier models. We know that the composer used them or at least tried to use them based on a report from the portrait artist August von Kloeber who mentioned that he was wearing a hearing trumpet during a 1820 conversation [5]. Beethoven offered contradictory feedback of their effectiveness.

Conrad Graf, a master piano maker and manufacturer, also tried to inventively help by constructing a "resonating plate" for use with his piano. This sound conductor was placed on the pianoforte and helped convey tones to his ear. Now lost, some question its effectiveness. Graf also constructed a special piano for the composer, now at the Beethoven House in Bonn, Germany. This instrument had quadruple strings and originally had an additional sounding board. No doubt its value was limited since Beethoven did not receive it until 1826, after he had already completed his major works [5].

An oft repeated and quite tantalizing anecdote attributed to the French physician Joachim-Agathon-Adjutor Rattel (1856–19-) claims that while working at the piano Beethoven used a wooden rod with one end clenched between his teeth and the other stretched to rest upon the sound box of the instrument. If true, this stick would have helped him perceive the instrument's vibrations [4, 5, 8]. Although unverified, it is a convincing story, and one that has a great deal of resonance with activities used within the Deaf Community as far back as the sixteenth century. Girolamo Cardano, an Italian physician who worked with the deaf, wrote about using bone conduction for low-frequency sounds in his 1521 book *De Subtilitate* [6, 9] and John Bulwer, as noted in Chap. 2, described it in *Philocophus* (1648, Figs. 2.2 and 2.3).

On the day after his death on March 27, 1827, Beethoven's associates discovered the Heilgenstadt Testament document quoted earlier, and other miscellaneous documents stored within a hidden compartment in his writing desk. The Testament is written to "my brothers Carl and [name left blank] Beethoven."[1] Dated October 6 and 10, 1802 [2], it is also addressed to the world at large, beginning "O ihr Menschen" ("O you people"). In this private testament the composer narrated his suffering in detail 25 years before his death and before he had lost all of his hearing. He notes his determination to overcome suicidal impulses and he confesses his desire to fulfill his artistic destiny.

> [W]hat a humiliation for me when someone standing next to me heard a flute in the distance and I *heard nothing* or someone heard a shepherd singing and again I heard nothing … Such experiences brought me close to despair; a little more of that and I would have been at the point of ending my life. The only thing that held me back was my art. Oh, it seemed to me impossible to leave the world until I had produced all the works that I felt the urge to compose (p. 119, [2]).[2]

What a double-edge situation, to lose one's hearing and to live only to produce musical work! Similarly paradoxical is that the music he subsequently produced

---

[1] It is not known why brother Johann's name is omitted in this testament to Beethoven's anguish.

[2] It is likely that Beethoven's references to suicide are linked to Goethe's *The Sorrows of Young Werther* [10], a book that was popular at this time. Goethe's book portrayed a hero who takes his life over the loss of his love.

was not therapeutic in a way that had the capacity to restore his hearing. Nonetheless, the activity of composing music was psychologically therapeutic.

Our knowledge that Beethoven composed part of his repertoire despite his hearing loss adds mystique to his legend, as do the abundant materials he left behind that demonstrate he retained his musical imagery. Since Beethoven retained his musical imagery, we can presume there was auditory agnosia, but no music agnosia; the former referring to the inability to recognize environmental sounds while the latter refers to the inability to recognize music (see [11]). Moreover, his writings and his compositions allow us to conceptualize how he composed and to perceive that for him composing music was an activity that included visual, tactile, and motor behaviors.

> His proverbially messy handwriting means that his music was taking shape on paper, before his eyes and ours. He didn't simply remember the sound of music, "hear it in his head" and write it down. He worked it out in intricate detail in writing because this was the way it increasingly made sense to compose. He also continued to work at the piano long after his ears had stopped receiving much benefit. The keys were still responsive, the hammers still rose and receded, and the instrument still vibrated to his touch (pp. x–xi, [8]).

Beethoven requested that following his death his favorite physician, Dr. Johann Adam Schmidt (1759–1809) describe and make public his condition [12, 13]. In 1827 Karl von Rokitansky and Johann Wagner performed the autopsy and analyzed his auricular system. Signed by Dr. Wagner, this report notes his very emaciated body and gives the evidence-based explanations for the abnormal sizes of the acoustic and facial nerves [14]. The doctors removed the tiny bones in his middle ear for future analysis, but unfortunately they were lost before any analysis was done [6].[3]

Based on this information and later research, Beethoven's death and his loss of hearing has been attributed to Cogan syndrome, Paget disease, lead toxicity, otosclerosis, syphilis, sarcoidosis, sensorineural hearing loss associated with inflammatory bowel disease, and other conditions [5–7, 13–17]. A chart covering diagnostic proposals from 1920 to 2020 shows the top choices in this one-hundred-year timeframe were otosclerosis, syphilis, and Paget's disease. Systemic lupus erythematosus (SLE), trauma, labyrinthitis and inner ear disease had the fewest supporters [16].

A sharp turn toward lead toxicity emerged in recent years. These studies relied heavily on his hair as a genomic data source, an evidentiary option that exists because many clipped locks of the composer's hair when he died as mementos. One 2007 study by Christian Reiter used chemical analysis of one of those purloined locks and a symptomatically similar case to argue for lead poisoning [15]. In 2020, Brotto et al. pointed toward a progressive sensorineural hearing loss due to lead intoxication as the most probable cause of not only Beethoven's hypoacusis but his overall health status as well [4]. If this is correct, and I am not convinced it is, the lead would add an environmental component to his development of the neurological condition.

---

[3] The short autopsy report is available in Thayer's *Life of Beethoven*, Part II (see pp. 1059–1060, [14]).

In 2021 Reiter and Prohaska concluded that medical treatment might have hastened Beethoven's demise by worsening his lead poisoning [18]. Four months before his death in March 1827, Beethoven began suffering from excessive abdominal swelling, possibly due to cirrhosis. Andreas Wawruch, his physician, decided to drain the fluid by puncturing his abdomen with a needle. According to these authors, the lead salts used to clean the wound were likely a culprit in the death, speculating that the lead worsened the composer's cirrhosis and hastened the composer's death [15]. Their analysis, also based on hair length data, established that the lead in the composer's hair spiked after each of the four abdominal punctures he received on his deathbed, draining between 7.7 and 14 liters of fluid from his body each time. In essence, their correlation suggests that the palliative measure worsened his overall situation.

A recent genomic analysis published in 2023 has added new wrinkles. Researchers first looked at eight independently sourced locks of hair attributed to Beethoven and determined five originated from a single European male. After concluding the matching samples were authentic Beethoven locks [13] the subsequent study, while unable to identify a genetic explanation for Beethoven's hearing disorder or gastrointestinal problems, did find that he had a genetic predisposition for liver disease. Furthermore, metagenomic analyses revealed that Beethoven had a hepatitis B infection in the months prior to his death.

This team concluded his genetic predispositions and his broadly accepted alcohol consumption presented plausible explanations for asserting that Beethoven's severe liver disease culminated in his death. Perhaps the most unexpected outcome of the 2023 study was finding the occurrence of an extra-pair paternity event in Ludwig van Beethoven's patrilineal ancestry. This was revealed as a part of the analysis of Y chromosomes sequenced from five living members of the Van Beethoven patrilineage [13].

Beethoven's case is among those that remind us how difficult it is to analyze someone born over 250 years ago. The details we have from his correspondence and other documentation further remind us that people who knew the world aurally before losing their hearing are more keenly aware of the sensory loss than those with congenital deafness who have little or no actual first-hand comprehension of life in a hearing world. Since hearing loss was not Beethoven's only health complaint, diagnostic speculation is additionally complicated.

Cases that intersect with aspects of Ludwig van Beethoven's case remind us that his case was more about the loss of hearing than music per se. Maurice Ravel (1875–1937) and Paul Wittgenstein (1887–1961) accentuate that creative musicians, like all cohorts, acquire a range of brain disorders. Unlike Beethoven, Ravel could hear music. He suffered aphasia, apraxia, agraphia and amusia [19]. Wittgenstein (1887–1961), who had a right arm amputated during the First World War, had various postamputation complications such as phantom limb, stump pain, and especially moving phantom [20, 21]. Nonetheless, he continued to perform. All three of these cases also speak to human creativity, first expressed through their relationship with music and then affirmed in their desire to continue musically despite obstacles their health placed in their way.

4.2 Maurice Ravel: Aphasia, Apraxia, Agraphia, Amusia

Maurice Ravel (1875–1937) was born 50 years after Beethoven's death. His musical gifts were noticed early in his life. It is also reported that his neurological health was unstable throughout his life due to neurasthenia, an anxiety neurosis marked by chronic fatigability, lack of energy and feelings of inadequacy, depression, and insomnia.[4] Before World War I his career thrived. Then Ravel's name was often coupled with that of the French composer Claude Debussy (1862–1918), an elder contemporary. Despite his neurasthenia, Ravel volunteered to serve in the war. The military deemed him disqualified due to a hernia and general weakness, but he did manage to serve in the ambulance corps.

Around 1916, the combination of war, sickness, and the emotional shock of his mother's death slowed him down, as did the subtle signs of a developing aphasia [19]. During one neurasthenic episode, in 1920, he complained of a terrible sadness [23]. This led him to leave Paris and retire from the world. In 1926, symptoms of confusion, insomnia, cloudy sensorium and intellectual inertia surfaced, possibly due to a transient ischemic attack [23].

According to Hélène Jourdan-Morhange, a violinist who often interpreted the composer's work, "at the beginning of his illness, Ravel flew into a passion when he could not find the word he sought" [24, 25]. Marguerite Long who often accompanied him on concert tours observed:

> He was constantly losing his luggage, watch, wallet, train tickets. He would put letters in his pocket and forget about them. Twice his neglecting to read or pass on mail caused unhappy situations, one when he did not acknowledge (let alone attend!) an important invitation by a prime minister. In Berlin, Ravel forgot to get off the train, leaving [Long] stranded, alone.... In the process...he had lost his train ticket to Holland. But what caused the most disruption was that he almost always forgot to pack his black leather shoes (p. 98, [26]).

Ravel's first symptoms became quite evident in 1927, when friends reported he had difficulties writing and speaking. French neurologist Théophile Alajouanine (1890–1980), who was Ravel's neurologist, and published an account of the composer's aphasia wrote up his case in 1948 [27]. He noted that his patient had lost his musical language, but not his inventiveness. Ravel's own statements suggest that his creative mind was still able to conceive musical patterns and ideas at this point. The problem was that he could no longer express the music he still heard; he could no longer put it on paper. One writer described it as "music trapped forever in his aphasic mind, Maurice Ravel, frustrated beyond words, could only cry" (p. 62, [28]).

Nonetheless, in 1928 Ravel declared himself well enough to tour America, against his physician's advice. The composer's first significant episode of amnesia resulted from travel exhaustion. This was followed by others, often lasting several

[4] Neurasthenia is an older term that makes reference to a constellation of neurological and psychological symptoms, including cognitive deficits; memory loss; insomnia; local, regional, and hemispheric weakness; fatigue, and lethargy, and so forth [22].

days. One noticeable incident was when he lost his place during a concert, while playing his *Sonatine*. He jumped from the first movement to the finale [19].

In the 1930s, as his health problems were increasing, so did his fatigue and lassitude. In October 1932 Ravel suffered a blow to the head in a taxi accident. Some teeth were broken, his head and chest were bruised, and he suffered shock. There is no evidence of any resultant neurological damage within the head, but it is common knowledge that an ostensibly minor head injury can be linked with deterioration of brain disease already present (as many assume was the case with Ravel) [19]. From 1932 until his death in 1937 his health continued to decline.

In 1933, at the age of 58, Ravel's aphasia was quite evident. By the end of that year, he was unable to sign his name, read, or even write a letter. He could spell words in his head but had forgotten how to form the letters [23, 28–30]. In November 1933, he made his last public appearance, performing *Boléro*, although the literature is undecided on whether his condition influenced the relentless repetitions in *Boléro* (1928) [19, 23–25, 28, 29]. As noted, it is known that Ravel had developed a semantic aphasia, an inability to deal with representations and symbols, abstract concepts, or categories.

We can also identify one notable exception to his general apraxia and agraphia. This was recorded and given context by his biographer Orenstein and took place during a trip to Morocco in 1935.

> The anguish and tragedy of Ravel's final years were perhaps best explained by the composer when he told Ernest Ansermet that his mind was replete with ideas, but, when he wished to write them down, they vanished. In January, 1935...Ravel was enchanted by the exotic scenery in Morocco, and unexpectedly found himself able to notate several sketches of *Morgiane*. These were to be the very last notes to come from his pen (p. 107, [29]).

Two years later Ravel's anguish was starkly exposed following a 1937 performance of *Daphnis et Chloé*, one of the last concerts Ravel attended. He began to sob: "I still have so much music in my head." Then he told Jourdan-Morhange, "I have said nothing. I have so much more to say" (p. 108, [29]).

The next year, in the autumn of 1937, Ravel's health declined sharply. On December 17 he was admitted to the clinic on rue Boileau in Paris. Two days later, a delicate brain operation, a craniotomy, was performed by the celebrated neurosurgeon Dr. Clovis Vincent, possibly to treat a subdural hematoma [19]. Ravel appeared to rally initially, but soon after he lapsed into a semi-coma. He died in the early morning hours of December 28, 1937 [29].

Like Beethoven, Ravel had a condition of uncertain etiology. Hearing loss was not a factor in Ravel's illness, rather he felt as if his music was bottled up inside of him [19] because, sadly, although he could hear music, he could no longer translate it. Even with his aphasia, apraxia, agraphia and amusia Ravel, like Beethoven, preserved most or all his auditory imagery even after his health deteriorated. But, unlike Beethoven, Ravel could hardly decipher musical signs analytically. He had no trouble recognizing them visually and was able to select a requested score at first glance without error, thus demonstrating the difference between pattern-object recognition and symbolic comprehension. This also shows he did not have agnosia

[23]. Nonetheless, it was as if he were two people: The artist who could hear the music in his head and a technician who had forgotten how to write, play the piano, and notate a score [23].

Finally, as with Beethoven, researchers have proposed several speculative diagnoses. In this case no autopsy was obtained and no records of blood pressure, urine analysis, and other laboratory tests have been found [19]. Baeck, Sacks, and Henson have asked if Ravel's illness belongs to the so-called Pick complex, which includes frontotemporal dementia, primary progressive aphasia, and corticobasal degeneration [19, 25, 31]. Early onset Alzheimer's has also been discussed [19, 27] since the slow evolution of his condition suggests a progressive degenerative complaint. Initially Alajouanine thought he suffered a form of stroke, but he later concluded that the complaint was a cerebral atrophy or degeneration [19, 27]. Both Valery Radot, Ravel's personal physician, and Alajouanine excluded systemic disease, specifically or by inference [19, 27]. One physician, noting his unusually large head considered hydrocephalus as a cause of his ailments [28, 30]. Given the absence of a necropsy we will never have a confirmed diagnosis.

4.3 Paul Wittgenstein: Phantom Movements

Paul Wittgenstein (1887–1961), a colleague of Ravel's, had a right arm amputated during the First World War and went on to become a famous left-handed concert pianist. Wittgenstein himself claimed the phantom movements of his right hand helped him to develop the dexterity of his left hand [21].[5] Erna Otten-Attermann, one of his piano students, affirmed aspects of this hypothesis during the 1930s:

> I had many occasions to see how involved his right stump was whenever we went over the fingering for a new composition. He told me many times that I should trust his choice of fingering because he felt every finger of his right hand. At times I had to sit very quietly while he would close his eyes and his stump would move constantly in an agitated manner. This was many years after the loss of his arm . . . His finger choice was always the best! (p. 298, in [21]).

Wittgenstein was one of nine children born into a wealthy, musical family. His brother, Ludwig, the philosopher, was born 2 years after him. They kept their distance. This family seems to confirm the old adage that money does not buy happiness. Three of his brothers committed suicide and two of his three sisters married men who went insane, one of whom also committed suicide. In terms of music, the story has more appeal. Visitors to the family home included Brahms, Mahler, and Richard Strauss. Indeed, Paul sometimes accompanied these illustrious composers when they visited.

[5] In the many videos online showing Wittgenstein playing, his extraordinary pace and dexterity are noticeable. Because he was filmed at an angle a viewer is apt to believe she is seeing someone using two hands.

After he was shot in the elbow during World War I and lost his right arm Paul Wittgenstein resolved to continue his career as a performer. In order to do so, he needed appropriate material, and so he asked others to write pieces for the left-hand only. His most enduring commission was Maurice Ravel's *Piano Concerto for the Left Hand*. Featuring march-like motifs and heroic themes, some view this work as echoing the darkness of war and see it as a dedication to the bravery of veterans. Wittgenstein premiered the Ravel concerto with the Vienna Symphony Orchestra on January 5, 1932, changing large portions of Ravel's work, to the composer's dismay. The two never spoke again [20].

Wittgenstein also composed works of his own [20]. Others who wrote for him included Benjamin Britten (1913–1976), Erich Korngold (1897–1957), and Sergei Prokofiev (1891–1953).

Thus, with commitment and the help of his colleagues, Wittgenstein was able to continue to perform despite various postamputation complications such as phantom limb and stump pain. His case, obviously, offered different kinds of challenges than those faced by Beethoven and Ravel but, as with them, his passion to continue musically was not muted.

An Elbert study of string musicians found that the cortical representation of the digits of the left hand of the string players they studied was larger than that in controls in a way that mirrored how these violinists used their digits while playing their instruments. These authors point out changes in the somatosensory cortex were strengthened due to how they practiced their craft. They also correlate their findings to a previous study showing humans with upper extremities amputations showed extensive cortical reorganization related to the amount of phantom limb pain experienced [32]. Although phantom limb pain is a maladaptive result of nervous system injury, and the Elbert musicians were neurotypical performers, the results nonetheless suggest that Wittgenstein's case many have involved cortical reorganization as well.

One element that sets Wittgenstein apart from Beethoven and Ravel is the degree to which he paved the way for others. Later pianists with one arm or only the use of one arm who desired to perform benefitted from the works Wittgenstein's commissioned to address his situation. Thus, Wittgenstein built an enduring place for music for the left hand on the concert platform. Among those who took this path are the German pianist Siegfried Rapp (1915–1982), who lost his right arm in World War II, and Leon Fleisher, (1928–2020), who continued a successful career as a soloist after losing the use of his right hand due to a neurological condition eventually diagnosed as focal dystonia [20].

In summary, each of these musician/composers adopted therapeutic strategies to meet his unique situation. Their ability to continue professionally also shows the degree to which available rehabilitation therapies intersect with individual aspirations and the aid of the community in developing tools to help them to circumvent challenges. Beethoven tried to find technologies to help him hear and soothe his depression, although he was unsuccessful, and these technologies did not cure his auditory loss. Still, his life indicates that his sensorium evolved as his loss of hearing increased. Ravel's combination of aphasia, apraxia, agraphia and amusia seemed to provide the fewest remedies. The case of Howard Engel, a writer who acquired *alexia sine agraphia* after a stroke (Chap. 9) is comparable to Ravel's in some ways. Notably, unlike Ravel, Engel's aphasia was sine agraphia, meaning that despite losing his ability to read, he retained the ability to write. Like many creative artists discussed throughout this book, Engel's resolve led him to develop strategies that subsequently made it possible for him to write several books despite living with aphasia. Returning to his creative profession, despite an inability to read, made a tremendous difference in his recovery.

4.4 Performing Art and the Deaf

The band Beethoven's Nightmare offers a perspective on neuroatypical challenges and environmental adaptations. Comprised of three deaf individuals, Bob Hiltermann, Steve Longo, and Ed Chevy, this band's name obviously alludes to how devastated Beethoven felt upon learning he was becoming deaf in his thirties. Many of the group's self-written songs reflect the emotions of deaf people, as well as how the band members experience living in a hearing world [33]. This band tours in the United States and has been together for over 50 years [34].

When performing, reverberation, or feeling the music, is paramount for both the audience and the band [34]. Much like Beethoven's rod may have helped him feel sound vibrations when composing, subwoofers aid their shows by reproducing low-pitched audio frequencies known as bass and sub-bass into a frequency that is lower than those which can be generated by a woofer. This sensory impact is highlighted by shining red, green, and yellow lights that convey the low, medium, and high frequencies [35]. A hearing listener at a concert recalled that it was obvious that the members of Beethoven's Nightmare kept rhythm and tempo together most effectively when they stared at each other and rocked out.

The group's drummer is German born Bob Hiltermann who went to Gallaudet University at the age of 18. There he learned Sign language and met the other two band members.[6] He was the only deaf member in a family of classical musicians. Spinal meningitis took his hearing at the age of four, although his deafness was not diagnosed until he was ten. His family simply concluded he was "slow." His repertoire now includes drums, percussions, harmonica, piano, and electric keyboard. Hiltermann is also well known for his pioneering work on stage and in film.

Guitarist Steve Longo was born profoundly deaf. Although he was unable to hear the words, it was upon "hearing" a Beatles record that he became hooked on music. He was able to hear most of the frequencies, just not the higher ones and letters like "S" and "Z." He read the lyrics to know what John and Paul were singing. After teaching himself to play the guitar, Longo plugged his headphones into his amp and used a device called an illuminator to further aid his musical education. This device

[6] John Carlin, who was discussed in Chap. 3, was instrumental in founding this school. It is a bilingual school that aims to ensure the intellectual and professional advancement of deaf and hard of hearing individuals through American Sign Language (ASL) and English.

synchs light with music: the light pulsates with the different parts and sounds of a track.

The third member, Ed Chevy, grew up in an all-deaf family. His father listened to country music and his mother used castanets to keep time. At seven, he got a hearing aid that allowed him to hear slow country rhythms and the clicking of castanets. He learned to count out beats and measures, taught himself to play the bass and started writing songs. Chevy's parents also taught him to perceive art visually and this influenced his song writing. Chevy believes that playing for deaf audiences ultimately comes down to three important elements: sound, sensitivity, and sight. He never stands still, rather he moves to the music.

There are, of course, challenges when one is a deaf musician, as Ludwig van Beethoven's case so clearly articulated. Beethoven's Nightmare, as a group, has found that one they confront in composing a new song is making sure the guitar and the bass are in the same keys. Chevy said he also has trouble hearing the pitch and the melody. To follow the changes in a song, he watches the light show and visual cues from his bandmates [35].

Just as many of Ludwig van Beethoven's colleagues worked to design hearing devices to aid Beethoven, new technologies are also helping Beethoven's Nightmare. One developer, Richard Burn, is working with them on a musical interface that will combine haptic and visual forms of feedback and will create a more inclusive experience for deaf people in general. Alongside vibrations, visual indictors will appear on a digital display that collectively forms a "sonic fingerprint" when an instrument is played, highlighting the different components that make up the sound. As he puts it:

> Traditional waveform representations are unable to truly describe what music actually sounds like, there are much more subtle characteristics of sound that may be better represented visually, for example, by using simple shapes and symbols to describe harmonic content [35].

Beethoven's Nightmare reminds us that performance and the deaf have a long history based in part on the vibration and the expressive hand and body movements that often accompanies Sign language. Pantomime, or the use of expressive movements, also speaks to a tradition within clinical neuropsychology where movement is used to assess aphasia, apraxia, kinetics, gestures, tool uses, mental rotation, to study various cortical areas, and so forth [36–39]. We also know that music acts as a mnemonics aid in therapeutic settings and may cognitively benefit both neurotypical and neuroatypical individuals. These kinds of activities have additionally aided individuals with memory impairment (e.g., from Alzheimer's) to learn and remember verbal information [40].

The National Theatre of the Deaf (NTD), founded in 1967 is also a part of the tradition of deaf individuals participating in the performing arts. One unique element of NTD's work is that both deaf and hearing performers are involved, and they strive to appeal to both deaf and hearing audiences. Deaf performers most often portray main characters, with hearing actors located on the stage's periphery voicing the main characters' lines, a technique called shadowing. Overall, deaf theater emphasizes the spatial and dancelike qualities of human communication through blending the spoken word, ASL, sign-mime, invented theatrical sign language, music, and movement that is both stylized in individual and coordinated ways.

NTD in some ways mirrors some of the known health benefits of performance overall, as mentioned earlier. To add to the above, dance movement have an influence on psychomotor development. Performance preparation also offers a form of socialization and integration with the environment [41]. In a general sense, the development of a performance demonstrates how theater use in education can foster community [42].

While deaf theater groups help us understand how sound, hearing, and deafness are interwoven, film offers a societal counterpoint worth mentioning. During the era of silent films (~1893–1929) deaf actors were in great demand due to their ease with facial and body expression. Older citizens of the Deaf Community consider the early silent films the "golden era" of film [43] because when all films were silent the deaf and hearing had equal access to what was on the screen. There was no need for third-party interpreters, special captions, elaborate sound systems, or any other assistive devices. With the development of "talkies" the deaf and hard of hearing experience was separated from that of the mainstream until interpretative aids were added toward the end of the twentieth century. Thus, with the advent of talkies, until late in the twentieth century the deaf experience was diminished because one needed help from persons who could hear to participate.

4.5 Visual Art: Stroke

Visual art, like music, offers an array of information about neuroscience, art, and creativity, particularly in terms of how acquired conditions necessitate social, environmental, and behavioral adjustments. The effects of stroke vary from person to person based on the type, severity, location, and whether there were earlier events. Left-hemisphere damage commonly provokes alterations in verbal production and comprehension, right-hemisphere stroke often leads to left-sided visuospatial neglect (also see [44–48]). Neglect patients characteristically fail to orientate, to report, or to respond to stimuli located on the contralesional side. Pairing stroke painters with their visual artwork presents easily perceived evidence of neglect and any other modification in artistic skills and compensation.

One pioneer in combining visual art and neurology was Richard Jung (1911–1986) who began to pursue research in this area after he retired as clinical director of the Freiburg Department of Neurology. Freed of his clinical duties he had more time to combine his hobby—the study of drawings, etchings, and art history—with his knowledge of brain physiology. One of his seminal articles included the cases of four German professional artists—Anton Räderscheidt, Lovis Corinth, Otto Dix, and Johannes Thiel—all of whom had right hemisphere lesions [49]. This research has since become a classic piece in the neurology and visual arts literature.

Another focused study on artists, by Bäzner and Hennerici, looked at 25 painters who showed a mean age of 63 at stroke onset. This cohort survived their first stroke for more than 8 years and the stroke incident significantly influenced the paintings of the vast majority [44]. Whereas in some cases a stroke, often a repeated stroke, ended artistic careers the majority reported resumed artistic production. In making adjustments that enable them to continue to work their cases mirror the musicians discussed earlier. Nine suffered a left hemisphere stroke and of these four had to switch from their prestroke dominant right hand to the left hand. The research also confirmed other research finding that visuospatial neglect is far more common in right hemisphere stroke (13 cases) than in left hemisphere damage [44].[7] Of the three anomalous cases reported, two were not painters before their strokes: Tommy McHugh (b. 1950) and Jon Sarkin (b. 1953). Their artistic pursuits began following the stroke damage.

McHugh's stroke was bilateral, with predominantly right hemisphere damage. He claimed that the brain injury left him obsessed with making art [44]. Sarkin's stroke occurred after he was operated on for tinnitus and vertigo. Severe postsurgery hemorrhaging led to a coma and necessitated an immediate craniotomy and removal of the hematoma and partial resection of his cerebellum. During recovery, the personalities of both men changed and their cases are discussed in more detail in Chap. 9.

Findings on visual artists who suffered strokes show that their histories and outcomes, like art itself, do not follow simplistic rules. One particularly useful element of these studies is how what is produced before and after stroke events robustly articulates the impact of functional loss. Another noteworthy element is that the artwork demonstrates that artists do not present a singular picture either before or after a devastating cerebral disability. Nonetheless, their work and experiences communicate effectively about both their neurological challenges and the desire to try to resume artistic production if possible.

A visually striking part of the stroke painter repertoire are artist self-portraits that depict their own physiological changes, and illustrate their own loss of self-resemblance. Painters such as Corinth, Dix, and Räderscheidt often show a certain degree of prosopagnosia (face blindness) [44, 51]. (Faces and prosopagnosia are discussed in more detail in Chap. 11.) Another particularly fascinating component of these studies is that despite how their artistic uniqueness remains intact, the creative work nonetheless documents their striking neurological changes.

[7] Right-sided spatial neglect is rare because there is redundant processing of the right space by both the left and right cerebral hemispheres, whereas in most left-dominant brains the left space is only processed by the right cerebral hemisphere. Although it most strikingly affects visual perception ("visual neglect"), neglect in other forms of perception can also be found, either alone or in combination with visual neglect [50].

4.6 Kurt Schwitters

One artist who captures how the complexity of a stroke patient's health profile extends beyond a single event or even repeated events is Kurt Schwitters (1887–1948). Schwitters, born in Hanover, Germany, suffered multiple strokes, multiple stresses, and epilepsy as well. Known as the founder of Merz, an art form associated with Dadaism, Schwitters suffered his first epileptic fit in 1901, at fifteen, signifying the start of a recurring condition that the artist felt continually impacted how he related to the world.

Schwitters married in 1916 and in that year was also declared unfit to serve in the military due to his epilepsy. The next year, 1917, his first son, Gert, was born and then died before his first birthday. When conscription was extended to a wider portion in of the population, in 1918, Schwitters enlisted and served as an illustrator in the army.

After the war, Schwitters began to develop his Merz style, which was dada-like in expression. In 1923 he also initiated the building of his architectural installation called the Merzhaus. He worked on this architectural piece for well over a decade. Built from found objects, one of its initial components was the death mask of his deceased son on a pedestal in a central room.

With the ascendency of the Nazi's in the 1930s, Kurt Schwitters was among the artists deemed degenerate—a label applied to various German artists who produced non-representational artworks. Schwitters was also a political target due to his severe epilepsy. In 1933 Germany passed the Law for the Prevention of Hereditarily Ill Offspring legalizing the sterilization of those considered to have hereditary diseases [22]. This legislation further endangered Schwitters' life because the law specifically included epilepsy [52]. All of these tribulations led the artist to flee to Norway in 1937, just before the onset of the second World War.

Ironically, although Schwitters' epilepsy had become a legal problem for him, as he aged his epileptic seizures decreased, and the attacks ceased around 1941 [52]. Despite this, his situation remained perilous. After the German troops invaded Norway in 1941, Schwitters became a refugee again, spending time in various internment camps in Scotland and England. During his exile in Norway, and later in Great Britain, he increasingly produced figurative drawings and paintings (primarily landscapes and portraits) to support himself [53], a style noticeably unlike his Dadaist-like Merz style. He also tried to rebuild the Merzhaus he had left behind in Hanover, Germany. However, more loss entered his life in 1943 when he learned that his original Merzhaus was destroyed by an incendiary bomb. Leaving Norway, he had to abandon a second Merzhaus and in 1943 was still intent on developing a third iteration in England.

In 1944, Schwitters suffered a right-hemisphere stroke with a subsequent paresis of his left arm during a severe bout of influenza. Following his recovery, he worked in smaller formats. By December 1945 the artist was making paintings, new collages, and sculptural objects that he described to his long-time friend Käte T. Steinitz as "better than ever before" [54]. A second stroke occurred in February 1946, due to vascular congestion. This event caused him to lose his sight for a period of 4 days and necessitated a move. Later that year he was confined to bed for several weeks due to a hip injury, the break of the neck of the femur. In July of 1947 he suffered a hemorrhage and died 6 months later in January 1948, of acute pulmonary edema and myocarditis [53].

As Schwitters' history underscores, our neurological profiles are complex and often compounded. For example, it is hard to evaluate to what degree the accumulated stresses and his history of epilepsy contributed to his strokes. What we do know is that none of his health problems diminished his will to create. We also know that the relationship between cerebrovascular disease and seizures is clearly illustrated by poststroke epilepsy. Seizures can also be the first manifestation of cerebrovascular disease and case-control studies have demonstrated that seizures carry an increased risk of subsequent stroke as well [55]. Whether epilepsy was a signal for Schwitters later health events is hard to say. Nonetheless, his case offers another example illustrating that people bring a broad range of social, environmental, physiological, and neurological spectrums to their art.

A quite different kind of case is that of the painter Jonathan Isaacson who acquired a form of cerebral achromatopsia, after a small stroke that he initially did not realize he experienced. His case is discussed after a few words about color and color-blindness.

4.7 Color-Blindness: Historical Notes

John Dalton (1766–1844), an English chemist, physicist, and meteorologist, is better known for introducing the atomic theory into chemistry than his theory on anomalous vision. He first presented the anomalous color theory in 1794 to the Manchester Literary and Philosophical Society in a paper titled "Extraordinary facts relating to the vision of colors" [56]. In it he postulated that a shortage in color perception was caused by discoloration of the liquid medium of the eyeball. His hypothesis was based on his own color-blindness and included the idea that the condition was hereditary since his brother was also color-blind.

The misleading term "color-blindness" was the invention of Sir David Brewster (1781–1868). It is said that Brewster introduced the term as a replacement for "Daltonism," a name that had taken hold as a result of Dalton's research in this area. Brewster's rationale was that no person wants to be immortalized by his imperfections [57].

Over time, Dalton's theories about color vision deficiency were disproven. A genetic analysis of his preserved eyeball in 1995, some 150 years after his death, nonetheless demonstrated that he was indeed color-blind [58]. He had *deuteranopia*, a type of congenital red-green color-blindness in which the gene for medium wavelength sensitive (green) photopsins is missing [59]. As he described the condition in his seminal paper:

> That part of the image which others call red, appears to me little more than a shade, or defect of light; after that the orange, yellow and green seem one color, which descends

> pretty uniformly from an intense to a rare yellow, making what I should call different shades of yellow [56].

Although the "discovery" of color-blindness is attributed to Dalton, we can date descriptions of the condition back to the sixteenth century. Swiss anatomist Felix Platter (1536–1614) was the first to suggest the retina and optic nerve's importance in the direct transmission of visual signals to the brain, further advancing the understanding of the human eye. He noticed instances in which people with apparently normal vision sometimes did not see red, yellow, or blue [60]. Nicolas Malebranche also noted that not everyone saw colors the same [57]. In 1684 Dawbenry Turbervile reported a case, the oculist of Salisbury, that suggests cerebral color-blindness:

> A Maid, two or three and twenty years old, came to me from *Banbury*, who could see very well, but no colour beside *Black* and *White*. She had such Scintillations by night, (with the appearances of Bulls, Bears, etc.) as terrified her very much; she could see to read, sometimes in the greatest darkness for almost a quarter of an hour (p. 1, [61]).

Robert Boyle (1627–1691), best known for Boyle's Law of gaseous electricity, also studied eye diseases and visual disturbance. His commentary on a man with blindness for unspecified colors [62, 63] was one of 14 unusual ophthalmological cases he published shortly before his death in "Observations about Vitiated Sight" [62], attached to a separate tract, *A Disquisition about the Final Causes of Natural Things* (1688) [64]. According to John Farquhar Fulton's *Bibliography of the Honourable Robert Boyle* this was one of the first instances in which case history was used in an ophthalmological treatise [65]. As noted in Chap. 2, Louis Verrey 1888 wrote on hemiachromatopsia) [66] and his work was validated through Semir Zeki's discovery of specific cells sensitive to wavelength and color toward the end of the twentieth century [67, 68].

4.8 Achromatopsia: Color-Blindness

Since John Dalton's presentation of color-blindness our knowledge has grown significantly, as has our understanding of how it manifests. More specifically, typical color-blindness, or achromatopsia, is not directly keyed to brain functionality, while cerebral achromatopsia is a brain deficit. Because not all visual disorders are fully centered within the brain, something Dalton would not have known, color vision, or its lack, offers an interesting window through which to think about how we see and understand vision and color, as well as how biological and acquired cases differ.

Earlier discussions on seeing noted that the retina is not a part of the brain's structure and essentially the optics of the eye create a focused two-dimensional image of the visual world on the retina, which then processes that image within the retina and sends nerve impulses along the optic nerve to the visual cortex to create visual perception. Color vision largely comes about due to photoreceptors in the retina of the eye known as cones. Found in the macula (the central part of the retina), these cones have light-sensitive pigments that enable us to recognize color. Cone recognition is based on wavelengths and each cone is sensitive to either red, green, or blue light (long, medium, or short wavelengths).

Normally, the pigments inside the cones register different colors and send that information through the optic nerve to the brain. This enables us to distinguish countless shades of color. Cones that lack one or more light-sensitive pigments are unable to see all colors and the result is that the relevant information is not sent to the brain. This "typical" kind of color-blindness arises from a retinal defect and is almost always partial. With this form of color-blindness, one has the inability to distinguish certain shades of color and some deficiency in differentiating colors. Typical forms of color-blindness are red-green or blue-yellow. Those with typical color-blindness find it difficult to do things like selecting ripe fruit, choosing clothing, and reading traffic lights. Red-green color-blindness is the most common form, and it affects up to 1 in 12 males (8%) and 1 in 200 females (0.5%) [69].

Cerebral achromatopsia, by contrast, is caused by damage to the cerebral cortex, as the names infers, rather than abnormalities in the cells of the eye's retina. Damage is generally localized in the visual area (V4) of the visual cortex (the major part of the color center in the brain), which receives information from the parvocellular pathway involved in color processing [70–72]. With this condition, people are uncomfortable in bright environments and have decreased visual acuity. (With rod monochromacy, one also sees the world in shades of gray [72].) Although complete color-blindness (achromatopsia or rod monochromacy) is very rare, affecting one person in thirty or forty thousand [73, 74], researchers have studied both congenital and acquired cases.

Chapter 2 also noted that the idea that the brain might reconstruct what we see and would develop a vocabulary for modalities like color and shape seemed absurd historically. To reiterate, we see this in the reaction to a paper published by Louis Verrey, a Swiss neurologist, entitled *Hémiachromatopsie droite absolue* (absolute right hemiachromatopsia), introduced earlier, describing a case of a semi-achromatopsia, or the loss of color perception in only one half of the visual field. This 1888 publication recorded his examination of a patient's lesion [66, 75]. He noted that he was able to delineate the exact area of the visual cortex that had been damaged by her stroke, and reports of problems with color perception and the determination of the lesions that caused her problems [76].

At the time, there was great resistance to this work, for it seemed to violate the idea that we are given an image world we simply "know" with all its color, form, movement, and depth. The idea of a central color area in the brain seemed nonsensical. As also noted earlier, other factors suggesting a problem with the study were at work as well. One was that the legendary neuroscientist Gordon Holmes (1876–1965) and his team produced complete topographical mapping of the visual field in the primary cortical vision center from the cases of 200 war injuries. Holmes emphatically noted that none of these showed isolated deficiencies in color perception [75–77].

Again, it was in the late 1950s and early 1960s that David Hubel and Torsten Wiesel definitively pierced the idea that perception is given to us fully formed [78, 79]. Later, Semir Zeki's discovery of specific cells sensitive to wavelength and color

[67, 68] validated both Verrey's research and the neurological condition achromatopsia he described.

In terms of individual cerebral achromatopsia cases, one artist, Megham Sims who was born with this condition speaks of exploring the richness of gray tonality in her artwork [80]. Jonathan Isaacson, who acquired cerebral achromatopsia, found that losing the chromatic versatility of the world he had always known was traumatic. In both cases, when living with cerebral achromatopsia, the individuals saw the world in terms of shades of gray. The key difference was that the congenital case, Sims, led her to become enamored with the variations offered by gray tones, all she ever knew. Isaacson, by contrast, saw the monochromatic world as dirty and disgusting after acquiring the condition [72].

4.9 The Color-Blind Painter: Jonathan Isaacson

Jonathan Isaacson (1920–?) is better known as Mr. I. or the color-blind painter through Oliver Sacks'presentation of his case [70, 71, 80]. Sacks' involvement in this case began early in March 1986 upon receiving the following letter from the artist:

> I am a rather successful artist just past 65 years of age. On January 2nd of this year I was driving my car and was hit by a small truck on the passenger side of my vehicle.
>
> When visiting the emergency room of a local hospital, I was told I had a concussion. While taking an eye examination, it was discovered that I was unable to distinguish letters or colors. The letters appeared to be Greek letters. My vision was such that everything appeared to me as viewing a black and white television screen.
>
> Within days, I could distinguish letters and my vision became that of an eagle—I can see a worm wriggling a block away. The sharpness of focus is incredible.
>
> BUT—I AM ABSOLUTELY COLOR BLIND.
>
> I have visited ophthalmologists who know nothing about this color-blind business. I have visited neurologists, to no avail. Under hypnosis I still can't distinguish colors. I have been involved in all kinds of tests. You name it.
>
> My brown dog is dark grey. Tomato juice is black. Color TV is a hodge-podge. Etc., etc. [70, 71]

Prior to becoming color-blind, Jonathan Isaacson lived for color in his artwork and in his life generally. After his accident, he aided researchers in evaluating his case by drawing visual objects and sculpting items to help illuminate the damage to his brain. Semir Zeki, one of the neurobiologists who studied his case, explained that this helped isolate the problem to color. For example, his drawings and sculpted pieces showed he retained a nearly perfect ability to reproduce forms despite his color loss [72, 81, 82].

Mr. I.'s case demonstrates that the life of an individual has a tremendous impact on how the brain develops in a number of ways [70]. Blind to color relationships in his general environment and in conceiving his artwork, Mr. I's entire life changed. One major change was that he found he preferred to venture outside when the light made it easiest for him to distinguish forms in his "gray" world. He also changed his diet, preferring foods he "knew" were "naturally" black and white.

One fascinating element of this case, and one often found when artist's lose neurological function, as shown in many cases discussed throughout this book, his initial sense of helplessness started to give way to a sense of resolution. He resolved he would paint in black and white and live in his black-and-white world as fully as he could. This resolution was strengthened by a singular experience about 5 weeks after his accident. The breakthrough came one morning as he was driving to his studio. He saw the sunrise over the highway, the blazing reds all turned into black, and he recognized that he had a perception of the world that was unique to him. As he described it:

> The sun rose like a bomb, like some enormous nuclear explosion, he said later. 'Had anyone ever seen a sunrise in this way before?' Inspired by the sunrise, he started painting again — he started, indeed, with a black-and-white painting that he called Nuclear Sunrise…'I felt if I couldn't go on painting,' he said later, 'I wouldn't want to go on at all' (p. 14, [70]).

After a year or more of experiment and uncertainty, Mr. I. moved into a strong and productive phase, as strong and productive as anything in his long artistic career. People applauded the black-and-white paintings he now produced and commented on what they characterized as a creative renewal when seeing this new "phase" he had "moved" into. The excitement people felt when seeing his monochromes is perhaps most meaningful when we reflect on the fact that very few people knew that this new phase was anything other than an expression of his artistic development. They failed to recognize that it was brought about by a calamitous neural loss [70].

Four aspects of his case are particularly noteworthy neurologically. First, since he continued to paint, researchers were able to use his work to explore how a brain could adapt to radically new modes of expression and ways of seeing. Second, although Sacks believes this kind of case allows us to assess how the cerebral "mapping" may be drastically reorganized and revised in cases of the special use or disuse of individual parts, we need not necessarily draw this conclusion. Clearly his brain did adapt. Yet, too, these adaptations did not bring color back into his memory or visual experience. For example, he had lost his color synesthesia, meaning that musical notes that had formerly evoked particular colors in him prior to the accident no longer did so. The synesthesia never returned.

Third, Mr. I.'s cerebral achromatopsia stretched beyond the definition of color agnosia. Color agnosia concerns the inability to recognize colors despite intact color perception, semantic memory for color information, and color naming. In Mr. I.'s case, he could no longer see, remember, or even imagine colors. "He knew all the names of the colors in his favorite paintings, but could no longer see them, either when he looked at them physically or in his mind's eye. He knew them, now, only by verbal memory, much like John Carlin's poetry 'explained' the sounds of phenomena he had never heard" (p. 7, [70]).

Fourth, after working with monochromatic vision for several years, surgery was proposed to Mr. I. Although it appeared surgeons may have developed techniques that could restore his color vision, he declined the opportunity to possibly return to a colorful world. He said that after becoming comfortable living monochromatically, he did not want to undergo another period of transition [70].

4.10 Conclusion

All of these cases give us insight into neuroscience and art, as well as the ongoing creative process artists use to adjust as their lives continue. These cases show a commitment to continue producing and remind us that art is not simply about aesthetic objects. Henri Matisse (1869–1954) offers a brief closing example that further confirms the desire to continue one's practice is evident in all stages of life, and frequently remains even when accompanied by physical and emotion challenges.

Matisse's health problems plagued him from childhood, and he decided to become a painter while in the hospital with a violent abdominal pain attack in 1889, at the age of 20. His medical history shows that his life-long somatic and psychosomatic health problems were multifaceted. They influenced his art, and his art helped him cope with these problems [83, 84]. For example, after he was partially disabled by surgeries at 71, he spent the rest of his life in bed or a wheelchair. Unable to stand at an easel, he produced delightful paper cut-outs with scissors [83, 84] and painted with a specially extended brush [85]. The brilliantly colorful paper cut-outs that Matisse made in his last decade offer another example of how artistic vitality aided him in overcoming his physical handicap. Indeed, Matisse developed a new artistic style in response to health problems [83].

As this chapter showed, artists, like all people, bring a range of skill sets and backgrounds to their efforts to cope with neurological events. In terms of neuroscience and art we find examples where new tools were developed and efforts were made to understand physiology and psychological process through their varied experiences. Part II presents a series of individual and thematic case studies. It opens with two chapters that contrast the work and conditions of two storytellers, Iris Murdoch and Oliver Sacks. Murdoch, a novelist/philosopher and a cultural icon, invented stories. Sacks, a neurologist, introduced the narratives of his patients to a larger public. Their health challenges and derivatives of their work are used to frame each in terms of their inspirations, aspirations, and accomplishments. The case studies presented in Chaps. 7, 8, and 9 are thematic in tone. Each offers a neurocultural examination of a topical area. These topics include technology, war, and therapeutics.

References

1. World Health Organization. Deafness and hearing loss. Geneva: World Health Organization; 2021.
2. Lockwood L. Beethoven: the music and the life. New York, NY: W. W. Norton; 2005.
3. Binns C, Low WY. Hearing loss in the 21st century and the legacy of Beethoven. Asia Pac J Public Health. 2020;32(5):224–5.
4. Brotto D, Fellin R, Sorrentino F, Gheller F, Trevisi P, Bovo R. A modern case sheds light on a classical enigma: Beethoven's deafness. Laryngoscope. 2021;131(1):179–85.
5. Thayer AW, Forbes E, Deiters H, Riemann H, Krehbiel HE. Thayer's life of Beethoven. Princeton, NJ: Princeton University Press; 1967.
6. Perciaccante A, Coralli A, Bauman NG. Beethoven: his hearing loss and his hearing aids. Otol Neurotol. 2020;41(9):1305–8.
7. Beethoven LV. Beethoven's letters: a critical edition: with explanatory notes. New York, NY: J.M. Dent; 1909.

8. Wallace R. Hearing Beethoven: a story of musical loss and discovery. Chicago, IL/London: The University of Chicago Press; 2018.
9. Cardano G. The De subtilitate of Girolamo Cardano. Tempe, AZ: Arizona Center for Medieval and Renaissance Studies; 1521/2013.
10. Goethe JWv. The sorrows of young Werther. Oxford: Oxford University Press; 2012/1774.
11. Vignolo LA. Music agnosia and auditory agnosia. Dissociations in stroke patients. Ann N Y Acad Sci. 2003;999:50–7.
12. Anderson E. The letters of Beethoven, vol. 3 vol. New York, NY: St. Martin's Press; 1961.
13. Begg TJA, Schmidt A, Kocher A, Larmuseau MHD, Runfeldt G, Maier PA, et al. Genomic analyses of hair from Ludwig van Beethoven. Curr Biol. 2023;33:1–17.
14. Wagner J, Rokitansky KV. Thayer's life of Beethoven, part II, appendix B, autopsy. In: Forbes E, editor. Thayer's life of Beethoven, part II. Princeton, NJ: Princeton University Press; 1967. p. 1059–60.
15. Reiter C. The causes of Beethoven's death and his locks of hair: a forensic-toxicological investigation. Beethoven J. 2007;22(1):2.
16. Thomas JP, Dazert S, Prescher A, Voelter C. Aetiology of Ludwig van Beethoven's hearing impairment: hypotheses over the past 100 years—a systematic review. Eur Arch Otorrinolaringol. 2021;278(8):2703–12.
17. Oiseth SJ. Beethoven's autopsy revisited: a pathologist sounds a final note. J Med Biogr. 2017;25(3):139–47.
18. Reiter C, Prohaska T. Beethoven's death—the result of medical malpractice? Wien Med Wochenschr (1946). 2021;171(15–16):356.
19. Henson RA. Maurice Ravel's illness: a tragedy of lost creativity. Br Med J (Clin Res Ed). 1988;296(6636):1585–8.
20. Boller F, Bogousslavsky J. Paul Wittgenstein's right arm and his phantom: the saga of a famous concert pianist and his amputation. Prog Brain Res. 2015;216:293–303.
21. Tatu L, Bogousslavsky J, Boller F. Phantoms in artists: the lost limbs of Blaise Cendrars, Arthur Rimbaud, and Paul Wittgenstein. J Hist Neurosci. 2014;23(4):355–66.
22. Scull A. Cultural sociology of mental illness: an A-to-Z guide. Thousand Oaks, CA: Sage Reference; 2014.
23. Cytowic RE. Aphasia in Maurice Ravel. Bull Los Angel Neurol Soc. 1976;41(3):109–14.
24. Vitturi BK, Sanvito WL. Maurice Ravel's dementia: the silence of a genius. Arq Neuropsiquiatr. 2019;77(2):136–8.
25. Baeck E, editor. The terminal illness and last compositions of Maurice Ravel. Basel: Front Neurol Neurosci; 2005.
26. Dunoyer C. Marguerite Long: a life in French music, 1874–1966. Bloomington/Indianapolis: Indiana University Press; 1993.
27. Alajouanine T. Aphasia and artistic realization. Brain. 1948;71(Pt. 3):229–41.
28. Seddon AM. Music inexpressible: the tragedy of Maurice Ravel. Med Probl Perform Art. 1995;10(2):62–5.
29. Orenstein A. Ravel: man and musician. New York, NY: Dover Publications; 1991.
30. Kanat A, Kayaci S, Yazar U, Yilmaz A. What makes Maurice Ravel's deadly craniotomy interesting? Concerns of one of the most famous craniotomies in history. Acta Neurochir. 2010;152(4):737–42.
31. Sacks O. Musicophilia: tales of music and the brain. New York, NY: Alfred A. Knopf; 2010.
32. Elbert T, Pantev C, Wienbruch C, Rockstroh B, Taub E. Increased cortical representation of the fingers of the left hand in string players. Science. 1995;270(5234):305–7.
33. Lawrence Livermore National Laboratory. "Beethoven's Nightmare"—a dream turned reality for Longo. Livermore, CA: LLNL News; 2007.
34. Perry A. No music, no life: deaf rockers in Beethoven's Nightmare feel the beat. Westword Insiders; 2019. https://www.westword.com. Accessed 23 March 2023.
35. Dose. New visions: how the world's first deaf rock band feels music. Medium [Internet] 2017 April 25, 2023.

36. Emmorey K, McCullough S, Mehta S, Ponto LLB, Grabowski TJ. Sign language and pantomime production differentially engage frontal and parietal cortices. Lang Cogn Process. 2011;26(7):878–901.
37. Goldenberg G. Facets of pantomime. J Int Neuropsychol Soc. 2017;23(2):121–7.
38. Helmich I, Meyer C, Voelk M, Coenen J, Mueller S, Schepmann J, Lausberg H. The pantomime of mental rotation: left-handers are less lateralized. Neuropsychologia. 2022;176:108385.
39. Niessen E, Fink GR, Weiss PH. Apraxia, pantomime and the parietal cortex. NeuroImage Clin. 2014;5:42–52.
40. Derks-Dijkman MW, Schaefer RS, Kessels RPC. Musical mnemonics in cognitively unimpaired individuals and individuals with Alzheimer's dementia: a systematic review. Neuropsychol Rev. 2024;34(2):455–77.
41. Pelc Z. Therapeutic values of dance movement and its influence on psychomotor development of deaf persons as a form of socialization and integration with the environment. Wiad Lek. 2002;55 Suppl 1(Pt 2):845–9.
42. Meath-Lang B. Dramatic interactions: theater work and the formation of learning communities. Am Ann Deaf. 1997;142(2):99–101.
43. Schuchman JS. The silent film era: silent films, NAD films, and the deaf community's response. Sign Language Studies. 2004;4(3):231–8.
44. Bäzner H, Hennerici M. Stroke in painters. International review of neurobiology, vol. 74. San Diego, CA: Academic Press; 2006. p. 165–91.
45. Chatterjee A. The neuropsychology of visual artistic production. Neuropsychologia. 2004;42(11):1568–83.
46. Chatterjee A. Neuropsychology of art and aesthetics. In: Huston JP, Nadal M, Mora F, Agnati LF, Cela Conde CJ, editors. Art, aesthetics, and the brain. Oxford: Oxford University Press; 2015. p. 341–56.
47. Piechowski-Jozwiak B, Bogousslavsky J. Painting in neurology. In: Colombo B, editor. Brain and art: from aesthetics to therapeutics. Cham: Springer International Publishing; 2020. p. 41–52.
48. Zaidel DW. Neuropsychology of art: neurological, cognitive, and evolutionary perspectives. 2nd ed. Hove England/New York: Psychology Press; 2016.
49. Jung R. Neuropsychologie und Neurophysiologie des Kontur- und Formsehens in Zeichnung und Malerei. In: Wieck HH, editor. Psychopathologie musischer Gestaltungen. Stuttgart/New York, NY: Schattauer; 1974. p. 29–88.
50. Giuseppe VM. Spatial neglect. J Neurol Neurosurg Psychiatry. 2015;86(9):e3.
51. Bäzner H, Hennerici MG. Painting after right-hemisphere stroke—case studies of professional artists. Front Neurol Neurosci. 2007;22:1–13.
52. Webster G. Kurt Schwitters' Merzbau. PhD dissertation. Open University, Milton Keynes; 2007.
53. Kurt and Ernst Schwitters Foundation. Kurt Schwitters biography. Hannover: The Sprengel Museum; 2019.
54. Steinitz KT. Schwitters and America. In: Schmalenbach W, editor. Kurt Schwitters: a retrospective exhibition. Dallas, TX: Dallas Museum of Fine Arts; 1967. p. 60.
55. Zelano J, Larsson D, Kumlien E, Åsberg S. Pre-stroke seizures: a nationwide register-based investigation. Seizure. 2017;49:25–9.
56. Dalton J. Extraordinary facts relating to the vision of colours: with observations. Memoirs of the Literary and Philosophical Society of Manchester; 1798.
57. Wade NJ. A natural history of vision. Cambridge: MIT Press; 1998.
58. BBC News. In pictures: the life and work of John Dalton (1766–1844): colour blindness. Boyle R. A disquisition about the final causes of natural things: wherein it is inquir'd, whether, and (if at all) with what cautions a naturalist should admit them? H.C.BBC News; 2011. http://news.bbc.co.uk. Accessed 12 Apr 2023.
59. Regan B. John Dalton's colour vision legacy. Br J Ophthalmol. 1998;82(2):203.
60. Koelbing HM. Renaissance der Augenheilkunde 1540–1630. Bern, Stuttgart: Huber; 1967.

61. Turbervile D. Two letters from the great, and experienced oculist Dr. Turbervile of Salisbury, to Mr. William Musgrave S. P. S. of Oxon, containing several remarkable cases in physick, relating chiefly to the eyes. Philos Trans R Soc Lond. 1684;14(164):736–8.
62. Boyle R. Some uncommon observations about vitiated sight. J. Taylor; 1688.
63. Hunter RA, Rose FC. Robert Boyle's "uncommon observations" about vitiated sight (London 1688). Br J Ophthalmol. 1958;42(12):726–31.
64. Boyle R. A disquisition about the final causes of natural things: wherein it is inquir'd, whether, and (if at all) with what cautions a naturalist should admit them? H.C.; 1688.
65. Fulton JF. A bibliography of the Honourable Robert Boyle. Oxford: Printed at the University press by J. Johnson; 1932.
66. Verrey L. Hémiachromatopsie droite absolue, conservation partielle de la perception lumineuse et des formes, ancien kyste hémorrhagique de la partie inférieure du lobe occipital gauche, par le Dr Verrey. G. Steinheil; 1893.
67. Zeki SM. Representation of central visual fields in prestriate cortex of monkey. Brain Res. 1969;14(2):271–91.
68. Zeki SM. Cortical projections from two prestriate areas in the monkey. Brain Res. 1971;34(1):19–35.
69. National Eye Institute. At a glance: color blindness. Washington, DC: National Institute of Health; 2019.
70. Sacks O. The case of the colorblind painter. An anthropologist on Mars: seven paradoxical tales. New York, NY: Vintage; 1995. p. 3–41.
71. Sacks O, Wasserman R. The case of the colorblind painter, vol. 34. New York Review of Books; 1987. p. 25–34.
72. Zeki S. Inner vision: an exploration of art and the brain. Oxford: Oxford University Press; 1999.
73. Sacks O. The island of the colorblind and cycad island. New York, NY: Alfred A. Knopf; 1996.
74. Thiadens AA, den Hollander AI, Roosing S, Nabuurs SB, Zekveld-Vroon RC, Collin RW, et al. Homozygosity mapping reveals PDE6C mutations in patients with early-onset cone photoreceptor disorders. Am J Hum Genet. 2009;85(2):240–7.
75. Zeki S. Improbable areas in color vision. In: Chalupa LM, Werner JS, editors. The visual neurosciences, 2-vol set. Cambridge: The MIT Press; 2003.
76. Sacks O. The river of consciousness. New York, NY: Alfred A. Knopf; 2017.
77. Fishman RS. Gordon Holmes, the cortical retina, and the wounds of war. Doc Ophthalmol. 1997;93(1):9–28.
78. Hubel DH, Wiesel TN. Receptive fields of single neurones in the cat's striate cortex. J Physiol. 1959;148(3):574–91.
79. Hubel DH, Wiesel TN. Receptive fields, binocular interaction and functional architecture in the cat's visual cortex. J Physiol. 1962;160(1):106–54.
80. Meghan S. Meghan Sims: Achromat. 2023. https://meghansims.ca/. Accessed 3 March 2023.
81. Zeki S. Art and the brain. Daedalus. 1998;127(2):71.
82. Zeki S. Art and the brain. J Conscious Stud. 1999;6(6–7):160.
83. Lacayo R. What powers artists who reach old age: the work. The Washington Post. December 6, 2022.
84. Lacayo R. Last light: how six great artists made old age a time of triumph. New York, NY: Simon & Schuster; 2022.
85. Zeidler H. Henri Matisse's medical history: multiple health problems and impact on creativity. J Med Biogr. 2021;29(2):63–70.

Part II

Case Studies

A Luminous Presence: Iris Murdoch

5

Novelist/philosopher Iris Murdoch died on February 8, 1999, at the age of 79 after becoming ill with a chest infection, having suffered for many years with Alzheimer's disease. After she died, her obituary by Richard Nicholls appeared on the front page of *The New Times*, a placement that underscored that her death was seen as a noteworthy cultural event [1]. In his tribute to her, Nicholls described Murdoch as a prodigiously inventive and idiosyncratic British writer whose 26 novels offered lively plots, complex characters, and intellectual speculation. He then added that despite her training in philosophy she did not see fiction as another and lesser way of dealing with philosophical questions. Rather, he noted, she argued that literature was meant for enjoyment and that the art of the tale was "a fundamental form of thought" in its own right [1].

In the *Guardian's* obituary, Murdoch's biographer Peter Conradi likewise underscored her cultural impact. He said that her best novels combine Dostoevsky with Shakespearian romance and love-comedy. He added that her achievement was not a modest one. According to Conradi, Murdoch had kept the traditional novel alive, and in so doing changed what it is capable of achieving. Conradi described her as "one of the best and most influential writers of the 20th century" [2].

Her relationship with art extended beyond the fictional work she published. For example, she taught at the Royal College of Art for 5 years. She also frequently included music and paintings in her books. During an interview Murdoch explained that she used painting more often than music because she knew more about it and did once very much want to be a painter [3]. Murdoch's case also adds to our understanding of brain functionality because her brain was studied after her death and its degeneration from Alzheimer's disease was correlated with the syntax in handwritten manuscripts she penned for her novels over the course of her career. From this research it was evident that her mental capacity was on the decline while writing *Jackson's Dilemma*, her final novel. Her philosophical writings, particularly her

A. Ione, *Neuroscience and Art*, Neurocultural Health and Wellbeing,
https://doi.org/10.1007/978-3-031-62336-3_5

academic expertise on Plato's views of art, are also included below. These ideas build on the historical linkages among neuroscience, the arts, and neuroculture discussed earlier.

5.1 Dame (Jean) Iris Murdoch

Anglo-Irish novelist/philosopher Dame Iris Murdoch (1919–1999) was born in Dublin in 1919 and her family moved to London a few weeks later. After reading classics, ancient history, and philosophy at Somerville College, Oxford, she worked as a civil servant in wartime London, as a United Nations refugee worker in postwar Europe, and joined the Communist party. In 1947 and 1948 she studied philosophy at Cambridge and then, in 1948, she returned to Oxford as a fellow of St Anne's College, where she taught philosophy until 1963. According to her biographer, it is said that the people she met during this time frequently became characters in her novels, although she denied her portraits were directly drawn from life [4].

Throughout Murdoch's life her achievements were in many ways anomalous. Even after she stopped teaching philosophy, Murdoch continued to publish in this area. Succeeding in philosophy was a rare accomplishment for a woman of her time. Moreover, whereas writing philosophy and fiction were pursued in France by illustrious figures like Jean-Paul Sartre (1905–1980) and Albert Camus (1913–1960), this was not the case in England. Perhaps Iris did so because she knew from a young age that she wanted to write. She started writing stories at about nine or ten, although she assumed she would eventually do "something else" as an adult. Philosophy was unforeseen and became the something else for a time, and then continued in tandem with her career as a novelist. Once established in both domains she said that "it is more fun to be an artist than a philosopher. Literature could be called a disciplined technique for arousing certain emotions" [5].

Iris Murdoch met John Bayley (1925–2015), her future husband and a Professor of English at Oxford, around the time she was crafting her first published novel, *Under the Net* in 1954 [6]. They married in 1956. One early health challenge entered her life around this time as well. She had been diagnosed as partially deaf in March 1954, in her mid-thirties. Murdoch's reaction was recorded in her journal: "It is 'real' deafness, failure of reception. Likely to get worse & no cure except amplifiers & lip-reading. I went for my first lip-reading lesson yesterday with a comic woman in the Radcliffe" (p. 482, in [4]).

Whether her loss of hearing impacted the later onset of her dementia is hard to know. Studies do suggest that subjects with hearing loss are more likely to develop Alzheimer's disease [7–9] but there is no consensus. Moreover, one recent study found that older adults with greater severity of hearing loss were more likely to have dementia, and the likelihood of dementia was lower among hearing aid users compared to non-users [10]. We also know from her husband John Bayley's memoirs that Iris Murdoch's mother had Alzheimer's, although neither Bayley nor Iris gave much thought to the genetics. Apparently, the specialists they consulted during her mother's decline did not mention the possibility either [11].

It is difficult to pinpoint exactly when the onset of Alzheimer's disease was evident in Iris Murdoch's life. It appears she showed no symptoms in 1987 when Tom Phillips painted her portrait. During their sittings and subsequent meetings Phillips and Murdoch discussed everything from socialism to school songs [12]. When Murdoch sat for Phillips, he described her as "a luminous presence…an electric light-bulb in that gloomy corner, glowing, casting out darkness" [13]. The final meeting between Iris Murdoch and Tom Phillips took place at a St Catherine's Feast in the late 1990s a decade or so later [12].

> Although Murdoch was, by this time, increasingly affected by Alzheimer's disease and often struggled to recognize her friends, she told Phillips straightaway: "Of course I know you, you are in a famous painting … in a portrait." Phillips was deeply touched; he has described how her face lighted up with 'all the old radiance' and he saw once again her "luminous beauty." Her reversal of the roles of artist and sitter indicates the confusion brought about by Alzheimer's, but it also signals her enduring preference for the role of scrutinizer, rather than that of the scrutinized [12].

Phillips included a detail from Titian's *The Flaying of Marsyas* in the background of the Murdoch portrait because it was her favorite painting. *Marsyas* was one of the final paintings created by this late Italian Renaissance artist and it was likely painted between about 1570 and his death in 1576, when he was in his nineties.[1] Throughout her life, Murdoch said that she was drawn to the ambiguities of this artwork that depicts the consequences incurred by Marsyas, the loser in a famed music contest with the god Apollo, as recorded by Ovid. According to the story, after Marsyas lost the contest, the flaying was his punishment. It was also among the subjects she and Phillips discussed during their conversations as he painted.

Iris saw the actual Titian painting when it was exhibited in London in 1984, and she had called it "the greatest in the Western canon" on many occasions. It was a part of an exhibition on Venetian art of the sixteenth century. At that time, the art critic Lawrence Gowing wrote that it was clear that Murdoch was not the only one in London under the spell of this ambiguous Titian masterpiece. He added: "At most hours on most days there is a knot of visitors riveted and fairly perplexed in front of it. It is a controversial picture…how it is possible that a horribly painful subject should be the occasion of beauty or greatness in art?" (pp. 13–4, [15]).

While we know of her regard for ambiguity and that she was taken by the ambiguity of the beauty juxtaposed with cruelty that Titian inscribed in this work, we do not know precisely what drew her to include *The Flaying of Marsyas* in her last book, *Jackson's Dilemma*, penned during the early onset of her Alzheimer's. In this book, the painting comes up during a conversation on Jackson's suffering, when one of the characters, Owen Silbury, says that artists know about suffering and then he

[1] It is hard to accurately say what Titian's age was. He liked to present himself as an elderly painter and gave contradictory statements about his age and date of birth [14].

adds, "How Titian must have felt it, when he was very old, *The Flaying of Marsyas*—the pain, the pain, the old man must have felt it deeply at the end..." [16].[2]

Very public signals of Murdoch's memory loss began to appear in the 1990s, even before she wrote *Jackson's Dilemma*, and before her Alzheimer's was officially diagnosed in 1997. Her biographer notes that even by 1993 her diary entries were reduced to heart-rending simplicity [4]. A 1994 episode mentioned in John Bayley's first memoir [11] recalls one memory lapse as her cognitive impairment was coming to the fore. She and Bayley were invited to Israel by the University of Negev to participate in a conference. Rather than present a paper, Iris opted to take part in a discussion of her novels and philosophical writings. Everything went all wrong. She was completely unable to summon the words she wanted. Many in the audience walked out and it quickly became paralyzing and embarrassing for all involved—except for Iris who seemed not to notice. Bayley thought she would explain what happened afterwards but, to his surprise, she seemed completely unaware of the fiasco [11].

The following year, 1995, while working on, *Jackson's Dilemma*, her last novel, she often mentioned writer's block, an unusual trait for her. Bayley's memoir tells us that this led her to change the book's name. She had initially intended to call it *Jackson*. Due to difficulties she was having getting a hold of his character, Iris ultimately decided on *Jackson's Dilemma* [17], in effect underscoring her own struggles.

Upon publication many respectfully reviewed her book, possibly in response to the longevity of her impressive career. A few critics expressed disappointment with the strangely altered quality of the language used. These reactions offer additional commentary on the influence of Alzheimer's on her mental faculties:

> A. S. Byatt, for instance, compares the structure of the novel to "... an Indian Rope Trick ... in which all the people have no selves and therefore there is no story and no novel," while Penelope Fitzgerald noted that the economy of the writing made it appear "...as though Murdoch had let her fiction wear through almost to transparency." Other critics were less oblique: Kate Kellaway found *Jackson's Dilemma* "... not a perfect novel: the narrative itself is, at times, a little distrait: like Jackson [a central character], it often moves with scant explanation," and Hugo Barnacle described it as reading " ... like the work of a 13-year-old schoolgirl who doesn't get out enough" (p. 251, in [18]).

Murdoch's biographer Peter Conradi recalled that the next year, 1996, she was intensely depressed and again lamented her inability to write. At one point she remarked that "singing *rescues* me" (p. 590, in [18]). He also notes that when she

---

[2] She had also referred to the work in earlier novels: "Murdoch never describes the *Flaying of Marsyas* explicitly, knowing that the attempt to do so would risk dangerously simplifying its meaning. She weaves a web of allusions which permit the painting to retain its complexity and ambiguity, so that it can point beyond itself towards a truth which can only be articulated indirectly. In *A Fairly Honourable Defeat*, for instance, Simon Foster sees the painting as emblematic of his relationship with his caustic lover Axel Nilssen. Later on it is referenced in *The Good Apprentice,* when the psychiatrist Thomas McCaskerville likens his patient Edward Baltram to Marsyas, and in *Jackson's Dilemma* when the painter Owen Silbury observes that the *Flaying of Marsyas* reveals Titian's remorse and great pain. Most significantly, *The Flaying of Marsyas* permeates *The Black Prince,* whose protagonist Bradley Pearson is flayed and goes through a process of redemptive suffering" [12].

got anxious, they would sing all kinds of songs together—Anglican hymns, Irish airs, folk songs, ballads, French *chansons*, and songs from the 1930s [4]. This calming effect correlates with studies that show people with Alzheimer's are able to participate in group singing and they likely derive benefit from the activity [19, 20].

Later, in May 1997, Conradi relates how her memory loss began to have a greater impact on how he saw his task as her biographer. To his surprise, Conradi realized he was asking himself if he was becoming her receding memory bank. As she was gradually forgetting her past, he was discovering and recording her history through her writings, journals, and correspondence. In addition, Conradi mentions that she was aware that her state of mind had diminished. In December 1997, a year or so before her death, she mentioned, "with grim exactitude, a friend she had heard nothing from 'since I began sailing away into darkness'" (p. 589, in [4]). Conradi also tells us that at this point, when language and memory deserted her, she would kiss the hands of friends or bow with her hands together, as in prayer [4]. Robert Weil similarly wrote that her abundant smiles always communicated her emotions when her words became unintelligible [21]. The idea of sailing into the darkness arose again at the end of John Bayley's second memoir on Iris, *Elegy for Iris,* when he described her death: "She is not sailing into the dark. The voyage is over, and under the dark escort of Alzheimer's she has arrived somewhere. So have I" (p. 266, [11]).

5.2 Studies of Iris Murdoch's Brain

As the progressive nature of her Alzheimer's condition became obvious Iris Murdoch underwent neurological and neuropsychological assessment in the period between November 1996 and June 1997, during which her condition deteriorated markedly. In early November 1996, she achieved 20/30 on a Folstein Mini-Mental State Examination (MMSE), with points lost mainly on tests of attention and orientation.[3] By the following summer, her MMSE score had dropped to 10/30. She could identify her surroundings only by the name of the city (Oxford), was unable to perform serial subtraction or reversed spelling, and could register but not recall a three-item word list [18].

An MRI brain scan performed in June 1997 (Fig. 5.1) showed changes of global atrophy with profound bilateral hippocampal shrinkage and a suggestion of temporal lobe predominance [18]. After she died in February 1999, at the age of 79 years, the presumptive diagnosis of Alzheimer's disease was confirmed by post-mortem as Braak stage 4–5 [18].[4] Microscopical examination demonstrated features of

[3] The MMSE is a 30-point questionnaire that is used extensively in clinical and research settings to measure cognitive impairment. It is commonly used in medicine and allied health to screen for dementia.

[4] Staging in Alzheimer's disease was described by Braak in 1991. Braak stages 1 and 2 are used when neurofibrillary tangle involvement is confined mainly to the transentorhinal region of the brain. Stages 3 and 4 denote involvement of limbic regions such as the hippocampus. Stages 5 and 6 imply extensive neocortical involvement [22].

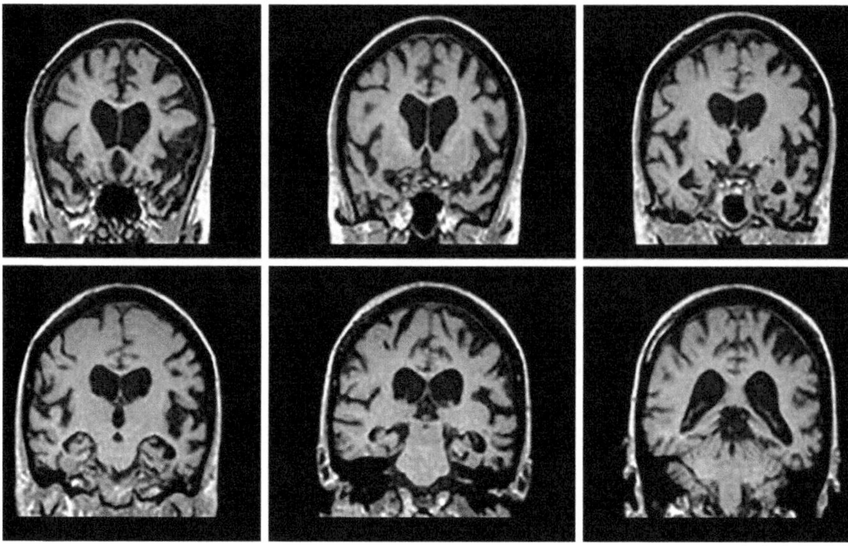

Fig. 5.1 An MRI brain scan performed on Iris Murdoch's brain in June 1997. Sample T1-weighted images from I.M.'s MRI (left hemisphere on right) showing bilateral hippocampal atrophy, moderate to severe global volume loss, with a temporal and parietal predominance. From *Brain*, Volume 128, Issue 2, February 2005, pages 250–260, https://doi.org/10.1093/brain/awh341 [18] with permission

Alzheimer's disease throughout the brain, with disproportionate deposition of plaques and tangles, as well as gliosis and spongiosis, in the temporal regions [18]. In conjunction with the physical analysis, research on her writing was conducted to determine whether her use of written language in her last novel, *Jackson's Dilemma*, also revealed signs of her Alzheimer's, although it was published in 1995 before her Alzheimer's diagnosis.

Garrard, who led the research team, explained that Murdoch was an excellent candidate for this type of study because she was known to write only in longhand, with few revisions of passages. Upon completion she sent the longhand manuscripts to her publishers with little allowance for editorial interference [18, 23]. She outlined her process to Jeffrey Meyers in an interview:

> Iris explained that she planned her novels in great detail—characters, scenes, the overall structure. Only when she had the shape of the book and the people in it very clear in her mind did she begin to write. She said, "I find inventing a novel very difficult. I can write it quite easily once I've invented it." …I make two complete drafts and then revise it, paying extra attention to every word. I like writing. I enjoy the English language, and constructing sentences. I like telling stories (p. 15, [3]).

Garrard's team analyzed three books, aided by special software capable of converting books into a digital format. One was *Under the Net* (1954), Murdoch's first published novel. *The Sea, The Sea* (1978) was the second of this group, written in the prime of her career. It received the prestigious Booker Prize. The third book was her final publication, *Jackson's Dilemma* (1995), completed a year before the Alzheimer's diagnosis. Even without analysis it was noticeable that *Jackson's Dilemma* was much shorter than her earlier books [16, 18, 23].

After digitization the researchers transformed each book into word lists that allowed them to evaluate the frequency of each word by word-type. The number of word-types allowed them to measure the variety of the vocabulary. When word types are examined at successive points in a text, they reflect the rate at which an artist, in this case Murdoch, introduces new words. Her spontaneous novel writing was also compared with her cognitive diagnostic tests, correspondence, and diary entries.

Overall, the study determined that the structure and grammar of Murdoch's writing remained roughly consistent throughout her career, but her last novel showed her vocabulary had dwindled, and the language was simplified. The enrichment of vocabulary between the early and middle stages of her career was mirrored in the word use as well. The award-winning *The Sea, The Sea* [24] was the most unusual and showed the largest introduction of new word types. Her first novel, *Under the Net* [6] was in the middle, while *Jackson's Dilemma* [16] contained the smallest number of word-types and the most commonplace words. It was nevertheless apparent that the grammatical characteristics of the three books remained largely unchanged.

Garrard notes that the *Jackson Dilemma's* results are consistent with the early stages of Alzheimer's, in which many sufferers experience word-finding difficulties, particularly for less commonly used abstract terms such as "equanimity" or "discretion," while retaining the ability to produce impeccably well-formed sentences [18]. This result also corresponds with research documenting that the temporal lobe degeneration in Alzheimer's disrupts the brain's semantic system. This can happen subtly before anyone has the remotest suspicion of intellectual decline. It may explain her problems with the last book, as well as her ability to ultimately finish it. Even still, as noted above, Murdoch experienced an intense and unfamiliar feeling of writer's block during this period [11, 17]. This uncommon behavior not only further reinforces the UCL study, it also confirms perceptions of Bayley and Conradi, who both noted there were already signs the disease was beginning to disrupt her cognitive abilities while writing *Jackson's Dilemma.*

5.3 Iris Murdoch on Film

The World Health Organization (WHO) estimates that in 2023 more than 55 million people will have dementia worldwide, 60–70% of these cases being Alzheimer's, the most common form [25]. The prevalence of it demonstrates that there is a large audience for a movie like the award-winning film *Iris* [26], produced after Iris Murdoch's death and drawn from Bayley's two memoirs of their life as Alzheimer's altered it.

In this film we meet two versions of Iris Murdoch. On the one hand, we meet her older self, the somewhat unpredictable woman in her seventies who is fading away from a combination of mid-life hearing loss and the onset of Alzheimer's, as discussed in the last sections. This fading woman is interspersed with the young, vibrant communicator who was a cultural icon revered by so many. Observing the smart, independent, youthful Iris Murdoch who became a venerated figure in twentieth century London as well as the woman who lost touch with her community, her husband, herself, and her memories as her dementia progressed, we understand how much her faculties diminished. We also feel how deeply the transformation of the woman in his care affects her husband, John Bayley. Repeatedly we sense that someone once so familiar to him and others now seems both close and far away. Additionally, we grasp how difficult it was for her at the end of her life, and even during the early stages of the disease, to write and to do the things that were intrinsically tied to her identity.

One poignant aspect of the juxtaposed renderings is that the presentation is as much a love story as it is a chronicle of how the onset of Alzheimer's altered Murdoch's life, her ability to continue to author novels, and the life of her husband, who was her caregiver. The film also adeptly highlights his efforts to address her needs as her health declined. Although it includes sequences showing clinical interventions, it is the love story behind the memoirs that the film emphasizes. It is through Bayley's words [11, 17], transferred into the creative film script, that we meet the various faces of the Iris he knew over the course of their lives together. This commentary both allows the audience to gauge the progression of her Alzheimer's and the impact it had on his life, her life, and their life as a couple. The audience also comprehends how her condition extended outward into their larger community interactions.

Because the film combines the two Bayley memoirs into one piece, it somewhat mutates the progressive nature of the disease. In his first biographical volume, Iris is fading but still somewhat present in her husband's day-to-day life. The second book, a much darker volume, is more about his own history: Bayley's reflections on his own life are brought to the fore as he tries to make sense of the many ways his relationship to Iris' deteriorating condition has changed both of them. Through his words we understand that when she had mild cognitive impairment; during the earlier more coherent stage of Alzheimer's, he could still find her. Yet, as the days pass, he is coping with her increasing silence and fluctuating behavior as she slips more and more out of reach. At the end, we comprehend his anguish as he finally places her in an institution. Clearly, Bayley is trying to come to terms with the range of psychological responses her illness evokes in him; she, by contrast, increasingly recedes as her decline thwarts emotional communication.[5]

[5]Arthur Kleinman's book, *The Soul of Care: The Moral Education of a Husband and a Doctor,* offers an excellent counterpoint from a physician's perspective. Kleinman's wife had a rare form of Alzheimer's that affects only 5% of the Alzheimer's population. It began in her occipital lobes, which meant that she became blind first. In his book, Kleinman talks about how he spent his entire career studying illness and caregiving only to realize how much he did not know about what it meant to be an intimate caregiver for a loved one on a daily basis [27].

A particularly affective sequence in the memoirs, also dramatized in the film *Iris*, illustrates that at times the challenges posed by her health expanded beyond their household. At one point Iris wanders out of the house after John Bayley inadvertently leaves the front door open. Once he realizes she is gone, Bayley spends hours frantically searching the neighborhood looking for her. He also contacts the police, all to no avail. Finally, an old colleague, Valentine Cunningham, knocks on the door with Iris by his side. Cunningham had found her aimlessly wandering unaware of her environment and brought her home [17, 26].

Key here is that the cinematic *Iris* is presented more as an artwork than a medical case study. It is intended to both touch and teach the audience. When it was released on DVD, supplemental educational segments more fully addressed Alzheimer's disease. In one, the actor David Hyde Pierce speaks on behalf of the Americans with Alzheimer's Disease Association. He focuses on the growing need to understand the condition. He tells us that his motivation to become a spokesperson for the association came about because his father and grandfather battled dementia. Pierce offers a heartfelt explanation of the disease and the efforts to find a cure. He emphasizes that the disease has a cultural impact as more and more people deal with it within their families.

In another educational segment, Jim Broadbent, who played John Bayley in the film, talks about being drawn to the movie because his mother had died of Alzheimer's. His recollection of how confused the family felt and their desire to learn more at that time mirrors the reactions of many when Alzheimer's enters their lives. Broadbent explains how he was drawn to the film's script as a way for those impacted by Alzheimer's to share their experiences with a broader public. This actor felt that the presentation through the film medium was the best way to capture some of the commonalities among Alzheimer's patients and caregivers as they deal with the ravages of the disease.

This affective emphasis underscores the film's neurocultural intent. The emotional sequences are juxtaposed and built upon Iris's life, her brain's neurodegeneration, and her humanness. We feel her early difficulty in finding the right words; and, as the diseases progresses, we perceive her changing moods and personality. What stands out is how cogently the script conveys that the disease does not just impact a person, but the wider circle of families, caregivers, and the community in general. Transmitting this information into an accessible and cultural art form, it provides a container for educating the broader community impacted by Alzheimer's disease, a large cohort.

5.4 Iris Murdoch on Art

It is hard to know how Iris Murdoch's interest in the ambiguities of art would align with the type of film *Iris* is, one that shows how much the younger Iris differed from her Alzheimer's self. We do know that her views on art co-existed with her academic background, which included an expertise on the ambiguity within Plato's views on art. Perhaps the younger Iris would have found the conflicting reactions to the film invigorating to debate?

For example, would she have argued with the film critic Roger Ebert who started his review of the film saying that after seeing it he asked himself why he disliked it so intently. He concluded that the film erred in not honoring her work. In his view, *Iris* mourns her life instead of celebrating her accomplishments.

> It's like a biopic of Shakespeare that cuts back and forth between his apprentice days and his retirement in Stratford. Alzheimer's is especially tragic because it takes away the person while the presence remains. The character of Bayley, meanwhile, is presented as a befuddled and ineffectual man who contends with the baffling Murdoch ... but there is much more to Bayley. He is one of the most brilliant of literary critics, whose essays grace the New York Review of Books and the Times Literary Supplement, but on the basis of this film, you would think of him, frankly, as a fond old fool [28].

Ebert concluded that because the film is well-acted and written with intelligence, it might be worth seeing, despite his objections to how it tells her story. One wonders if Ebert read Bayley's books, for the film accurately depicts Bayley's descriptions of their final years and how he thought about her in relation to their more than forty years together, including how confused he was about how to deal with the challenges her mental decline posed.

The neurologist Eelco F. M. Wijdicks, in contrast to Ebert, gave the film *Iris* his highest rating in his *Neurocinema* books [29]. He applauded its accuracy and characterized it as one of the remarkable films portraying dementia in the elderly. Coupling these two reactions we see the ambiguity often present in responses to art and, by extension, the range of neuroculture.

A similar albeit more indirect criticism that some expressed after seeing the film centered on their fear that people would come to conceptualize Murdoch only in terms of her final days with Alzheimer's and her achievements as a novelist would become lost as a result [30]. We do not know where Murdoch would have landed in this discussion. Murdoch researcher Glliam Dooley does not believe she would have approved of using the word "tragic" to describe her Alzheimer's since it was such a contingent state of affairs [31]. Robert Weil similarly believes that despite being a private person, Iris would have approved of John Bayley's quite public announcement of her illness, particularly in light of his decision to record his experiences in caring for her in spite of the late twentieth century society's fear of old age and death [21]. Bayley said that in light of her valuation of education, she would have supported having her brain researched after her death and therefore she might have welcomed the discussion. In addition, since one thing she valued in art was that it was open to interpretation, the spectrum of views the film generated would no doubt strike her as a positive and educational outcome. This range speaks to how cultural attitudes about health, aging, well-being, and community education are another part of the neurocultural paradigm, an emphasis throughout this book.

In terms of her own fiction, we can see that she wrestled with the questions concerning art's ambiguity in terms of human life and behaviors. In her work, this was usually through paradoxical situations. For example, in *The Black Prince*, the name of the book alludes to Shakespeare's *Hamlet*, she presents the stories of two artists, Bradley Pearson and Arnold Baffin, both of whom had different aesthetic standards [32]. Bradley is obsessed with the idea that he is a great artist, although he is not. This points to a theme discussed throughout this book: there is more than one view on what art is. In addition, Bradley Pearson is also a trifle deaf, like Iris. Here we do not know to what degree she is drawing on her own experiences. Similarly, in the book Murdoch notes that Bradley has an unusually acute sense of smell [33]. It is impossible to know if this an insight gained through her personal experience, but surely worth pondering since quantitative investigations showing sensory compensation were more speculative when she wrote the book [34–36].

She also explained that she did not write philosophical novels, although of course philosophy imbues them. Rather, she said, she includes philosophy because it is what she knows. In other words, someone who knew cooking would likely include this expertise in their work. She also said that she wrote literature because it is open to interpretation and is a versatile form: "I think a novel is allowed to be ambiguous, I mean it's not surprising if people interpret a novel in different ways" [37]. She also noted that she believed in the literary tradition:

> It's a marvelous versatile form; within what looks like—and I suppose is—a conventional novel you can do anything under the sun. You can investigate anything, you can use any mode of thought you like, you can use language almost any way you like. It's a very big house as it were, the novel, within which all sorts of things can happen and a lot of experiment can take place without the reader being necessarily disturbed (p. 34, [38]).

5.5 What Is Art?

This embrace of art's openness to ambiguity raises the question: What is art? Given that Murdoch saw art as a big house, it is unsurprising that this was a question she wrestled with broadly, and not only in her fictional writing. We see this in 1975, when she presented the Romanes Lecture[6] at the University of Oxford, later published as *The Fire and the Sun: Why Plato Banished the Artists* [39]. In these lectures she presented a multi-faceted critique of Plato's contradictory views on art, in effect speaking about his own ambiguity on the subject. One focus in her study was why Plato himself, whose dialogues demonstrate his awe-inspiring artistry, spoke out against art so passionately—except when he did not. As Murdoch points out, his banishment of the artists from the *Republic* creates several conundrums when evaluating his views on art. These are useful to consider here for they still pertain when evaluating what art is, and, by extension, what art is in relation to the neurosciences. We see this definitional ambiguity clearly in research surrounding neuroscience and art because definitions of art vary so widely in this literature, where studies range from art as beauty to art as a therapeutic tool.

[6] The Romanes Lecture was created in 1891. Each year they invite a distinguished public figure from the arts, science, or literature. The first lecture was given in 1892 by William Gladstone.

Perhaps most noteworthy in terms of Plato's views, conundrums, and artistry is that he, himself, encompasses many of the perspectives we find in the multi-faceted literature. He wrote so lyrically and on such a broad range of subjects that people across disciplines continue to quote (or misquote) him. In his works we find discussions of beauty, morals, political philosophy, cosmology, education, ethics, medicine, a tripartite mind, and more. Moreover, as Murdoch points out, Plato was not offended by all the arts all the time, although he often was. Sometimes his poetic language admonishes the poets who mislead us by portraying the gods as undignified and immoral. We see another perspective when he advocates for art that teaches us values, respects religion, admires those who are good people in his view, and so forth. Similarly, we see his moralistic framing when he presents his views on music and the theater: they should encourage stoical calmness, not boisterous uncontrolled emotions [39]. Also, as she notes and as discussed in Chap. 1, according to the historical evidence the Greeks did not view art as we do.

We can also see that his alluring allegories understate nuance. Plato's *Allegory of The Cave* is an elegant metaphor of early film technology and is discussed in Chap. 7. Another example is his Charioteer analogy in *Phaedrus* where Plato compares the soul to a person driving a chariot pulled by two flying horses; one white, one black. The white one is moral and the other immoral. The moral horse, of finer spirit, is beautiful and noble. This mare wants to soar into heaven. For Plato, the other—ugly, base, and immoral—represents human nature when driven by passions and irrationality. Plato's paints his poetic picture so elegantly that it is easy to miss how overly black-and-white [sic] his construction is. Putting aside Plato's urge to position artists in terms of the black mare—and how this positioning had an influence on later attitudes—it is not hard to see that many human realities are not black-and-white. An obvious example explored in Chap. 8 is that the tragedy of war has fostered brain injuries as well as an impressive range of medical advances to address these insults to the brain regardless of which side one fought for in the field. War also leads to heartfelt art from those on all sides of a conflict.

His views on art are also interwoven with commentary on one class of entities, his well-known Forms, which he postulated were accessible to the mind alone. In other words, the Forms were inaccessible to the senses. Given Plato's influence on later thinking related to the natural sciences, psychology, psychiatry, and neurology, as discussed earlier, it is helpful to keep in mind here that the inaccessible Forms demote how our sensory system gives us information about the world external to ourselves and how we construct this knowledge in our brains. Moreover, how we construct knowledge in our brains is influenced by both our biology and education, as discussed in chapters throughout this book. Also, our senses are involved in monitoring the state of our own bodies [40–45].

Plato's most alluring conundrum, as Murdoch points out, is his own artistry, as inscribed in the dialogue form he invented. His writing is itself strongly enticing, affective, and effective. Filled with wit and anecdotes about the nature of human life, Plato's expressive talent and originality reveals an inquiring, imaginative mind that created a spirited questioning form filled with words that postulate and elegantly assert his philosophical views. In the *Republic* (607b) he speaks of "an old quarrel between philosophy and poetry" and even dubiously allows that a defense of poetry might one day be made by a poetry-lover who was not a poet (as indeed it was by Aristotle) [39]. Plato later notes that technique alone will not make a poet, for poets may intuitively understand things of the greatest importance (*Laws* 628a). As Murdoch puts it:

> [Plato] was impressed by the way in which artists can produce what they cannot account for (perhaps this suggested certain ideas to him), and although he sometimes, for instance in the *Apology* and the *Ion*, holds this against them, he does not always do so" (p. 2, [39]).

A final point Murdoch makes on Plato's view of art, and because it is still debated this point has neurocultural importance, is that while Plato is proposing an ideal society, and gives beauty a crucial role in his philosophy, he defines it to *exclude* art. Murdoch notes that he constantly and emphatically accuses artists of moral weakness or even baseness. "One is tempted to look for deeper reasons for such an attitude; and in doing so to try (like Plotinus and Schopenhauer) to uncover, in spite of Plato, some more exalted Platonic aesthetic in the dialogues" (p. 2, [39]).

Plato's exclusion of art stands out even more in light of how later thinkers, beginning with Plotinus's Neoplatonism, have re-interpreted his views, as Murdoch notes. Over the years, following Plotinus' lead, others began to equate art with beauty and aesthetic sensibility. As this equation took hold, art was more and more wedged into a metaphysical or spiritual domain that mitigated the brain's involvement with the making and experience of art. Words like spiritual, unconscious, and intuitive were used to characterize art and this characterization obscures that art includes the use of motor skills, memory, active visualization, the development of specialized skills, and so forth [46, 47]. This broader perspective on art is adopted in this volume. In other words, people turn to art for a range of reasons and various areas of the brain support the making and viewing of it. Many examples throughout this book demonstrate arts breadth and variety.

5.6 Conclusion

Given Murdoch's passion for language, communication, and education, it is hard not to feel saddened that with the onset of Alzheimer's she lost what was most precious to her: her ability to communicate and use language. One striking aspect of this in relation to her fiction is that her novels mostly examined human dynamics through the same sorts of themes explored in her philosophy: the Good, morals, ethics, sexuality, the unconscious, and so forth. Neurological degeneration does not come up as much in her novels although aging and disabilities do on occasion, but these areas do not seem to capture her imagination. In some ways, her emphasis on the body and mind reflects twentieth century sensibilities. One wonders how her work would have changed if she had lived into the twenty-first century and seen how imaging technologies have advanced our understanding of the mind, the body, the brain, emotional experience, and so forth.

In terms of her commentary on Plato, his often-bizarre comments on science and medicine are largely outside of her scope as well. Others have recognized that his work has retained its appeal and readability despite its biological and medical inaccuracies, an area in which he and Aristotle have much in common, despite their philosophical differences.

One who wrote in this vein was Sir William Osler (1849–1919), discussed further in Chap. 8, in reference to shell shock. Osler was the Canadian physician now known for creating the first residency program for specialty training of physicians and a founding professor of Johns Hopkins Hospital. In his position there he was primarily interested in clinical practice and urged young doctors to read extensively in the humanities. His urge to bring medical students out of the lecture hall included introducing them to bedside clinical training, at a time when some thought medicine was turning to technology at patient expense. His was an effort to emphasize our shared humanity [48].

Osler used Plato to speak about the importance of the humanities, a sentiment this book shares. In a paper he read before the Johns Hopkins Hospital Historical Club on December 14, 1892, titled "Physic and Physicians as Depicted in Plato" [49], he began by pointing out Plato's medical errors. Then, like Murdoch, he praised the philosopher's "melodious voice" [49]. He also echoes Murdoch in pointing out how the philosopher is often inconsistent and frequently wrong. Yet, he concludes that despite the crude and imperfect nature of Plato's anatomy and physiology, his work has value because he affirmed that striving to build a more perfect society remains a worthy ideal [49]. Or, as Osler notes, from a humanistic perspective, it is illuminating to consider the inspirational and aspirational resonance within Plato's artistry, despite how clear it is that his ideas derive from another time:

> More than two thousand two hundred years have passed away since [Plato] returned to the place of Apollo and the Muses. Yet the echo of his words continues to be heard among men, because of all philosophers he has the most melodious voice…He is the inspired prophet or teacher who can never die…Other teachers of philosophy are dried up and withered,—after a few centuries they have become dust; but he is fresh and blooming and is always begetting new ideas in the minds of men (p. 20, [49]).

This book discussion of Plato's art similarly is intended to acknowledge his artistic legacy rather than his ideas, for his inspirational and aspirational spirit has resounded across centuries. Neurologist Oliver Sacks, the subject of the next chapter, offers a contrasting perspective on how the inspirational and aspirational spirit comes together in a neurocultural manner.

References

1. Nicholls R. Iris Murdoch, novelist and philosopher, is dead. The New York Times. 1999 February 9; 1999, Sect. A.
2. Conradi PJ. Iris Murdoch obituary. Guardian; 8 February 1999.
3. Meyers J. Remembering Iris Murdoch: letters and interviews. London: Palgrave Macmillan; 2013.

4. Conradi PJ. Iris Murdoch: a life. New York: Norton; 2001.
5. Murdoch I. Existentialists and mystics: writings on philosophy and literature. London: Penguin Publishing Group; 1999.
6. Murdoch I. Under the net, a novel. New York: Viking Press; 1954.
7. Bakhos D, Villeuneuve A, Kim S, Hammoudi K, Hommet C. Hearing loss and Alzheimer's disease. Geriatr Psychol Neuropsychiatr Vieil. 2015;13(2):195–204.
8. Loughrey DG, Kelly ME, Kelley GA, Brennan S, Lawlor BA. Association of age-related hearing loss with cognitive function, cognitive impairment, and dementia: a systematic review and meta-analysis. JAMA Otolaryngol Head Neck Surg. 2018;144(2):115–26.
9. Bosmans J, Jorissen C, Cras P, Van Ombergen A, Engelborghs S, Gilles A, et al. Impact of hearing loss and vestibular decline on cognition in Alzheimer's disease: a prospective longitudinal study protocol (Gehoor, Evenwicht en Cognitie, GECkO). BMJ Open. 2020;10(9):e039601.
10. Huang AR, Jiang K, Lin FR, Deal JA, Reed NS. Hearing loss and dementia prevalence in older adults in the US. JAMA. 2023;329(2):171–3.
11. Bayley J. Elegy for Iris. New York: St. Martin's Press; 1999.
12. Moden R. Tom Phillips, Iris Murdoch, and the Flaying of Marsyas. The Iris Murdoch Society; August 23, 2023.
13. Phillips T. Iris Murdoch. London: National Portrait Gallery; 1987. https://www.npg.org.uk. Accessed 3 Mar 2023
14. Sohm PL. The artist grows old: the aging of art and artists in Italy, 1500–1800. New Haven: Yale University Press; 2007.
15. Gowing L. Human stuff. London Rev Books. 1984;6(2):13–4.
16. Murdoch I. Jackson's dilemma. London: Chatto & Windus; 1995.
17. Bayley J. Iris and her friends: a memoir of memory and desire. New York: W.W. Norton; 2000.
18. Garrard P, Maloney LM, Hodges JR, Patterson K. The effects of very early Alzheimer's disease on the characteristics of writing by a renowned author. Brain. 2004;128(2):250–60.
19. Koger SM, Chapin K, Brotons M. Is music therapy an effective intervention for dementia? A meta-analytic review of literature. J Music Ther. 1999;36(1):2–15.
20. Bannan N, Montgomery-Smith C. `Singing for the Brain': reflections on the human capacity for music arising from a pilot study of group singing with Alzheimer's patients. J Royal Soc Promotion Health. 2008;128(2):73–8.
21. Weil R. Memories of Iris, 1999. In: Dooley G, editor. From a tiny corner in the house of fiction: conversations with Iris Murdoch. Columbia: University of South Carolina Press; 2003. p. 251–3.
22. Braak H, Braak E. Neuropathological stageing of Alzheimer-related changes. Acta Neuropathol. 1991;82(4):239–59.
23. News UCL. Iris Murdoch's last novel reveals first signs of Alzheimer's disease. London: University College London; 2004.
24. Murdoch I. The sea, the sea. London: Chatto & Windus; 1978.
25. World Health Organization. Dementia: key facts. Geneva, Switzerland: World Health Organization; 2023.
26. Eyre R. Iris: Miramax films; 2001.
27. Kleinman A. The soul of care: the moral education of a husband and a doctor. New York: Viking; 2019.
28. Ebert R. Iris Roger Ebert's movie yearbook 2003. Kansas City: Andrews McMeel Publishing; 2002. p. 292.
29. Wijdicks EFM. Neurocinema: when film meets neurology. Boca Raton: CRC Press, Taylor & Francis Group; 2015.
30. Altorf M. After cursing the library: Iris Murdoch and the (In)visibility of women in philosophy. Hypatia. 2011;26(2):384–402.
31. Dooley G, editor. From a tiny corner in the house of fiction: conversations with Iris Murdoch. Columbia: University of South Carolina Press; 2003.
32. Bove CB. Understanding Iris Murdoch. Columbia: University of South Carolina Press; 1993.

33. Brans J. Virtuos dogs and a unicorn: an interview with Iris Murdoch. In: Dooley G, editor. From a tiny corner in the house of fiction: conversations with Iris Murdoch. Columbia: University of South Carolina Press; 2003. p. 155–93.
34. Alencar CDC, Butler BE, Lomber SG. What and how the deaf brain sees. J Cogn Neurosci. 2019;31(8):1091–109.
35. Glick H, Sharma A. Cross-modal plasticity in developmental and age-related hearing loss: clinical implications. Hear Res. 2017;343:191–201.
36. Wettstein M, Wahl H-W, Heyl V. Visual acuity and cognition in older adults with and without hearing loss: evidence for late-life sensory compensation? Ear Hear. 2018;39(4):746–55.
37. Todd R, Fletcher J, editors. Encounters with Iris Murdoch: proceedings of an informal symposium on Iris Murdoch's work, held at the Free University, Amsterdam, on 20 and 21 October 1986. Amsterdam: Free University Press; 1988.
38. Glover S. Iris Murdoch talks to Stephen Glover. In: Dooley G, editor. From a tiny corner in the house of fiction: conversations with Iris Murdoch. Columbia: University of South Carolina Press; 2003. p. 33–43.
39. Murdoch I. The fire and the sun: why Plato banished the artists. New York: Viking; 1977.
40. Dissanayake E. What art is and what art does: an overview of contemporary evolutionary hypotheses. In: Martindale C, Locher P, Petrov VM, editors. Evolutionary and neurocognitive approaches to aesthetics, creativity, and the arts. Amityville, NY: Baywood Publishing Company; 2007. p. 1–14.
41. Warburton N. The art question. London and New York: Routledge; 2003.
42. Freeland CA. But is it art? An introduction to art theory. Oxford and New York: Oxford University Press; 2001.
43. Danto AC. What art is? New Haven: Yale University Press; 2013.
44. Tolstoy L. What is art? And essays on art. London and New York: Oxford University Press; 1932.
45. Beuys J. What is art? Conversation with Joseph Beuys. West Sussex: Clairview Books; 2007.
46. Ione A. Art and the brain: plasticity, embodiment, and the unclosed circle. Leiden: Brill; 2016.
47. Ione A. Innovation and visualization: trajectories, strategies, and myths. Amsterdam: Rodopi; 2005.
48. Bliss M. William Osler: a life in medicine. Oxford and New York: Oxford University Press; 1999.
49. Osler W. Physic and physicians as depicted in Plato (Johns Hopkins Hospital Historical Club, December 14, 1892). Boston: Damrel & Upham; 1893.

Biography and Biology: Oliver Sacks

6

The name Oliver Sacks (1933–2015) became a household word after he published his book *Awakenings* in 1973. This chapter introduces him as a doctor, a writer, and a neurologist who reported on his own neurological conditions. As the second chapter in Part II of this book, the chapter further emphasizes the value of case studies, a methodology Sacks used to introduce his patients as more than their pathologies. This approach is also a core aim of this book. As this chapter demonstrates, Dr. Sacks' case studies allowed him to distinguish between studying neurological disturbance in terms of how the nervous system is organized and learning from the actual people who were struggling to adapt and survive.

Awakenings had a broad appeal because it is an extraordinary volume that recounts the life histories of a group of patients who were victims of the 1920s encephalitis lethargica (sleeping sickness) epidemic. It also conveys Sacks' efforts to help them in the 1960s with the drug L-DOPA. Sacks later explained that it was after arriving at Beth Abraham Hospital (Mount Carmel in the book) and meeting these patients that he began to think about the difference between studying neurological disturbance in terms of how the nervous system is organized as compared to learning from the actual people who were struggling to adapt and survive. In a preface added to the 1990 republication of the volume Dr. Sacks tells readers that it was at Beth Abraham that he came to see his work in terms of identity and thus as at the junction of biology and biography, a phrase adopted for this chapter's title:

> There evolved a new concern: that of a commitment to the patients, the individuals under my care. Through them I would explore what it was like to be human, to *stay* human, in the face of unimaginable adversities and threats. Thus, while continually monitoring their organic nature—their complex, ever-changing pathophysiologies and biologies—my central study and concern became *identity*—their struggle to maintain identity—to observe this, to assist this, and, finally, to describe this. All this was at the junction of biology and biography (pp. xxviii–xxix, [1]).

A. Ione, *Neuroscience and Art*, Neurocultural Health and Wellbeing,
https://doi.org/10.1007/978-3-031-62336-3_6

Oliver Sacks also explained that his urge to see people in terms of their identity, biography, and biologically was a product of his upbringing. As noted earlier, he said that growing up in a household filled with doctors and medical talk led him from childhood to see patients in terms of their stories, and how they responded to illness or injury, stress, or misfortune. Referring to these conversations, Sacks said: "Perhaps it was inevitable that I myself became both a physician and a storyteller (p. ix, [2])."

Sacks' range and story-telling style firmly brings him into neuroculture. Rather than adopting a detached, dense form of medical reportage, his sensitive approach brought a literary flair to his pathological notations and reached out to readers because his patients became living beings. This encouraged—and continues to encourage—cross-disciplinary exchange. Indeed, adaptations of his work have further expanded the conversations between neurology and culture. In addition, his person-centered approach juxtaposed neurology with cultural modalities like music and art, and these practices were frequently interwoven into his work with patients.

Nonetheless, at times his ability to translate complex conditions into easy to follow language led medical practitioners to find his writings too accessible. For example, a pre-publication reviewer of his first book, *Migraine* [3] wrote that he found the text too easy to read and this might arouse suspicions [4]. This reviewer's reaction was not an isolated example. Later, in 1990, while Lewis Rowland was president of the American Academy of Neurology, there was a discussion about whom they might ask to speak at the Academy's annual meeting where they planned to announce the Decade of the Brain, an initiative launched by the United States President George H. W. Bush.[1] After some members of the group proposed they invite Dr. Sacks, one member dismissed the suggestion saying: "We need a prominent investigator, not a story teller" (p. 247, [6]). To this Rowland replied, "But this 'storyteller' happens to be the most famous neurologist in the entire world" (p. 247, [6]).

As this chapter will demonstrate, Sacks' own story and his relationship to his identity and humanness is as compelling as the stories he told. In terms of where the non-fictional stories he tells "fit" neuroculturally, one can look to historical medical case studies, discussed shortly, or to storytelling as an art form. As an art form, an excellent counterpoint is novelist and cultural icon Iris Murdoch, who was introduced in the last chapter. She saw literature as a way to craft identities that were resonant as living beings and so she, too, was interested in biography, human behaviors, and identity. In 1977, several years before Oliver Sacks' trailblazing revival of powerful neurological narratives as a conveyor of both biology and identity, she said in an interview that humans, as storytelling creatures, use language in the parallel arts of literature and living:

[1] United States President George H. W. Bush designated 1990–1999 "The Decade of the Brain." It was a part of a larger effort involving the Library of Congress and the National Institute of Mental Health of the National Institutes of Health. The purpose of this delineation was "to enhance public awareness of the benefits to be derived from brain research" [5].

> Literary modes are very natural to us, very close to ordinary life and to the way we live as reflective beings. Not all literature is fiction… When we return home and "tell our day," we are artfully shaping material into story form.… So in a way as word-users we all exist in a literary atmosphere, we live and breathe literature, we are all literary artists, we are constantly employing language to make interesting forms out of experience which perhaps originally seemed dull or incoherent [7].

Finally, Sacks was a man of contrasts. He was a motorcyclist, a swimmer, a weightlifter, a writer, a compassionate doctor who had a tremendous influence on others, and a lonely, closeted gay man for much of his adult life.[2] Thus, in Oliver Sacks, we see a melding of many pursuits and a complex person. This comes through in his 2007 dual appointments at Columbia University. He joined the faculty of the Medical Center as a professor of neurology and psychiatry and was also appointed as the first "Columbia University Artist" in recognition of his work in bridging the arts and sciences. This dual appointment underscores how seamlessly his persona fits within the neuroscience, art, and neurocultural theme of this volume, as does his person-centered approach and the urge to see the identities of the people behind their pathologies.

6.1 Beginnings: What Is It To Be Human?

Born in London, England, in 1933, Oliver Sacks was the youngest of four children. His parents were both physicians, as were his two older brothers. At the age of 6, in December 1939, he and his brother Michael were evacuated from London and sent to a boarding school in the English Midlands to escape the Blitz, where he remained until 1943. The trauma of this experience had a tremendous impact on his life and that of his brother Michael, who was sent to another boarding school in 1941 and eventually became psychotic.

In *Uncle Tungsten* [8], Sacks relates that after he returned home from boarding school he set up a chemistry lab that doubled as a darkroom to keep his world separate from Michael's. His laboratory work enthralled him, and his chemistry experiments often filled his home with noxious smells but, as he put it, he fortunately never burned his house down. These kinds of explorations accentuate that his inquisitive nature as an adult was already evident in childhood. He always desired to probe and was always attempting to discover how things work. At 12, one of his schoolmasters noted that he would go far, if he did not go too far [4].

Even once he established this work space to shield himself from his brother's psychosis, the reality of his brother's fragility nonetheless had an impact on his life. Sent to an institution at sixteen, Michael was given shock therapy in 1944, then a first line treatment for schizophrenia. He also received electroconvulsive treatment.

[2] His biker friends called him "Wolf," his middle name. Non-medical achievements early in his life included setting a California state record in 1961 doing a full squat with 600 pounds. This is a free weights exercise that primarily targets the quads.

Later, around 1953, new treatments were equally heartbreaking. Tranquilizers (Largactil/Thorazine) calmed him down but left him grossly Parkinsonian. Without the drugs, and even with a reduced dosage, his full-blown psychosis returned. Like Sacks' professional patients later, Michael was a person that helped formulate the question that resounded throughout his life: what is it to be human?

After Sacks completed his Oxford education in 1956, he began to sense that research was not his calling. A key event happened while working at the Laboratory of Human Nutrition. He was tasked with designing an animal model to study the jake paralysis that was causing the crippling neurological damage seen during Prohibition when drinkers denied alcohol turned to Jamaica ginger (jake). The jake included a very unpleasant tasting compound, triorthocresyl phosphate (TOCP), then freely available as a "nerve tonic." Sacks despaired when he saw his favorite hen, her number was 4304, die from the TOCP that was in the jake. His sadness in watching this unnamed bird, "an animal of unusually docile and sweet disposition" sink "to the ground on her paralyzed legs, chirping piteously" (p. 26, [4]) stayed with him his entire life.

After a brief sojourn at a kibbutz in 1955, Oliver Sacks began medical school, receiving his medical degree in 1958. He then interned and did residency work in California before moving to New York. where he spent most of his career. A fellowship at the Albert Einstein College of Medicine was followed by work with patients as a neurologist at Beth Abraham Hospital's chronic-care facility in the Bronx. Here he found the survivors of the 1920s sleeping sickness encephalitis lethargica, mentioned above, and employment that was a better fit for his talents.

Awakenings records how he found himself trying to find ways to help these patients in a state and situation that was not a part of the medical literature. Here, too, we see his interest in what it is to be human:

> This sense of the dynamics of illness and life, of the organism striving to survive, sometimes under the strangest and darkest circumstances, was not a viewpoint which had been emphasized when I was a student or resident, nor was it one I found in current medical literature. But when I saw these postencephalitic patients, it was clearly and overwhelmingly true (p. 171, [4]).

6.2 Case Studies: Seeing People As More Than Pathologies

Throughout his career Dr. Sacks explained that his attraction to the anecdotal case study stemmed from the belief that it added a greater understanding of a patient's condition. Within the detailed description of an individual's life the person is juxtaposed with his or her symptoms. To be sure, through his case study presentations Sacks in effect aligned patient histories with the biographies discussed over dinner in his childhood. His education additionally contributed to this approach as well for it generated an interest in historical medical literature more generally. As he tells it, even in medical school he was wondering who the patients behind the pathologies were.

> When I became a medical student … it was the patients I saw, their predicaments and their stories, that gripped my imagination, and these experiences imprinted themselves upon me indelibly. Lectures and textbooks, abstracted from living experience, left almost no impression. I was, however, strongly drawn to the case histories that abounded in the nineteenth-century medical literature—rich, detailed descriptions of patients with neurological or psychiatric problems (p. 17, [9]).

Of course, the case study approach was developed well before the nineteenth century. Hippocrates, discussed in Chap. 1, introduced it because he believed that diseases follow a course from first symptoms to their resolution, which could be a happy or a fatal one. This led Hippocrates to advocate for a description of the disease's progression and the patient's experience with it to comprehend how disease worked. He characterized this as an essential element in understanding and learning.

A more recent advocate who had a major influence on Sacks was Soviet neuropsychologist A.R. Luria (1902–1977). Luria spoke of his own books as neurological detective stories. After reading *Awakenings* Luria described this Sacks book as a major contribution to medical literature in a letter to the author:

> Unfortunately, the ability to describe which was so common to the great neurologists and psychiatrists of the nineteenth century is lost now, perhaps because of the basic mistake that mechanical and electrical devices can replace the study of personality. Your excellent book shows that the important tradition of clinical case studies can be revived and with great success (p. 202, [4]).

Biographies of Sacks bring his identify and his life into focus. One documentary by Ric Burns, *Oliver Sacks: His Own Life* [10], for example, is filled with interviews and vignettes that capture how Sacks' personality intersects with his work, his large circle, and the public. In one sequence a man relays that the head of Columbia University's neurology department had told him that 70% of the applicants who aspired to do neurology as a concentration mentioned Oliver Sacks as the reason that they wanted to become neurologists. We also see the attendees at a party in his home wearing name tags, a common practice when he entertained. His prosopagnosia—face blindness—meant that the doctor had difficulty recognizing faces, a problem the name tags solved.

Face blindness and Sacks' other neurological conditions reinforce another theme this book emphasizes, that congenital conditions are frequently not seen by those who have them in the same way as acquired ones. Writing in *The New Yorker*, Sacks revealed that he long assumed that his difficulties remembering people were the result of general absent-mindedness. It was not until halfway through his life that Sacks—despite specializing in neurology—realized that he had the neurological disorder prosopagnosia [11]:

> I have had difficulty recognizing faces for as long as I can remember. I did not think too much about this as a child, but by the time I was a teen-ager, in a new school, it was often a cause of embarrassment. My frequent inability to recognize schoolmates would cause bewilderment, and sometimes offense—it did not occur to them (why should it?) that I had a perceptual problem. … I have been accused of "absent-mindedness," and no doubt this is true. But I think that a significant part of what is variously called my "shyness," my "reclusiveness," my "social ineptitude," my "eccentricity," even my "Asperger's syndrome," is a consequence and a misinterpretation of my difficulty recognizing faces (p. 37, [11]).

This comment brings to light that he did not perceive of himself as someone whose medical "conditions" interfered with his health and well-being despite his prosopagnosia, migraines, and other conditions mentioned in his writings. Or, as he states in his autobiography: "As I entered my seventies, I was in excellent health; I have a few orthopedic problems but nothing serious or life threatening. I did not give much thought to illness or death" (p. 174, [4]).

6.3 Writings and Derivatives

Over the course of his career, Sacks' topical portfolio included Tourette's syndrome, agnosias, migraines, hallucinations, color-blindness, languages (the loss of language, Sign, etc.), and more. Even in his writings about his own neurological and medical conditions (such as migraine, prosopagnosia, Asperger's, etc.) we hear his complex, yet clear way of juxtaposing the person with the problem. We also see cultural adaptations that offer affective renderings of the patients he presents. Often, these derivative offer a window into human behaviors and pathologies and sometimes update his cases. Not all mirror Sacks' writings precisely, for on occasion a derivative alludes to neurology without presenting a factual rendering. The ways in which adaptors interweave Sacks patients' stories with other narratives further emphasizes that the biographies are human stories, and that Sacks had a talent for bringing people to life, even those who were at times suffering from what many would see as unimaginable challenges.

In some ways Dr. Sacks' storytelling life began when he started keeping journals as a young teenager. He wrote on all kinds of subjects and continued to record his thoughts throughout his life. Aside from giving him a familiarity with language, this practice aided him in finding the right word or expression. His journaling habit also enabled him to write a quite detailed and annotated autobiography in his later years [4]. One journal he kept in 1960, when he was 27, indicated his intention to write five novels and one memoir on his chemical boyhood. It took 45 years for Sacks to partially accomplish this youthful goal, with the publication of *Uncle Tungsten: Memories of a Chemical Boyhood* in 2001 [8]. A detailed second autobiography followed in 2016 [4]. Although the novels did not come to pass, he does refer to *A Leg to Stand On* as a sort of neurological novel. His patient journals, written during his recovery, served as a source for this book, as discussed shortly [12].

In addition, his book *Awakenings* [13] is the only science book to date to have received the Hawthornden Prize. It was awarded to Sacks in 1974. This British literary award was established in 1919 by Alice Warrender, who was born at Hawthornden Castle. It is given annually to an English writer for "the best work of imaginative literature" [14], and it is one of the few that Iris Murdoch never received. Obviously, awarding the prize to Oliver Sacks shows that the qualifications are liberally interpreted.

Four of Sacks works are discussed in the remainder of this chapter, although others are referenced to expand on this limited selection. The first is *Migraine: The Evolution of a Common Disorder*, published in 1970. This early work shows Dr.

Sacks as he was beginning his career. The *Migraine* section also integrates notations from later books in which he expands on the subject. Second, *Awakenings*, published in 1973, made him a worldwide figure. This book spawned many derivatives, also discussed below. Third, Sacks has described *A Leg to Stand On* (1984) as a more personal book. In it Sacks describes an accident he had in Norway that resulted in a disturbance of proprioception and how music proved therapeutic. The last of the four sections is his essay "The Man Who Mistook His Wife for a Hat," published in 1985. Originally presented as a case involving agnosias, the patient's autopsy showed atypical Alzheimer's Disease.

6.4 *Migraine: The Evolution of a Common Disorder* (1970)

Even if Dr. Sacks did not see his various neurological conditions as health problems, it is illuminating that he adapted to some of his conditions more easily than others—and sometimes he was fascinated by them. For example, Sacks has reported that he had visual migraines from an early age:

> I have had migraines for most of my life; the first attack I remember occurred when I was three or four years old. I was playing in the garden when a shimmering light appeared to my left, dazzlingly bright. It expanded, becoming an enormous arc stretching from the ground to the sky, with sharp, glittering, zigzagging borders and brilliant blue and orange colors. Then behind the brightness came a growing blindness, an emptiness in the field of vision, and soon I could see almost nothing on my left side. I was terrified—what was happening? My sight returned to normal in a few minutes, but these were the longest minutes I had ever experienced (p. 102, [3]).

His alarm was soothed by his mother, who too was a migraineur. She reassured him that everything would return to normal in a few minutes, as it did. His migraines were the classical visual presentation of patterns like twigs or geometrical structures: lattices, checkerboards, cobwebs, and honeycombs inside the scotoma [3]. They did not include headache, vomiting or painful sensitivity to light and noise. This may explain why his initial fear gave way to a fascination with the way his "vision could be unmade, deconstructed frighteningly but fascinatingly, in front of me, and then be remade, reconstructed, all in the space of a few minutes" (p. 13, [4]). He was not similarly fascinated by his prosopagnosia (face blindness), perhaps because the interpersonal discomfort of not recognizing faces brings others into the experience.

His first medical book, *Migraine* [3], was published in 1970, before the 1973 publication of *Awakenings*, his breakthrough book. In *Migraine* Sacks explains that when he saw his first migraine patient, he had thought of migraine as a peculiar type of headache, no more and no less. After seeing more patients, it became apparent that headache was never the sole feature of a migraine and was not even a necessary feature of all migraines. In line with his life-long approach, as he delved into the literature of the subject, he found his patients more instructive than any book. He also found himself astonished by the complexity of the histories. Eventually Sacks came to see the condition as an entire encyclopedia of neurology [3].

He also found that the kinds of treatments that worked differed from patient to patient. Drugs worked with some patients. Some patients only needed the magic of attention and interest. The most challenging were the severely afflicted patients who at times defeated his therapeutic endeavors until he probed into their emotional lives. This led him to envisaging migraine as a structure whose forms were implicit in the repertoire of the nervous system, and also something that might be employed to any emotional or biological end.

In the 1970 book Sacks portrays migraines as both physical and symbolic events. Part I describes the forms of migraine attacks as experienced by the patient and observed by the physician. Part II brings in the many circumstances—physical, physiological, and psychological—that may provoke isolated or repeated migraine attacks. Part III considers the physiological mechanisms of the migraine attack as well as the biological and psychological roles which migraines, and certain allied disorders, may fill. He ends in Part IV with a therapeutic approach to migraine.

Much of *Migraine* grew out of Sacks work at a Bronx Clinic headed by Arnold P. Friedman, an expert on the topic. Sacks has mentioned in a number of his writings that after reading Edward Liveing's 1873 *On Megrim, Sick-Headache, and Some Allied Disorders: A Contribution to the Pathology of Nerve-Storms* [3, 4, 15, 16] he resolved to write a comparable, updated book. The resulting *Migraine* is not as lively as some of his later works, but even in this early work Oliver Sacks' depth and reach is already clear. Its sweep includes historical contrasts and illustrative reports. For example, he mentions Hippocrates (c.460–c.370 BCE) and Hildegard of Bingen (1098–1180). He also acknowledges the clinical autobiography of the Italian physician Girolamo Cardano (1501–1576), mentioned in Chap. 1 in terms of his work with the deaf, Cardano also talked about the strange visions he experienced between the ages of three and six [3, 17].

Anonymous physicians are among the patients introduced. Knowing that Sacks was a migraineur, I wondered if his own experience of migraines was included. I also wondered how his history influenced his thinking and writings on the subject, particularly given how freely he talks about his own migraines later in life, once his career was more established. For example, 42 years later, in his 2012 book *Hallucinations* [16] he says:

> For most people with classical migraine, the scintillating scotoma is the chief visual effect, and things go no further. But sometimes, within the scotoma, there are other patterns. In my own migraine auras, I would sometimes see vividly with closed eyes, more faintly and transparently if I kept my eyes open—tiny branching lines like twigs or geometrical structures: lattices, checkerboards, cobwebs, and honeycombs. Unlike the scintillating scotoma itself, which had a fixed appearance and a slow, steady rate of progression, these patterns were in continual motion, forming and re-forming, sometimes assembling themselves into more complicated forms like Turkish carpets or complex mosaics or three-dimensional shapes like tiny pinecones or sea urchins....One has to call these hallucinations, even though they are only patterns and not images, for there is nothing in the external world that corresponds to the zigzags and checkerboards—they are generated by the brain. And there may also be startling perceptual changes with migraine. I might sometimes lose the sense of color or of depth (for other people, color or depth may intensify) (p. 103, [16]).

Many of the migraine patterns Sacks describes have a great deal of resonance with art and psychological studies of pattern and vision. In *Hallucinations* [16] Sacks correlates the migraine patterns with hallucinations at times, as he does above, because they do not correspond to anything in the external world. In the external world similar patterns do, however, adorn architecture, classical and medieval motifs, pottery, and are found in Islamic art.

Migraine Art by Klaus Podoll and Derek Robinson [18] offers a nice visual counterpoint to Sacks' study. Their neurological interest in this topic was in part spawned by a cultural event, a campaign initiated by Robinson while working for a pharmaceutical company. His company decided to sponsor a contest seeking migraine art they could use to promote a new drug for migraine patients. The migraine contests took place over several years and at the time the book was published the migraine art collection consisted of 562 pieces. The artwork in it offers another neurocultural example of artists using visual syntax to communicate a personal physiological experience with others. Both amateur and professional artists submitted work and their illustrations of individual pain and visions are enhanced by interviews that explain the art [18]. Historical information in the book complements Sacks work as well.

Among the migraineurs, Podoll and Robinson include a Sumerian poem written in approximately 3000 BCE in which the anonymous sufferer bewailed his "sick-headed" and "sick-eyed" condition as he prayed for relief in some painless afterlife [18]. Others examined are Hildegard of Bingen, Blaise Pascal, and Lewis Carroll. One particularly alluring monochromic image by a professional artist shows the visual field is concentrically contracted by black-and-white checkerboard patterns. According to this artist, his small events were more frequent than the full-blown migraines:

> The picture was an attempt to depict the full-blown effect of the visual disturbance during that phase of the attack.…The shimmering can start anywhere, but always expands clockwise. It appears predominantly on the edges of vision, but it is not as discrete as a tunnel; it is more like a frame. And while my vision remains in color, the actual disturbance does seem to be predominantly monochrome, with sharper definition from light to dark than what I'm actually looking at. …I have several of these a week (I think of them as *petit* migraine) while I only suffer from *grand* migraine about once a month, or after a specific trigger, which is usually a release of tension, and occasionally bright lights (p. 166, [18]).

Migraine Art, like Sacks in his *Migraine* book, notes that migraine patterns are linked to certain distinctive shapes introduced by German-American scientist Heinrich Klüver (1897–1979), now called Klüver forms or Klüver constants, discussed in Chap. 12. These form simple illusory patterns like spirals, fan shapes, or concentric circles, all of which correspond to oriented stripes of neural activity on the visual cortex, or they form illusory lattices that correspond to a superposition of cortical stripes. Klüver studied them broadly and in relation to mescaline, flickering lights, transcranial electric currents and magnetic fields, epileptic seizures, and even by binocular pressure on the eyeballs [19]. These forms are also linked with both art and synesthesia.

Finally, Sacks wrote his *Migraine* book in the summer of 1967 while in England. After completing it, he sent the manuscript to A. P. Friedman, the head of the migraine lab where he worked, in the hopes that Friedman would like it and write a foreword. Instead, Friedman told Sacks the book was awful and that he would fire the young doctor if he published the manuscript. Moreover, if he published it, Friedman said, he would make sure that Sacks could never get another neurologist job in America [4]. Sacks published it despite these threats. After it was released in 1970 puzzled colleagues asked Sacks why earlier versions of some chapters had appeared under the pseudonym A. P. Friedman [4].[3]

6.5 *Awakenings* (1973): Encephalitis Lethargica

As mentioned earlier, Sacks began seeing patients at Beth Abraham, a chronic disease hospital affiliated with the Albert Einstein College of Medicine in the Fall of 1966. There he found about eighty patients of the approximately five hundred residents were survivors of the *encephalitis lethargica* (or sleeping sickness) that broke out early in the twentieth century. When he arrived at the hospital, he was struck by how the post-encephalitic patients were frozen in odd postures and parked in wheelchairs around the hospital.

> Many were frozen in deeply parkinsonian states, some stuck in catatonic postures—not unconscious but with their consciousness suspended at the point where the disease had closed in on certain parts of the brain. I was amazed when I heard that some of the patients had been like this for thirty or forty years—indeed that the hospital had originally been opened in 1920 for these first victims of the *encephalitis lethargica* [4].

By the late 1930s most of those initially affected in the 1920s had died and the disease itself was largely forgotten by the world-at-large. Sacks found that several features stood out in the surviving group as a whole. One was that no two patients were alike. Another was that there were enormous disturbances occurring at every level of the nervous system. Sacks also found that the nurses at Beth Abraham who worked with them regularly believed that their minds and personalities were intact. They concluded this through periodic incidents that brought the patients to life. Music, for example, was particularly effective. Dance might briefly liberate a patient

---

[3]As Sacks puts it: "When *Migraine* came out, I got a couple of rather puzzled letters from colleagues asking why I had published earlier versions of some of the chapters under the pseudonym A. P. Friedman. I wrote back, saying I had done nothing of the sort and that they should address their question to Dr. Friedman in New York.…I think Friedman had delusions of ownership, a feeling that not only did he own the whole subject of migraine but that he owned the clinic and everyone who worked there and was therefore entitled to appropriate their thoughts and their work. This painful story—painful on both sides—is not an uncommon one: an older man, a father figure, and his youthful son-in-science find their roles reversed, when the son starts to outshine the father. This happened with Humphry Davy and Michael Faraday—Davy first giving every encouragement to Faraday, then trying to block his career. I am no Faraday, and Friedman was no Davy, but I think the same deadly dynamic was at work, at a much humbler level" (p. 124, [4]).

who could not walk. Some could occasionally sing, despite their inability to speak; or move spontaneously (*kinesia paradoxa*) [4, 13],

After the doctor spent a year and a half observing these patients, taking notes, and sometimes filming them, he asked the director of the hospital if he could move them onto a single ward in the hopes that a community would form [4, 13]. This did help them form a sense of community, one that was strengthened once the L-Dopa drug began to liberate them and then, at times, torment them.

In the 1950s it was established that the parkinsonian brain was deficient in the transmitter dopamine and that it might be "normalized" by raising the dopamine level. Sacks hypothesized that experimental drugs being used with parkinsonian patients might work with his patients as well because the encephalitic disorder resembled Parkinson's. He also had concerns because the encephalitic disorder was more complex and of greater severity. In an effort to test his hypothesis, Sacks applied and received a special experimental license from the United States Drug Enforcement Administration to test the L-DOPA drug. With this license Sacks began a double-blind trial with six patients. After experimenting with dosage amounts for a few weeks he found one that worked. Indeed, the results were so convincing that Dr. Sacks decided to make the drug available to all of the patients.

Despite the apparent initial success, however, there were ups and downs, particularly in the first years. It was also clear that the outcomes were unpredictable. Each patient had a different reaction although eventually some were able to continue to enjoy lives with possibilities completely out of the question before L-DOPA woke them, but not all. Sacks began to document their changes using a tape recorder, photography, a moving camera, and in his notebooks. Over time, as Sacks worked with these patients, he thought about why the struggle to survive that he found in his patients was not emphasized when he was a student or resident, or even in the medical literature of that time.

Awakenings, his writeup of this Beth Abraham work, was first published in 1973 with the British publisher Duckworth & Co. [13]. The title, *Awakenings,* comes from Ibsen's play, *When We Dead Awaken* [4]. The doctor was initially quite confused by the mixed reactions to the book. On the one hand, upon publication, Richard Gregory (1923–2010), a psychologist, wrote a wonderful review in *The Listener*. Another extraordinary outcome of the publication is that in 1974 the book won the Hawthornden Prize, as noted earlier. Also, prior to his death, his friend the poet W. H. Auden read the book and told Dr. Sacks that he thought it was a masterpiece. This was noted on the cover of subsequent releases and the 1976 edition of the book is dedicated to the poet's memory with an extract from Auden's 1969 poem "The Art of Healing," a poem that references the poet's father, a physician:

> 'Healing,'
> Papa would tell me,
> 'is not a science,
> but the intuitive art
> of wooing Nature.'

On the other hand, the reaction from the British medical community ranged from silence to outright condemnation. Their attitude was so surreal that Sacks decided to postpone publication in the United States [4]. He was particularly unsettled by a review that accused him of making up the patients, the hospital, and even the disease. In some ways these accusations were ironic because Sacks' had deliberately obscured the names and renamed the hospital to protect patient confidentiality [4].

The silence and accusations also led Sacks to work with filmmaker, Duncan Dallas on a documentary for Yorkshire Television about his patients [20]. This was in 1973, and he did the film with the permission of the patients. They urged Dr. Sacks to make the movie after the accusation that the book was fabricated. This documentary incorporates research films, still photographs, interviews, and the doctor's writings. It also offers a potent counterpoint to both the book and the 1990 commercial picture directed by Penny Marshall for Columbia Pictures, starring Robin Williams and Robert De Niro. One fascinating feature we see in comparing the two films is that at times there are similar behaviors represented in both. For example, some noteworthy scenes in both movies were taken directly from the doctor's films. One striking example is that both films include a sequence where we see the seemingly inert patients surprisingly catching balls thrown at them.

Largely unseen by the public now, the documentary did win a Red Ribbon at the 1978 American Film Festival. It also won first prize at the 1978 International Rehabilitation Film Festival. A particularly artful component in the documentary is the way it edits in the early footage Sacks took of the actual patients to record progress and behaviors, such as the ball catching scene mentioned earlier. In production, they also edited in newspaper articles and historical reports about the encephalitis epidemic. Because the actual patients are seen in the documentary their gaits and expressions offer a fuller description of the awakenings. Additionally, in the documentary, the young Dr. Sacks sometimes speaks in the episodes. At other times we hear an off-camera interviewer speaking to the patients. Another noteworthy feature is that this actual footage does not embellish or clean up their stories in the manner of the commercial movie.

While space does not allow me to introduce all of the patients, a few do show how widely they differed. One patient we meet is Sylvia (Sylvie) Schneider. The film's narrator introduces her as the most dramatic awakening. One night, in 1926, Sylvia dreamed the world had come to a stop, that she was a living statue of stone. By morning, the nightmare had come true. Then, after 35 years, frozen and inert in the hospital, she regained the enthusiasm for the life that she had known 40 years before.

Sacks noted in his diary that despite her enthusiasm, something in her manner was strange: "Is it possible that Sylvia has in fact never moved on from the past?" [20]. His speculative question illustrates her difficulty in adapting to the world in which she now found herself. It was clear from the start that she has retained memories, for in one frame we see her writing her name on a blackboard with the date July 10, 69.

> She was in a marvelous mood, for a while. She was talking away and singing and dancing, but almost everything she said and did had reference to 1926 or before. She referred to figures who were topical at that time... Some of her mannerisms, some of her slang was of a sort which had been obsolete for 40 years. She was, so to speak, a "flapper" who had come to life. Sylvia's mood was joyous, elated, and salacious. Alone in the room, she sang her favorite songs [20].

Sacks goes on to say that he felt, at the time, that it was the most extraordinary scene he had ever witnessed. Sylvie was a flapper in the 1920s before contracting *encephalitis lethargica* and, as it turned out, she wanted her memories more than the person she found upon awakening. Sylvie was one who was unable to make the adjustment. After the summer of 1969 she fell back into that inscrutable place where she had fallen after the terrible nightmare that had changed her life in 1926. Nothing they did would budge her. Her case helped Sacks recognize that they had not been literally asleep, but in abeyance for decades.

Sylvia, with Mary, inspired the character of Lucy in the commercial movie starring Robin Williams. The singing belonged to Mary; the nostalgia belonged to Sylvia who wanted to awaken in 1926 still. Sacks called her the "Sleeping Beauty" in his book.

Mary had a tragic first response to L-Dopa. It shattered her personality so badly that they doubted any possible happy future outcomes. Yet, by the early months of 1973, Mary started to put herself back together. Her lovely attitude was finally able to surface again, and so did her skills. Sacks writes in his last notes that the last 4 years of her life were the best she had experienced since she was stricken. For the first time she was freed from the most severe post-encephalic symptoms and from the roller-coaster of her early L-Dopa years, Mary would engage with the other patients, take care of the garden and, most of all, she would sing. Unfortunately, in 1976 Mary broke her hip. The surgery resulted in complications that left her in bed for months, once again. She did not complain and the day before she passed away, she asked to hear Mozart's Requiem.

Another patient well-articulated in the documentary is Lillian. At one point, the interviewer asks her about fixing her hair while her head was gyrating up and down. Lillian's manner is calm, and she seems somewhat amused by the activity of fixing her hair with comb and pins as she is coping with her head's gyrations. The sense of amusement, or perhaps bemusement, is particularly evident when the off-camera interviewer asks if it is hard to do her hair when her head is moving about so much. She explains that she has gotten used to it, "but it's the devil to make a part." Lillian also explains—and we see for ourselves—that when the drug wears off, she can hold her head still, but then her hands and legs become useless and tremble.

Lillian was still well enough to come to the set when the commercial movie was made in 1990. Unfortunately, the one scene in which she actually appeared, with Robert De Niro, was cut, leading Dr. Sack to quip, "I guess they thought real patients weren't authentic enough" [21]. She died in 1992, having outlived all the other original patients.

Lola, another patient, was subject to very extreme reactions. Sometimes she was "exploding" and at others "freezing." According to Sacks her management of these reactions was balanced and she was pretty much on top of things. She would engage in all available activities: Bingo games, gardening, poetry readings, even day trips. Despite the worsening of her dystonia and kyphosis this was possible with the help of another patient she had befriended. Sacks characterized this as a sort of symbiosis that developed between the two of them; they came to the rescue of each other. Examples include her friend pushing her chair so that she could move around. When her speech became too difficult for most people to comprehend, her friend was able to step in and "translate" her words. Because Lola's friend had become slightly brain-damaged, Lola would guide her. In some ways, according to Sacks, their relationship was like that of two friends helping one another and their connection inspired everyone.

The commercial film's primary patient was Leonard Lowe, played by Robert de Niro. His actual name was Ed, and he was awakened in 1969. One of the first awakened, the commercial film barely encompasses his experiences. For Ed things were up and down because the drug's side effects hit him the hardest and Sacks took him off medication almost immediately. Then the doctor tried multiple times and unsuccessfully to figure out the proper dosage, but the balance was never found. Both L-Dopa and amantadine would only trigger unbearable tics and exaggerated pathological responses. For Ed, the awakenings felt like a curse.

As the various accounts show, the commercial film combined some patients and re-wrote others. Of more importance, the disease went on even after *Awakenings* was filmed and published. Watching the two cinematic productions makes it evident that the actors in the Hollywood version studied the documentary in minute detail, which does add some legitimacy to their portrayals. Similarly, Robin Williams studied Oliver Sacks' mannerisms. Indeed, I can recall attending a Sacks talk many years ago in which he spoke about how Robin Williams mirrored him so well in the film that for months after its release people would come up to the doctor on the street and say, "You're Robin Williams, aren't you?"

After seeing the documentary in 2021, Wendy Lesser revealed that Dr. Sacks and his staff had thought about airing it on American television at about the same time the commercial film was scheduled for release. It was decided that this was not a good idea since Columbia Pictures said they did not want it competing with their movie. Therefore, it was not shown [21].

Finally, unlike the documentary, which tried to verify the accounts in the 1973 book, multiple embellishments in the commercial film skewed many aspects of the reality. Most notably it introduces a fictional doctor, Dr. Sayer, played by Robin Williams, a character based on Sacks. Using a fictional character allowed for modifications and the addition of dramatic sequences such as Sack's flirtation with a female nurse. The real Oliver Sacks was a gay man, and this kind of re-write is a common device used to enhance a movie's entertainment value. In the film they also toned down the more violent or sexually aggressive elements of the Beth Israel patients. Leonard Lowe, de Niro's character, was perhaps the closest to reality but even he has some scenes in the *Awakenings* movie that do not appear in the book. Among them are his flirtatious moments that were similarly added to inject more Hollywood drama and to keep audiences engaged.

6.6 *Awakenings* Derivatives

One of the first adaptations of *Awakenings* was *A Kind of Alaska* a play by British playwright and Nobel Prize winner Harold Pinter.[4] Prior to receiving Pinter's play Sacks had reservations about dramatic representations or any kind of fictional adaptation of his work. He felt this kind of production could distort the reality of the actual cases because the writer of the fictional derivation lacked direct contact with the patient. After Pinter's work premiered in 1982 the play convinced Sacks that this kind of project had an affective character that was not necessarily a dilution of reality. Rather, in terms of neuroculture, the people in an adaptation become interwoven into our lives in a way that compels us to ask about identity, pathologies, and human behavior [1]. Later Sacks claimed that it was the Pinter play that showed him how a great artist can re-confer, reimagine, reality. "I felt Pinter had given me as much as I gave him: I had given him a reality, and he had given me one back" (p. 393, [4]).[5]

Pinter's one-act play takes place over the course of a single afternoon. It opens with the awakening of Deborah, who has been in a strange, inaccessible, frozen state ("a kind of Alaska") for years. She is now 45, but feels she is 16. (She is much like Rose R. in the book, a patient inspired by Sylvia Schneider.) Curiously, Judi Dench, who portrayed the aged Iris Murdoch as an Alzheimer sufferer in the film *Iris* [23], had the starring role in Pinter's original production.

Another adaptation of *Awakenings* was an opera by Tobias Picker and Aryeh Lev Stollman. It premiered in St. Louis on June 11, 2022 [24]. Composed by Picker, it was commissioned by Opera Theatre of Saint Louis, with a libretto by Picker's husband Stollman, an acclaimed novelist who had never written a libretto before [24]. Picker himself is a successful composer of European classical music and knew Sacks for many years before writing the opera. He had contacted the doctor for help with his Tourette syndrome. Sacks wrote about Picker in two of his books where he theorized about a connection between Tourette syndrome and Picker's early composition style, which at times featured shifts in tone and dynamics.

This opera was the first derivation of the *Awakenings* book to portray Oliver Sacks as a gay man, in this case a shy, gay man struggling with his sexuality. Sacks had not revealed he was gay publicly until his 2015 memoir, where he spoke openly

[4] Pinter first sent his manuscript for *A Kind of Alaska* to Oliver Sacks in 1982, before the commercial *Awakenings* film was released. The play was later performed as part of a trilogy of Pinter's plays titled *Other Places Other places* was initially published as a trilogy in 1982, comprised of "A Kind of Alaska," "Victoria Station," and "Family Voices" [22]. In 1984 a fourth play, "One for the Road" was added to the group [22].

[5] In his autobiography Dr. Sacks notes that he felt similarly inspired by Peter Brook's theatrical presentations of *L'Homme Qui* in 1993 and *The Valley of Astonishment* in 2014, and by a ballet inspired by *Awakenings* with music by Tobias Picker.

about his sexuality [4].[6] Like the commercial film, the opera dramatizes the *Awakening* story rather than presenting actual events. This version has a triangle of unrequited love among Sacks, a doctor known as Mr. Rodriguez (Andres Acosta), and patient Leonard Lev.

With the *Awakenings* opera Picker was able to speak to Sacks' closeted period through introducing an incident Sacks referenced in his memoir in 2015. When Sacks was a teenager, his mother became enraged upon discovering he was gay. She called him an abomination and said she wished he had never been born [4, 24]. Part of this incident involved a larger and complex cultural reality. When she made her comments, homosexuality was illegal in England. This made the exchange about more than a personal condemnation and Sacks wrestled with what it meant to be a homosexual from a young age. By the age of 12 he had read many books on sexual pathology by authors like Richard von Krafft-Ebing, Magnus Hirschfeld, and Havelock Ellis [4]. After Sacks came out publicly it was clear that his homosexuality had had an influence on how the doctor's life unfolded and in his development of a neurological approach that had a keen eye for identity and humanness.

Sacks himself, after many years of celibacy, met the writer Bill Hayes, with whom he spent the last 6 years of his life. Before this, he had gone through many phases. Essentially, he found it difficult to feel he had a "condition" and that his identity could be reduced to a name or diagnosis. He also thought about how homosexuality aroused horror in some people. As Picker pointed out, the importance of including this episode with his mother is that Sacks treasured her and how he wrestled with her comments are key in understanding both Sacks and the empathy he had for people whose medical conditions made them outsiders or intersected with questions others raised about their identity [4, 24].

The various derivatives of *Awakenings*, as is the case with the opera, underscore the myriad of ways in which Sack's patients have become a part of an extended family, and when Sacks refers to them from publication to publication, his readers know them too. As he cross-references cases from book to book and project to project readers get a better grasp of related neurological conditions.

6.7 *A Leg to Stand On* (1984): Disturbance of Proprioception

A Leg to Stand On [12], one of Dr. Sacks' more personal books, is a rather slim volume that took 11 years to write.[7] He has said it may be regarded as a sort of neurological novel or short story (p. 11, [12]), and it does have the flavor of a fictional tale

[6] Earlier Picker had wanted to write an opera on Sacks' life, a project the neurologist encouraged, but there was a conspicuous hurdle since Sacks had not yet come out publicly as a gay man. Finally, even before the opera, in 2010, Picker teamed up with choreographer Aletta Collins to write a ballet inspired by *Awakenings,* 2010 for the Rambert Dance Company that premiered in the United Kingdom.

[7] As he often did, he also recorded the case in *The New York Review of Books* as well, under the title "The Bull on the Mountain" [25].

in some ways, despite how real the experience was for him. It also evokes a comparison with *Migraine* because, despite Sacks being a migraineur, his personal voice was not apparent in his early book on this subject. By contrast, his autobiographical *A Leg to Stand On* introduces the doctor as both a neurologist and a patient.

Leg records an accident in Norway that followed a sequence of trying professional events. First, Sacks was fired unexpectedly by Beth Abraham Hospital, where he worked with the encephalitis lethargica patients memorialized in *Awakenings.* With his firing, he also lost the apartment they provided him. This led him to take a part-time job at the Bronx State hospital where he worked with a variety of problems: autism, retardation, fetal alcohol syndrome, tuberous sclerosis, early-onset schizophrenia, etc. [4].

Appalled by the behavior modification policy adopted by his ward there, Sacks tried to bring out the potentials of his patients by engaging them through play and art. Several of the autistic patients were stimulated by music and drawing so he had his piano transported there. The young patients loved it, often performing elaborate dances as he played [4]. The hospital administration did not share the patients' enthusiasm for these new stimuli. They were dismayed by what they saw as Sacks' effort to undermine their program. This admonishment led Sacks to leave his position and take the trip to Norway discussed in *A Leg to Stand On.*

In Norway, Dr. Sacks' tragic encounter with a bull was followed by his miraculous rescue by local huntsmen and the flight from Norway to England for an operation. *Leg* offers an account of the ups and downs of patient Sacks' recovery, and his discovery of the therapeutic value of music in fostering his recovery.

> The leg incident taught me in a way which I could not, perhaps, have learned otherwise about how one's body and the space around one are mapped in the brain and how this central mapping can be profoundly deranged by damage to a limb, especially if this is combined with immobilization and encasement. It also gave me a feeling of vulnerability and mortality which I had not really had before (p. 218, [4]).

Briefly, a series of events left him with a complete rupture of the quadriceps tendon and marked impairment of conduction in the femoral nerve that supplies it. After the operation he could neither move nor feel his damaged leg. Despite the pain of his fractured femur and his miraculous survival, during his recovery he struggled to understand what had happened both physiologically and philosophically. What was both puzzling and disconcerting was that he experienced an odd lack of ownership of his deafferented limb, even after the quadriceps tendon came back to life. He was unable to walk because the leg felt alien to him, as if it were not his. He has described it as the opposite to a phantom limb:

> [I]t was an external limb with no internal image to give it presence and life.…I ruptured the quadriceps tendon in my left leg. Though the tendon was repaired surgically, there was damage at the neuromuscular junction, and additionally, the leg was hidden from sight and touch, immobilized in a long, opaque cast. Under these circumstances, where it was impossible to send commands to the injured muscle and there was no sensory or visual feedback, the leg disappeared from my body image, leaving (so it seemed to me) an inanimate, alien thing in its place. This continued to be the case for thirteen days (p. 222, [16]).

One of the more difficult parts of this entire ordeal was his inability to communicate his experience to the surgeon who operated on him, an encounter that influenced his approach to patients. Equally influential was finding the value of music in his eventual recovery. One day, after a futile session with the physiotherapist, he decided to listen to music while attempting to perform an exercise. Meditating on the music and without thought, he stood on both legs and began to walk with the music despite weakness. According to neurologist Sacks, the epiphany as patient Sacks was perhaps the most eventful and crucial 10 minutes of his life.

As he recalled, he had been listening to a Mendelssohn's Violent Concerto tape constantly in the hospital and he believes that the "kinetic melody"was the driving force. When the music stopped, he needed to return to it to keep going. Then, in his words: "within an hour, I regained fluent, automatic walking and no longer needed any imaginary musical accompaniment" (p. 217, [4]).

His loss of proprioception was essentially at the center of the event. Explaining what it seemed like to him to not "feel" his leg, Sacks wrote:

> When I closed my eyes, as a start, I had no feeling whatever of where the leg lay—no feeling that It was "here" as opposed to "there," no feeling that it was anywhere—no feeling at all. And what can be felt, what can be posited, about something that is 'not there'? It seemed, indeed, as if it was this profound disturbance of proprioception, which had only been discovered and revealed by a fluke, though then investigated carefully by Nurse Sulu and myself—it seemed as if this, in particular, was somehow the "last straw," Serious problems and questions had already arisen, relating especially to the injured and operated muscle: its great atrophy, its atonia, its apparent paralysis. And questions of a "higher" sort, just before I fell asleep—the apparent breakdown in "know-how" and "idea," so that I could no longer "think" or "recall" how to make muscular movements involving the muscle. There was already something odd going on at this point. But a total, absolute, "existential" breakdown immediately followed, and seemed precipitated by, the discovery of the breakdown of sensation and feeling, for it was then, and only then, that the leg suddenly assumed an eerie character—or, more precisely, if less evocatively, lost all its character—and became a foreign, inconceivable thing, which I looked at, and touched, without any sense whatever of recognition or relation. It was only then that I gazed at it, and felt I don't know you, you're not part of me, and, further, I don't know this "thing," it's not part of anything. *I have lost my leg* (pp. 74–75, [12]).

Sacks has explained that he had some difficulty completing *A Leg to Stand On* due to its personal nature [4]. One obstacle was that working on it forced him to re-live the traumatic accident and the "passivity and horrors of being a patient." Another obstacle was that the book required Dr. Sacks to move from "doctorly" writing to a form that would expose his own intimate feelings, a positioning that made him uncomfortable then. After working on the project between 1977 and 1982, *Leg* was eventually published in 1984 [12] with the editorial assistance of Kate Edgar, who became his life-long editor and collaborator. The book spent half a year on the New York Times best-seller list in 1985. Its broad success demonstrated that there clearly is an audience for this kind of therapeutic story.

6.8 *The Man Who Mistook His Wife for a Hat* (1985)

The Man Who Mistook His Wife for a Hat is one of Oliver Sacks' most inspiring essays and it was published in a book by the same name [26]. The title was derived from a scene in the essay where the patient, Dr. P., momentarily grabbed his wife's head thinking it was his hat. This behavior articulates how strikingly bizarre brain damage in individuals can appear to an observer.

Dr. P., a composer, singer, and music teacher visited Sacks about his visual agnosia, sometimes described as prosopagnosia in the literature. Initially he and his wife presumed that his diabetes has affected his vision. An ophthalmologist, suspecting a neurological problem, referred him to Sacks. Testing showed Dr. P. was able to perceive separate features of objects but could not correctly identify them or the whole objects of which they were a part. In the essay, Sacks, finding he was unable to treat Mr. P.'s problem, encouraged him to focus on his musical interests, which seemed at this point to remain intact.

In 1986, an opera of the case study was composed by Michael Nyman (b. 1944), an innovative English composer, pianist, librettist, musicologist, and filmmaker. When Michael Nyman first approached Oliver Sacks with the idea, the doctor felt that opera was the furthest thing from a case history even if some stories have an operatic quality. Much as was the case with Pinter's play, at the end, Dr. Sacks decided the project worked, it even seemed natural. Although it was Nyman's first opera, this mesmerizing work successfully met the challenge of depicting Mr. P.'s displacement of a visual world through a musical medium.

One dilemma Nyman needed to resolve in successfully putting the work together was how to construct and build a narrative structure. How does one capture an exploration into the inner world of a patient? How does one convey what the neurologist is asking and how it might be helpful? Nyman decided to address these requirements from the perspective that the patient himself has a need for a narrative. Using this approach, the opera is not a narrative with a plot but rather a presentation that retains its neurological underpinnings in looking to define the patient's situation.

Artistically fusing both visual and musical elements affectively and effectively captures the case despite its lack of a conventional story line. Of course, this would not work well for every case, but it does here because Mr. P.'s musical identify was so much a part of his person even before his visual challenges were noticeable. One splendid part of the work is how sequences build a coherent presentation despite how they range from a sense of randomness and confusion to musical sequences that evoke a magical wonder.

On stage the opera unfolds as a series of diagnostic events with each one making music the central hero of the story. Nyman wanted to create a world that would be logical on its own terms and the question he asked himself was how to create music that represented a face that would lose contact with the details of another face. Thus, we are initially introduced to the world of Dr. P., a man who first notices he cannot identify people (face blindness or prosopagnosia). The script then advances into a more general visual agnosia (or "mental blindness"). In these scenes he could see but was unable to see accurately. He saw colors, lines, boundaries, simple shapes, patterns, movement, and so forth—but he could not bring the elements of the world together coherently.

Staging the events dramatically allowed for creative license so the presentation allows us to grasp that Dr. Sacks recognized Dr. P.'s musical gifts and how Dr. P.'s musical persona used his music as an organizing principle for his behavior. We also perceive that Mr. P., as a gifted performer, could manage life once he put his actions into music.

Just as the documentary and commercial version of *Awakenings* present the case in different ways, the televised version of *Hat* differed from the performed opera in key respects. An important one is how editing and close-ups of the performers in the television production bring the audience more deeply into the details. Another effective use of editing was the way the director used it to seamlessly update information gleaned on the brain's pathology after Dr. P. passed away. Using parallel editing, a common technique that intercuts scenes from different locations, the editor added new information. Oliver Sacks and pathologist John Tighe explain that Dr. P.'s was a case of atypical Alzheimer's. It affected only one portion of his brain until its latter stages:

> It turned out that his illness was Alzheimer's, irreversible shrinkage of brain tissue, a form of pre-senile dementia. Yet in spite of its gradual advance, Dr. P's inner musical world sustained him in living and teaching for some years. But when, as it will, this tragic degenerative process went its course, rendering him speechless lacking all bodily control, apparently soulless, demented, and barely alive, music sustained him, found its way through the choked pathways of his mind, massaged his neural decay, fueled his spirit, defined his humanity to the end [27].

Mr. P.'s Alzheimer's was quite unlike that of Iris Murdoch although musical aspects of their world sustained both and were therapeutic. For Murdoch music input did not give her the same kind of operational and organizational power on a day-to-day basis. Mr. P., however, was able to live a fairly full and normal life outside of an institution because his music aided him in doing daily tasks.

A helpful and enthralling difference between the written case and the opera came about because of Nyman's interactions with Mr. P.'s widow. Before starting the work, Nyman visited her and decided that Sacks failed to do justice to her help in sustaining Mr. P. We see this in one expanded sequence in the opera during the discussion on Mr. P's paintings. His visual art was only summarily mentioned in the case study. This visual addition enhances the opera's overall presentations and is additionally compelling because we can perceive his neurological degeneration through his painted compositions. In this opera, after Sacks comments on the paintings on the wall of their home, Mrs. P. responds that her husband was a gifted painter as well as a singer. Because his canvasses were hung in chronological order, Sacks was able to perceive alterations within his style over the course of his life.

Mr. P.'s early work was naturalistic realism, with vivid mood and atmosphere, finely detailed and concrete. Years later, the paintings became less vivid, less concrete, less realistic, less naturalistic, more abstract, and even cubist. The last canvases looked like nonsense to the doctor—mere chaotic lines and blotches of paint. Commenting on this to Mrs. P., she exclaimed: "Ach, you doctors, you're such Philistines! ... Can you not see artistic development—how he renounced the realism of his earlier years, and advanced into abstract, nonrepresentational art?" (p. 17, [26]). In the essay, Sacks had explained that he concluded that,

> ...this was not the artist, but the pathology, advancing—advancing towards a profound visual agnosia, in which all powers of representation and imagery, all sense of the concrete, all sense of reality, were being destroyed. This wall of paintings was a tragic pathological exhibit, which belonged to neurology, not art (p. 17, [26]).

As Sacks often noted later, the latter days of Dr. P. were left out of the case study because he did not know the full details when he wrote the case. Years later he learned about his Alzheimer's. He also learned that Mrs. P. took him to a concert in a wheelchair shortly before he passed away. At that point, Mr. P. could not speak and was without many faculties. Nonetheless, in the middle of the concert he shouted NO, because he thought the music was begin badly played. This would suggest that his musical faculty was still alive deep in his center [28].

In an interview, Sacks spoke about Mr. P.'s reaction, saying that musicality is absolutely rooted in our being. One sees it in all sorts of patients, but it must be music with a particular significance for the person [28, 29]. Comparing musical perception with linguistic perception, Sacks added that the musical precedes linguistics and also survives it. People with aphasia who have lost understanding and use of language may be able to sing perfectly but in a song with words the words might be garbled [29].

6.9 Peter Brook: *The Man Who*

Among the derivatives of *A Man Who Mistook His Wife for a Hat* is one by English theatre and film director Peter Brook (1925–2022). Brook's adaptation used the entire book with a sequence that alludes to the Mr. P. essay. Like Pinter, Brook made his play a part of a trilogy on the human brain. The *Hat* play is titled "The Man Who" (1993). All three components of the trilogy explore various neurological conditions. Each is intended to capture the radically different ways in which people perceive the world. The second work, "Je suis un phenomene" (1998), is based on the book of the Russian neurologist Alexander Luria about Solomon Shereshevsky [30]. The third, "The Valley of Astonishment" (2014), presents synesthesia, a physical condition where two senses are inextricably linked; we meet a painter who sees colors when he hears music. In this play Kathryn Hunter played a woman with an almost limitless memory. The character of the synesthetic painter in Peter Brook and Marie-Hélène Estienne's play The Valley of Astonishment, 2014, was based in

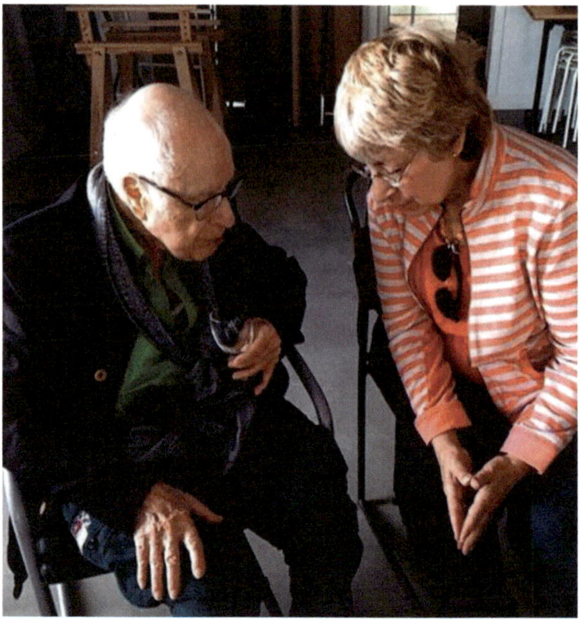

Fig. 6.1 Peter Brook and Carol Steen talking about synesthesia and hypnagogia above Peter's theatre Bouffe du Nord, Paris. July 13, 2019. Photograph courtesy of Carter Jones

part on the life and work of artist/synesthete Carol Steen's life, discussed in more detail in Chap. 12 [31] (Fig. 6.1).

"The Man Who" includes a succession of 17 scenes intended to show the myriad perceptions of reality of which the brain is capable. As with Brook's other experimental works. This work calls for wild leaps of imagination and leaves the narrative somewhat open to interpretation. In this it is unlike the Nyman opera's interpretation of Sacks' work, where the case is largely intact, and it is in the syntax of the narrative where we see the creator's hand. In "The Man Who" the cast takes turns playing the parts of doctor and patient, erasing the line between them, and raising the question of who is normal in the process.

One scene often mentioned by those who have seen the play performed introduces a young man sitting immobile in a hospital room. He has lost all feeling in his body from the neck down. His lost proprioception (kinesthesia) and the sense of self-movement, force, and body position brings to mind Sack's descriptions of his disturbance of proprioception as described in *Leg*. At one point in the play the man realizes that if he concentrates on his right arm, he can raise it and even make the right hand grasp the side of the chair. More concentration allows him to stand, and as long as he sees his feet, he is able to move around the room, lurching in the stiff-legged manner of Frankenstein's monster.

> "I can walk," he announces, "but nothing has been achieved." Each day the young man must again learn to do what he did the day before. "There are no habits," he says with resignation, "only tricks and stratagems." It's as if his body were a land devastated by war, a place without civil order where life's most basic activities have become endeavors worthy of an Odysseus [32].

Oliver Sacks' life, writings, and related derivatives illuminate the complexity of both neuroscience and art. The next chapter expands the technological tools Sacks used for research into a case study on technology. Indeed, technology has brought neuroscience and art together historically and this integration is also a part of the neurocultural milieu.

References

1. Sacks O. Awakenings. New York, NY: HarperPerennial; 1990.
2. Sacks O. The mind's eye. New York, NY: Alfred A. Knopf; 2010.
3. Sacks O. Migraine: the evolution of a common disorder. Berkeley, CA: University of California Press; 1970/1992.
4. Sacks O. On the move: a life. New York, NY: Alfred A. Knopf; 2015.
5. LC/NIMH Decade of the Brain Project. Project of the decade of the brain. Washington, DC: Library of Congress; 2022.
6. Rowland LP. In memoriam: Oliver Sacks, MD (July 9, 1933, to August 30, 2015). JAMA Neurol. 2016;73(2):246.
7. McGee B. Modern philosophy: philosophy and literature with Iris Murdoch and Bryan Magee. BBC; 1977.
8. Sacks O. Uncle Tungsten: memories of a chemical boyhood. New York, NY: Alfred A. Knopf; 2001.
9. Sacks O. Preface (2013). The man who mistook his wife for a hat and other clinical tales. New York, NY: Vintage; 2013.
10. Burns R. Oliver Sacks: his own life. Kino Lorber; 2019.
11. Sacks O. A neurologist's notebook: face-blind: why are some of us terrible at recognizing faces? The New Yorker; 2010:36–43.
12. Sacks O. A leg to stand on. New York, NY: Summit Books; 1984.
13. Sacks O. Awakenings. London: Duckworth; 1973.
14. University of North Carolina Asheville. The Hawthornden Prize.
15. Liveing E. On megrim, sick-headache, and some allied disorders: a contribution to the pathology of nerve-storms. J. and A. Churchill; 1873.
16. Sacks O. Hallucinations. New York, NY: Alfred A. Knopf; 2012.
17. Cardano G. The book of my life (De vita propria liber). New York, NY: EP Dutton and Co.; 1576/1930.
18. Podoll K, Robinson D. Migraine art: the migraine experience from within. Berkeley, CA: North Atlantic Books; 2008.
19. Billock VA. Hallucinations, Oliver Sacks. Optom Vis Sci. 2013;90(4):e124–e5.
20. Dallas D, Sacks O. Awakenings [documentary film]. United Kingdom: Yorkshire Television; 1973.
21. Lesser W. Seeing 'Awakenings' with its real-life cast. The New York Times; January 21, 2001.
22. Pinter H. Other places: three plays. London: Methuen; 1982.
23. Eyre R. Iris: Miramax films. 2001.
24. Goodwin JD. An 'awakenings' opera that shows Oliver Sacks as his friends knew him—a gay man. St Louis Public Radio; 2022. https://news.stlpublicradio.org/arts. Accessed 18 Jan 2023
25. Sacks O. The bull on the mountain. The New York Review of Books; June 28, 1984.
26. Sacks O. The man who mistook his wife for a hat and other clinical tales. New York, NY: Perennial Library; 1987.
27. Rawlence C, Morris M, Nyman M. Michael Nyman—The man who mistook his wife for a hat (1987 TV Opera). 1987:1:12:46.
28. Nyman M. The man who mistook his wife for a hat: one act chamber opera. London: Libretto by Christopher Rawlence, Michael Morris, & Michael Nyman; 1986.

29. Institute of Contemporary Arts Oliver W Sacks, Michael Nyman, Michael Norris and Christopher Rawlance, In Conversation. Lecture. London: Institute of Contemporary Arts; 1987.
30. Luria AR. The man with a shattered world; the history of a brain wound. New York, NY: Basic Books; 1972.
31. Audience TfaN. The Valley of Astonishment (Theatre for a New Audience 360° Series). In: production TdBdN, editor. 2014.
32. Canby V. Theater review: the man who; Peter Brooks charts twists and turns of the human mind. The New York Times; March 15, 1995.

Neuroculture and Image Technology

7

The last chapter on Oliver Sacks offered a glimpse of how film and other technologies have both aided neuroscientific study and brought its work into the cultural sphere. The impact of film, photography, and other media on both neuroscience and art correlates with the technological revolution of the nineteenth century. Then, photography, moving images, and X-rays impacted popular culture, art, and therapeutics while also advancing studies of physiology and living systems [1, 2]. How these new modalities, particularly film, radically altered the neurocultural landscape is the subject of this chapter. One reason to examine these technological developments neuroculturally is that each showed that the brain is creatively involved in activities that alter the world, art, and culture. Another, in neuroscientific terms, is that just as earlier adjustments to our knowledge base furthered our grasp of physiological functions but did not fully replace the value of engaging with gross anatomy and individuals; these new technologies advanced our knowledge base and cultural possibilities without replacing other ways of exploring the human brain, human psychology, art, and culture.

Photography—the ability to capture and chemically fix images of the external world on a light sensitive surface—offered a means to expand our memory through visually recording discrete moments. Mechanized moving images (film) added to our understanding of dynamics and, like photography, memorialized what was recorded. X-rays led to applications that could perceive what is invisible to the naked eye. X-ray images additionally led to ways of accessing and transcribing invisible radiation [1, 2]. With the entry of computers, it became possible to combine X-rays and other non-optical imagery with expansive data visualization options, a fascinating subject beyond the scope of this volume.[1]

[1] The best source for material on how X-rays changed both science and culture is still *Naked to the Bone: Medical Imaging in the Twentieth Century* by Bettyann Kevles [1]. Also see my *Innovation and Visualization: Trajectories, Strategies, and Myths* [2] and *Art and the Brain: Plasticity, Embodiment, and the Unclosed Circle* [3].

A. Ione, *Neuroscience and Art*, Neurocultural Health and Wellbeing,
https://doi.org/10.1007/978-3-031-62336-3_7

Tools like photography and film have a long and overlapping history dating back to Paleolithic times. Cave art and artifacts show that humans were exploring representations in visual forms even before writing was invented. One excavated object of animals drawn on a spinning disc articulates one way that people long ago were also toying with combinations that added motion to representation [4–6]. Photography's most immediate precursor was the *camera obscura*, a darkened room or box with a pinhole. Light entering through the small hole projects an image onto a surface opposite. Knowledge of this type of projection system dates back at least to the fifth century BCE, when the Chinese philosopher Mozi mentioned it in his writings. Shortly thereafter Aristotle noticed the phenomenon as well [7]. A precursor to film, the magic lantern, was the *camera obscura's* cousin. It is an antecedent of the movie projector, and the now ubiquitous slideshow presentation. As shown shortly, this neurocultural technology was used in research, education, and as entertainment.

Remarkably, the use of light to build the projection apparatus of early cinema is somewhat akin to Plato's (c.427–c.347 BCE) imagery in his famous Allegory of the Cave (*Republic*, 514a–520a) [8]. Plato's story posed philosophical questions about education, what we see, and what knowledge is. Although philosophical perspectives are largely beyond the scope of this book, it is helpful to periodically note their existence because they show how people use what they know to create metaphorical analogies about the world and how some come to influence later ideas and resemble later technologies. The Cave Allegory also reminds us that each innovative technology butts against older philosophical ideas.

Figure 7.1 depicts key features of Plato's Cave allegory. According to the allegory, chained prisoners are watching shadows projected on the wall in front of them. These images are projected by a fire hidden behind the wall in back of them by walking figures holding objects. The prisoners do not realize that their perceptions are merely shadows or that they are inaccurate representations of the actual world. They are also not cognizant of the people hidden behind the back wall carrying the

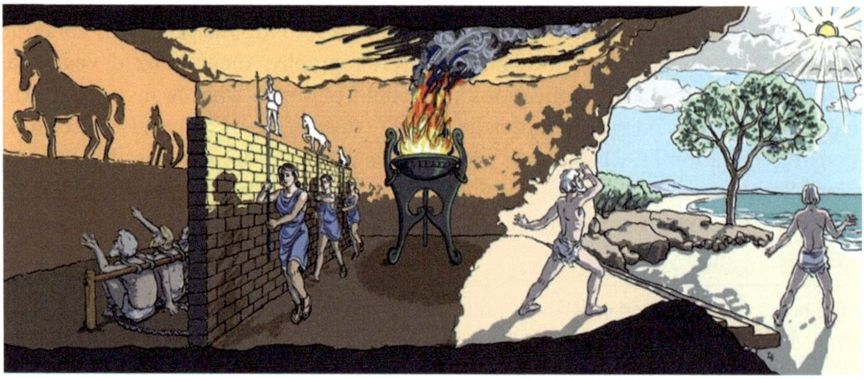

Fig. 7.1 An illustration of the Allegory of the Cave, from Plato's *Republic*. This file is licensed under the Creative Commons Attribution-Share Alike 4.0 International license

projected objects. As Socrates points out in the dialogue, why would they think the shadows were not real since they have never seen anything else? [8] Indeed, in Plato's parable, a prisoner escapes and sees the world of sunlight. When he returns and tells the others, the prisoners conclude that leaving the Cave makes one crazy.

Unlike the chained prisoners, we absorb new data, technologies, and ideas on an ongoing basis. When the various mechanized imaging technologies discussed in this chapter were introduced, people understood that they did not depict reality. Some embraced them, concluding they revealed more about the world than the eye could perceive normally. Others were critical of how they altered human living or the mechanical facets of their creation. Among the most fascinating cases are those with ambivalence. An artistic example is the painter Edgar Degas (1834–1917) who used photography secretly and privately [9].

7.1 The Magic Lantern

The earliest recorded magic lanterns are largely associated with the Dutch scientist Christiaan Huygens (1629–1695) and the German polymath Athanasius Kircher (1602–1680). These were simple devices that used a box into which a light was inserted. A single lens would project the image outwards, through an opening, onto a wall. The first magic lantern images were bright, hand-painted images on glass, like a single frame of film. Each was comparable to a single work of art and thus the painting (or preparation) process was quite time consuming.

Huygens used an optical lens for projecting his painted glass slides. Candlelight brought them to life. Kircher suggested projecting live insects and shadow puppets from the surface of a mirror in his Steganographic system around 1645. These were intended for dramatic scenes 1645 [10].

Camera obscura images had already shown that this kind of setup inverted projected images. Magic lantern boxes resolved the inversion problem by inserting the glass slides upside down, using lenses, or using a square prism [12]. Figure 7.2, a page from the third volume of Gulielmo Jacobo Gravesande's 1720 book *Physices Elementa Mathematica*, reproduced Jan van Musschenbroek's magic lantern. The top rectangular box in the illustration shows the lantern projecting a monster. The wooden lantern case, on the left, stands on an adjustable base. A tin chimney on the top of the box allows the smoke and heat of the oil burning lamp to escape. The bottom rectangular image from the reproduced page shows the inside of the box: there is an oil burning light source with four flames, a concave mirror, and an ingenious lens arrangement. This setup allowed for a projection of up to 10 meters (about 33 ft.) [13].

Figure 7.3 shows a nineteenth century counterpoint. The slide in this photograph is right-side-up, so in a proper pictorial orientation. If this box has a mechanism inside to correct the image projection orientation, then this is fine. If not, when this slide is inserted into the box it will project upside down [12].

By 1713 magic lantern slides were used for lectures [14]. In this form, they were generally shown sequentially, and without enhanced movement. Operationally, they

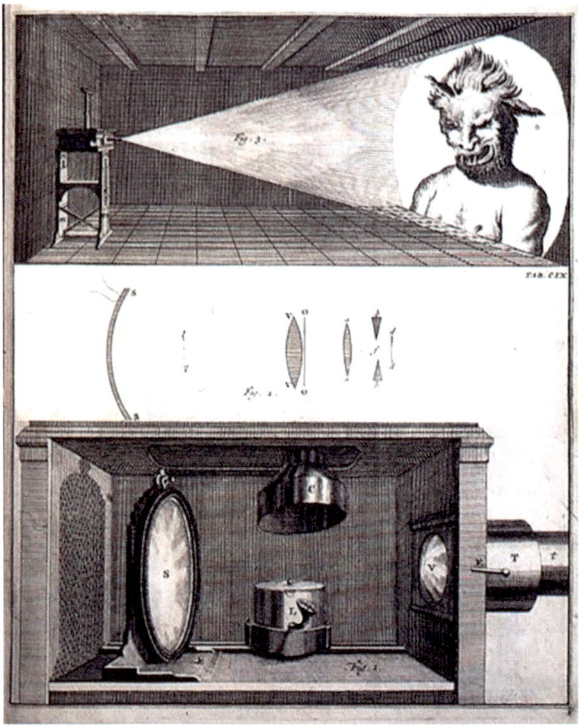

Fig. 7.2 Jan van Musschenbroek's magic lantern in Gulielmo Jacobo Gravesande's 1720 book *Physices Elementa Mathematica* [11]

worked much like a manual twentieth century 35-mm slide projector that advanced celluloid slides one at a time, or a contemporary unenhanced digital PowerPoint presentation. It was possible to create motion by adding cranks, gears, and other technologies [15], an option used more for entertainment purposes. For example, lever mechanisms might move the glass side-to-side or up-and-down. Cranking a handle attached to the slide mounting could create circular motion. The most common way to convey a scene in motion was to use two pieces of glass. On one there might be a stationary scene of a shoreline, for example, and on the second a boat might appear to pass by due to a hand-operated pulley wheel that moved the slide to create the sense of motion [16].

Hermann Vogel gave a lecture in 1874 on using magic lanterns as an instructional aid for scientific lectures. He explained that 2-inch square small slides projecting pen sketches, woodcuts, photographs, charts, graphs, and so forth could be enlarged to 4 or 5 feet on projection. Vogel prepared his slides by varnishing his material—paper photographs, woodcuts on silk paper, or drawings, etc.—onto the glass [17]. This size is somewhat smaller than the slide shown in Fig. 7.3, and about the size of a slide used in a 35-mm carousel slide; these were generally 2-inch square mounts.

In the seventeen and eighteenth centuries "magic" became a part of the instrument's name. Magic alludes to the "ghost-like" nature of the projections produced [18]. "Ghosts" were particularly evident in the pre-electricity stages when light sources were typically candles or oil lamps. Limelight, a type of staging light, was

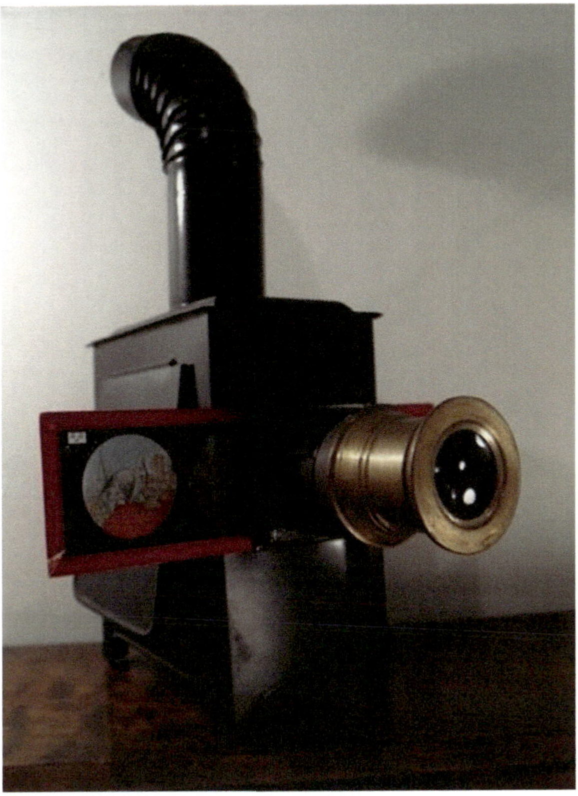

Fig. 7.3 Nineteenth century magic lantern. Andrei Niemimäki, CC BY-SA 2.0 <https://creativecommons.org/licenses/by-sa/2.0>, via Wikimedia Commons

also used before electricity. It provided the means to make the projections even larger and brighter. When presented to audiences as entertainment, stage technologies were sometimes incorporated to increase dramatic possibilities. One was projecting onto smoke [15].

Part of the magic lantern's value in its time was its portability. This made it particularly well suited for medical and art history venues where visual illustrations are more informative than language alone. Although not as instantaneous or as simple as a colleague sharing aspects of a surgical procedure with a cellphone movie recording, these devices nonetheless offered a congruent means for sharing charts, graphs, drawings, and instructional images and thus aided information dissemination.

Figure 7.4 conveys the magic lantern technology as a form of entertainment, a popular form culturally until cinema replaced it. This cut paper and wash example was crafted by the French artist Auguste Edouart in about 1835. As the figures within the image articulate, he was a proficient silhouette artist in his time. The showman, center left, is projecting the image using a mounted magic lantern box [19]. Clearly, as the children show best, the affective and expressive nature of this picture gives it vitality and makes a viewer feel as if she is a part of the projected performance.

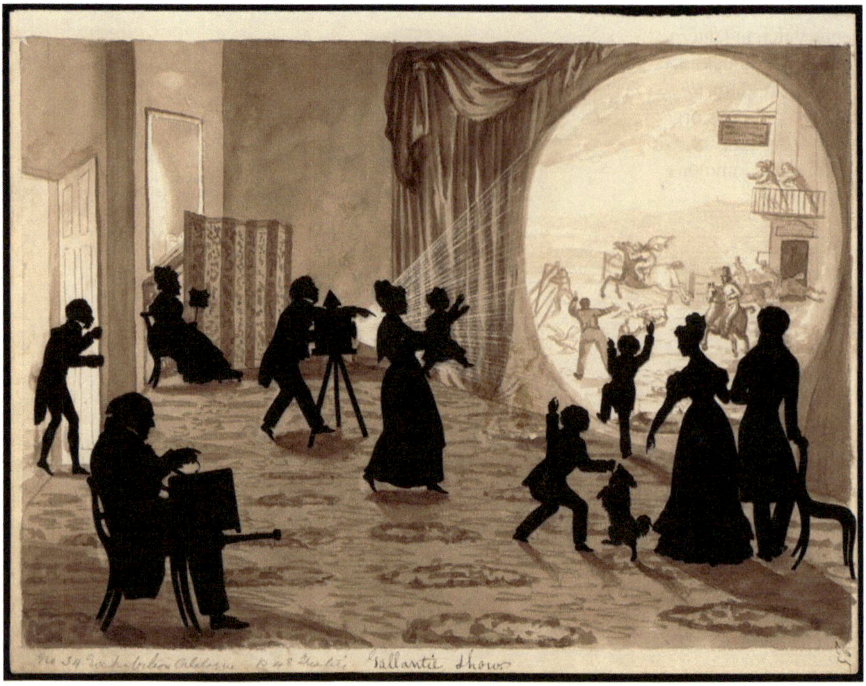

Fig. 7.4 *The Magic Lantern*. Artist: Auguste Edouart (French; 1789–1861); Date: ca. 1835; Medium: Cut paper and wash; Dimensions: image: 9 1/2 × 13 3/8 in. (24.2 × 33.9 cm); sheet: 10 1/4 × 13 1/2 in. (26 × 34.3 cm). Courtesy of the New York Metropolitan Museum of Art, Bequest of Mary Martin; 1938; Accession Number: 38.145.392

7.2 Nineteenth Century Magic Lantern and Photographic Therapeutics

In the nineteenth century, Dr. Thomas Story Kirkbride (1809–1883) used the magic lantern in his practice before he transitioned to photography. Like Francis Dercum, who later in the century collaborated with Eadweard Muybridge (1830–1904) on neurological studies of movement photography, Kirkbride graduated from the Perelman School of Medicine, commonly known as Penn Med today. Founded in 1765, it is the oldest medical school in the United States.

Originally. trained as a surgeon, Kirkbride wrote his medical thesis on neurology. He was offered his first surgery appointment at the same time as a job opened up to oversee the Pennsylvania Hospital for Mental and Nervous Disorders and he opted for the latter [20]. As the first superintendent and physician-in-chief of this psychiatric hospital, Kirkbride had an opportunity to construct the institution and implement his ideas about humane methods of treatment. This hospital, also known as the Institute of the Pennsylvania Hospital and Kirkbride's Hospital, operated from 1841 through 1997. Kirkbride is now remembered for the Kirkbride Plan he implemented and is also seen as one of the fathers of the modern American practice of psychiatry.

The Kirkbride Plan was a type of mental asylum construction system the doctor designed that was widely used during the mid-to-late-nineteenth century in the United States. Kirkbride defined his hospitals as curative buildings, as healing environments. The Plan stipulated exposure to spacious landscapes (sometimes maintained by patients), natural light, air circulation, and an overall layout that promoted patient privacy and comfort to the degree possible [21, 22].

A Quaker, Kirkbride's underlying requirements were based on the philosophy of moral treatment and a belief that environmental factors could cure impaired human minds [22, 23].

> Moral treatment de-emphasized traditional medical therapies and regarded the total environment of the hospital as therapeutic; stress was placed on the psychological and emotional state of the patients. The use of shock, terror, harsh punishment, and restraint was modified and eventually partly abandoned. Kindness, consideration, freedom of movement, attention to individual needs and feelings, supervised occupation and recreation, and a family or home atmosphere were all considered vital aspects of treatment and indispensable to the cure as well as the comfort of patients. The physician, whether or not he served also as asylum superintendent, would live on the premises and see each patient daily (pp. 183–184, [24]).

Kirkbride's thesis was that a daily regimen, complete with exercise and interactions with nature, would free his patients from the over-stimulation and stress of the modern world, which was thought to be the cause of their insanity [25]. Landy points out that in Kirkbride's history on the use of the magic lantern, written in 1863, he stated that photographic slides were shown in his facility as early as 1849 [23]. Lectures were available to both staff and patients. The magic lantern was also employed to abate patient boredom [18, 24, 26].

In his 1856 report, Kirkbride described the magic lantern shows as an "indispensable means of treatment" and added that all patients showed excellent results, even those in "the most excited wards" (p. 37, in [18], also [27]). In 1869 Kirkbride's Committee of Managers report also alludes to their success, saying "one of the prominent causes of our success in the treatment of disease as well as the general estimation in which we are held has long been owing to the great pains taken to vary both the day and evening amusements" (pp. 262–263, in [28]).

We do not know exactly when Kirkbride transitioned from expensive hand-painted images to the cheaper photographs. What is known is that around 1851, shortly after the invention of photography, he began to collaborate with two local Philadelphian photographers, William and Frederick Langenheim [24]. Between 1849 and 1865, the Langenheims produced over 90 stereoviews of the hospital (Fig. 7.5) and its staff. Their preserved collection, now at the University of Pennsylvania, contains over 3500 magic lantern and photographic slides that were used for patient therapy, to document the facilities, and other activities [29].

One noteworthy aspect of Kirkbride's integration of photography into his hospital routine is that his moral philosophy emphasis led him to use this technology in a

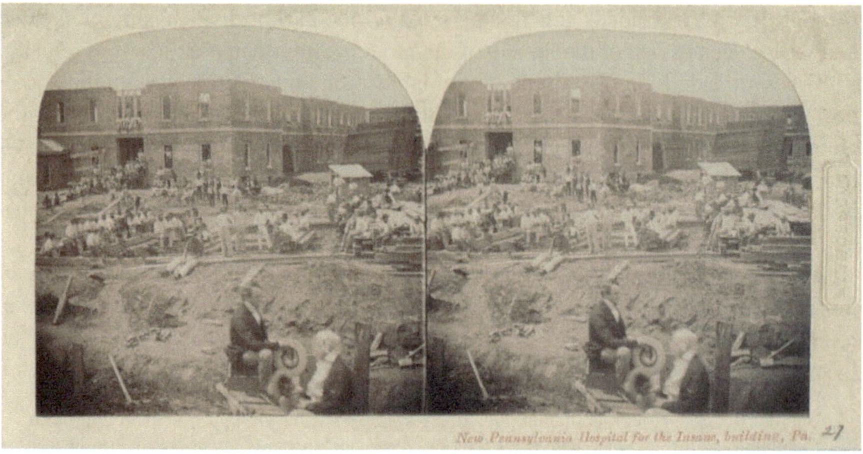

Fig. 7.5 Stereogram of the New Pennsylvania Hospital for the Insane building (Langenheim Collection). Courtesy of the Library Company of Philadelphia

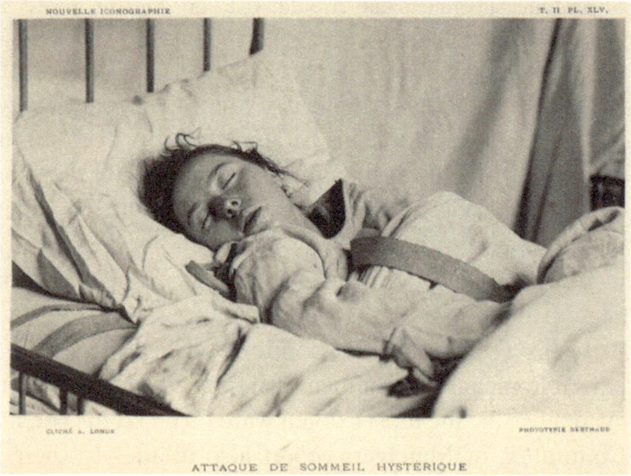

Fig. 7.6 Attaque de Sommeil Hysterique. Albert Londe, *La Photographie Médicale: Application aux Sciences Médicales et Physiologiques* [29], Paris 1893, Plate VIII. Courtesy of the Wellcome Collection

way that differed from many other nineteenth century doctors who were using photography in their work. For one, Kirkbride prohibited photographs of his patients, unlike many who sketched, exhibited, and recorded patient appearances for study (e.g., Jean-Étienne Esquirol (1772–1840), Jean-Martin Charcot (1825–1893), Alphonse Bertillon (1853–1914), Hugh Diamond (1809–1886, etc.). Figure 7.6 is an example of a photograph from Charcot's Saltpêtrière showing a female patient with hysteria-induced narcolepsy. It was taken by Albert Londe, the director of photography there. By all accounts it does not seem likely that Kirkbride's hospital would not have produced this kind of patient photograph. Not only did Kirkbride discourage photographing patients, his facility, as noted, also discouraged the use of restraint devices like the kind of leather restraining strap we see in this Londe picture. Kirkbride also refused to use patient images to determine a person's "type" [18].

Much like Oliver Sacks wanted to work with patients as individuals, Kirkbride did as well. He saw the magic lantern and later photography as tools that aided in stimulating patient minds. In other words, he did not see his patients as objects to record and study only in terms of their psychopathologies. Rather, he sought to enhance their human experience. Moreover, while Kirkbride discouraged photography of patients, the patients themselves were encouraged to learn how to make their own photographs, just as they were encouraged to care for the landscape [18].

Some in the medical community were critical of his approach [18, 28] Tomes describes it as a sales tool used for fundraising and points to a few problematic incidents that challenged the validity of the hospital's persuasive approach. She delineates incidents of patients confined against their will and patients who did not believe that their "insanity" diagnosis was correct [28]. According to Tomes' history of asylums:

> To the extent that families came to look upon institutionalization as a solution to their internal disruption, the asylum inevitably became the focus of familial conflicts....As Kirkbride envisioned it, the hospital's rationale depended upon a therapeutic persuasion, an agreed upon set of beliefs about mental disorders and their treatment. But however well executed this program was—and the Pennsylvania Hospital for the Insane represented the best of its type—asylum medicine rested upon an untenable premise: that the patient, as well as the patron, accepted the necessity for treatment. In truth, the vast majority of patients were confined and forced to submit to asylum care without their consent....To patients who did not believe in their own insanity, the hospital did indeed seem like a prison. Out of this disparity in perceptions grew the endless allegations of abuse (pp. 262–263, [28]).

7.3 Photography and Film

The move from the magic lantern to photography and then film included, first, developing a way to permanently affix something we can see in the external world to another surface. For film, the related challenge was conveying the appearance of natural motion.

In 1826, Nicéphore Niépce (1765–1833) fixed the first photographic image. This achievement was hardly a resounding success, but it was a move in the right direction. He needed over 8 hours of exposure time to print the crude final result now known as "View from the Window at Le Gras" ("Point de vue du Gras"). This heliographic image was captured with a camera obscura projected onto a 16.2 cm × 20.2 cm (6.4 in × 8.0 in) pewter plate thinly coated with bitumen of Judea, a naturally occurring asphalt [31].

Later results were more encouraging. Around 1839, William Henry Fox Talbot (1800–1877), a scientist, successfully demonstrated a negative photographic process that theoretically allowed unlimited positive prints to be produced from each negative. His multiple print processes are now seen as precursors to the photographic processes of the later nineteenth and twentieth centuries. Also in 1839, Niépce's associate Louis Daguerre (1787–1851), an artist, developed the single print daguerreotype process using a polished sheet of silver-plated copper. These positive prints only required minutes of exposure and produced clear, finely detailed results.

Early adopters and enthusiasts of still photography included many nineteenth century luminaries like Charles Darwin (1809–1882), Jean-Martin Charcot's (1825–1893), Étienne-Jules Marey (1830–1904), and Eadweard Muybridge (1830–1904). Darwin's use of photography in *The Expression of the Emotions in Man and Animals* [32, 33] is noteworthy because rather than mechanically recording the "emotional brain" in action he used innovative photography to create the emotional expressions he hypothesized [33]. In terms of neuroscience, Charcot, who established the psychiatric center at La Salpêtrière Hospital in Paris, was among the first neurologists to see the promise of photography and later film. His work built upon Esquirol's effort to classify neurological conditions.

Muybridge and Marey perhaps had a longer reach in terms of interweaving neuroscience and art. Marey's pioneering research into movement involved multiple exposures. Called chronophotography, his sequential images were useful in cardiology and aerodynamics. Marey's methodology entailed first taking a succession of photographs on a single circular plate rotating in front of a lens equipped with a shutter. He later devised a photographic gun that could take 12 consecutive frames a second with all the frames recorded on the same picture. Measurements of these successive phases of motion aided in studying biological motion procedurally [3, 34].

Muybridge's sequential image efforts were more artistic, and his sequential series operated like a flip book. Flip books, which were introduced in 1868,[2] included a series of slightly different pictures that seem to move when one frees the pages on the unbound edge of the book one-at-a-time in quick succession. When moving at this rapid pace, the images appear animated and simulate motion. A disadvantage of a flip book is that the animation stops rather quickly.

Muybridge's contributions are of particular interest in terms of neuroscience and art because it was a neurological injury sustained during a stagecoach accident that turned him toward photography [36]. Thrown from the vehicle, he hit his head on a boulder and was knocked unconscious. The events came to light when he was on trial for murdering his wife's lover. At the trial, he used this incident to plead insanity and was acquitted. As Shimamura writes: "The emotional changes that followed Muybridge's head injury are familiar to neurologists. Damage to the anterior part of the frontal lobe, known as the orbitofrontal cortex, disrupts the control and regulation of emotions" (p. 341, [36]).

Despite their different backgrounds, both Muybridge and Marey were also influential in fostering the celluloid age, which began at the end of the nineteenth century. One incentive to develop a celluloid product for photography and cinema was

[2] Although one would think that the flip books have a long, rich history, the oldest known documentation of it appeared when it was patented on March 18, 1868, by John Barnes Linnett under the name *Kineograph* (or "moving picture") [35]. The patent reads: "935. JOHN BARNES LINNETT, of Birmingham, in the County of Warwick, Printer, for an invention for — "Improvements in the means of producing optical illusions." *Letters Patent sealed*" (p. 65, [35]).

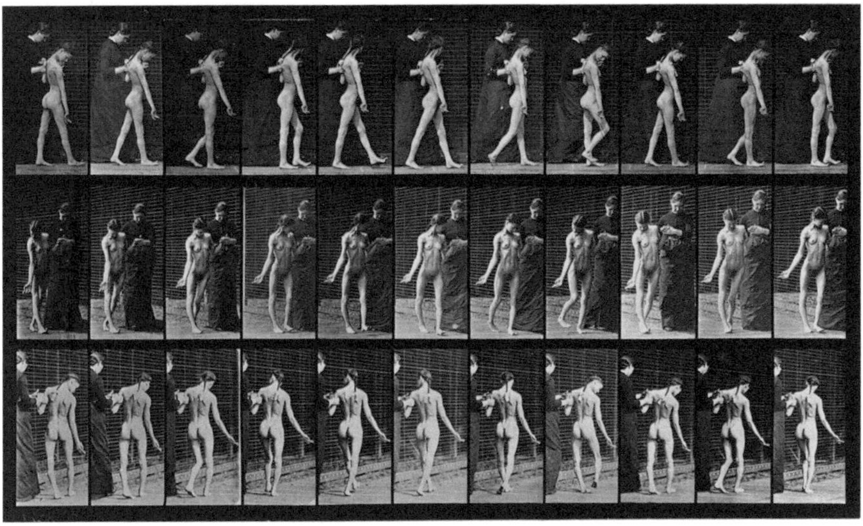

Fig. 7.7 Plate from *Animal locomotion: an electro-photographic investigation of consecutive phases of animal movements*. A girl with multiple cerebral-spinal sclerosis walking with a nurse. Collotype after Eadweard Muybridge, 1887. Courtesy of the Wellcome Collection

that the kind of series some photographers produced needed a flexible and durable material capable of holding complicated events and thousands of images, particularly Marey's work. Using celluloid as a recording medium solved the problem [5]. Early celluloid products tended to spontaneously ignite. Once Eastman Kodak resolved this, the material became popular, particularly once the film industry adopted it in the 1930s. Analog film was replaced in the twenty-first century by digital methods.

Muybridge's collaboration with Francis Dercum (1856–1931), a neurologist, began in 1872. As noted, Dercum, like Thomas Kirkbride, was a graduate of the Perelman School of Medicine, the oldest medical school in the United States. Dercum is best known today for treating President Woodrow Wilson in October 1919, after he suffered a stroke in the White House, and for his description of Adiposis dolorosa (or Dercum's disease), a rare condition characterized by multiple painful fatty tumors (lipomas) [37].

With Dercum's support, Muybridge obtained financing from the University of Pennsylvania in 1883 to study humans and animals in motion. Work with Dercum included patients with abnormal gait and pathological movements [30].[3] Figure 7.7 shows a girl with multiple cerebral-spinal sclerosis walking with a nurse. The

[3] The University of Pennsylvania has put much of this material online including over a hundred of these plates, animations of the still photographs, letters written by Muybridge, and letters written about Muybridge and the Animal Locomotion project. Images of the photographic equipment used is also displayed, accompanied by explanation of the technology and how it was used.

patient was from Blockley Hospital for the Poor, a site across the street from the university, and she is attended by a nurse from the hospital [38].

Published in 1887, the mainly neurological studies are in the eighth volume, titled *Abnormal Movements* [39]. Another volume, *The Human Figure in Motion* (1901) [40], also contained photographs of a number of neurologic movements diagnosed as "locomotor ataxia." It also included chorea-ballistic movements and a spastic, high-stepping gait caused by multiple sclerosis, poliomyelitis, and tabes dorsalis [41].

Notebooks recently found by Geoffrey Noble in the Hospital of the University of Pennsylvania's Dispensary provide a window into the Muybridge-Dercum project and their rigorous analyses of human locomotion, including walking, running, and jumping. Dercum's handwritten clinical records of nine subjects allow us to follow the clinical course of the individual patients. Some of the patients were seen for several years and in some case the records note they were photographed by Mr. Muybridge [42].

The majority of the subjects were white, working-class trade workers in the area who were already seeing Dercum through the Hospital of the University of Pennsylvania's Dispensary [42]. Patient "HR" (also "HB Rogers"), for example, was of particular importance in Dercum's analysis of the abnormal gait due to "locomotor ataxia," or the loss of control of one's bodily movements due to long-standing and untreated syphilis [42]. He initially came to the Dispensary in 1883 as a 34-year-old clerk, with a history of syphilis from age 22. HR first reported headaches and indigestion and was diagnosed with general nervousness secondary to overwork. He returned to the Dispensary due to additional difficulty buttoning his shirt and walking. His twenty visits over 6 years narrate his clinical decline through syphilis along with his physicians' largely unsuccessful treatment efforts.

A sample of those within the neuroscientific community who were turning to photography and/or film includes Gheorghe Marinescu (1863–1938), a Romanian neurologist. He perfected the use of cinematography as a research method from 1899 through 1902. He had studied with Marey and Charcot and adapted Charcot's method of lining up several patients with the same disorder, convinced that showing them together would allow for appreciation of archetypes and *formes frustes* (an atypical or attenuated manifestation of a disease or syndrome). These cinematographic studies led to several original contributions in clinical neurology [43]. Arthur Van Gehuchten (1861–1914) also had an inclination toward film He had worked with Santiago Ramón y Cajal (1852–1934) and actively participated in the formulation of the neuron doctrine [44]. Other experimenters included the Walter Greenough Chase, Theodore Weisenburg, Paul Schuster, Albert Londe (see Fig. 7.6), and Camillo Negro [41].

One particularly noteworthy early enthusiast was the physician, filmmaker Eugène-Louis Doyen (1859–1916), a neurosurgeon. When he began to show his films, he discovered that people had mixed views about seeing the medium in public settings. This was apparent after he presented the very first moving recording of a craniotomy in 1898 at the British Medical Association in Edinburgh [41]. While popular with attendees there and at international medical conferences in general, his decision to publicly air this work was harshly criticized by his contemporaries in France who felt that the films compromised the integrity of their profession [41, 45]. This response brings to mind similar comments made about Oliver Sacks' popularization of neurology, as discussed in the last chapter.

This incident also highlights that from the beginning photography and film were frequently greeted with suspicion. Despite their use in science and medicine, the media were often characterized as if they were lesser instruments. Artists and art historians were similarly conflicted as to whether mechanically made images were genuinely art. Many nineteenth century painters like Edgar Degas (1834–1914) and Thomas Eakins (1844–1916) kept their use of photography secret [9, 46, 47]. Some, however, like the pioneering American photographer and artist, John Moran (1831–1902), sought to reconcile photography and painting. Film had a similar reception and scholars have only recently begun to examine how film within the cultural milieu influenced artists such as Picasso, Braque, and Cubism [48].

7.4 Actualities

As noted, the challenge in transitioning from photography to film involved developing tools that would add movement to still images. In other words, the illusion of film is based on the optical phenomena known as persistence of vision and the phi phenomenon. The first of these causes the brain to retain images cast upon the retina of the eye for a fraction of a second beyond their disappearance from the field of sight. The latter creates apparent movement between images when they succeed one another rapidly. Together these phenomena permit the succession of still frames on a film strip to represent continuous movement when projected at the proper speed (traditionally 16 frames per second for silent films and 24 frames per second for sound films). Before the invention of photography, a variety of optical toys exploited this effect by mounting successive phase drawings of things in motion on the face of a twirling disk (the phenakistoscope, c.1832) or inside a rotating drum (the zoetrope, c.1834), which produced the illusion of motion by displaying a sequence of drawings or photographs showing progressive phases of that motion. A drawback of the zoetrope is that it displays the animation as a continuous loop.

Actualities is a name used for the first successfully mechanized films, and these productions were silent. The name comes from their ability to record whatever was in front of the camera, or the actual world. Due to their ability to capture motion and dynamics, scientists and clinicians found them beneficial in researching motion and movement. The general population took to this novelty immediately as well, although there were no dedicated theatres in which to view them initially. People watched them at fairgrounds or in parlors. Sometimes films were a part of vaudeville variety shows where they were shown for lower class clientele (not elites!). Film historian Tom Gunning has described this phase as the "cinema of attraction(s)" [49].

Overall, the earliest actualities did not try to tell a story. Rather, their intent was to display a marvelous medium that moved. Because early films only recorded what was seen directly—unmediated and unedited—they are often now considered precursors of educational and documentary films. Unfortunately, due to their fragility, we only have about 10% of these early, experimental films.

Auguste and Louis Lumière, who sold photographic supplies before they invented the Kinetograph (Cinématographe) motion picture system, were also early popularizers of actualities. The cinématographe name was chosen in 1893 by Léon-Guillaume Bouly, a French inventor who patented a design he called the cinèmatograph. Unable to pay the fees in 1894, Bouly sold his rights to the name and design to the Lumière Brothers. It provided the inspiration for the Lumière multiple-function device they invented in 1895 with the ability to record, develop, and project motion pictures [5, 50].

The first Lumière films premiered before an audience in Paris on December 28, 1895 [51]. Among them was the "Train arrival in the station of La Ciotat" *(L'Arrivée d'un train en gare de La Ciotat).* It shows a train pulling into a station and the passengers disembarking, complete with the use of a diagonal train track that perceptually enhances the viewers experience of both the depth and the motion effect [51]. This actuality is among those that allowed early filmmakers to perceive the value of diagonals compositionally. Diagonals then and still serve as an effective device in affectively and effectively accentuating temporal and spatial characteristics. This perceptual accent operates similarly to how Renaissance painters used three-dimensional perspective once they recognized that convergent lines deepen the illusion of depth on a flat, still surface. In a film, the diagonal is even more striking compositionally because the movement presented is not lateral and, as noted, with the diagonal motion of the train entering the station the effect reinforces both the sense of motion and the sense of depth.

In 1894, a year before *L'Arrivée d'un train en gare de La Ciotat* was shown, Thomas Edison's company filed the oldest surviving motion picture copyright in the United States. The film "Record of a Sneeze" (Fig. 7.8) is colloquially known as "Fred Ott's Sneeze." (Fred Ott got the part because he was an Edison employee known to his fellow workers in the laboratory for his comic sneezing and other gags [52]). This less than 20 seconds short, made by W. K. L. Dickson, was initially produced for advertising purposes [52]. Its availability to us today is something of an historical accident. In 1894 there was no provision in the copyright law for the registration of motion picture film. The film survived only because the Copyright Office allowed the Kinetoscopic Record of a Sneeze to be registered as a photograph [52].

In 1896, Charles and Émile Pathé launched Pathé Frères, a company still in business today. By 1908 they were selling up to five hundred projectors a month. Their film catalog included entertainment as well as educational and scientific productions. Among the latter are many films showing prosthetics such as a 1914 film on making artificial limbs that ends with a disabled man walking back and forth with a cane in front of a workshop: presumably he is trying out an artificial leg. A case study of multiple personalities from 1923, discussed shortly, records a case of what is now commonly known as dissociative identity disorder (DID). One film from 1928 (that is not for the squeamish) explains blood circulation.

Although many early films are now lost, a few surviving actualities of artists offer entry into this period from both an art and medical perspective. One, a 1915 film of Pierre-Auguste Renoir (1841–1919), shows him painting at home (Fig. 7.9)

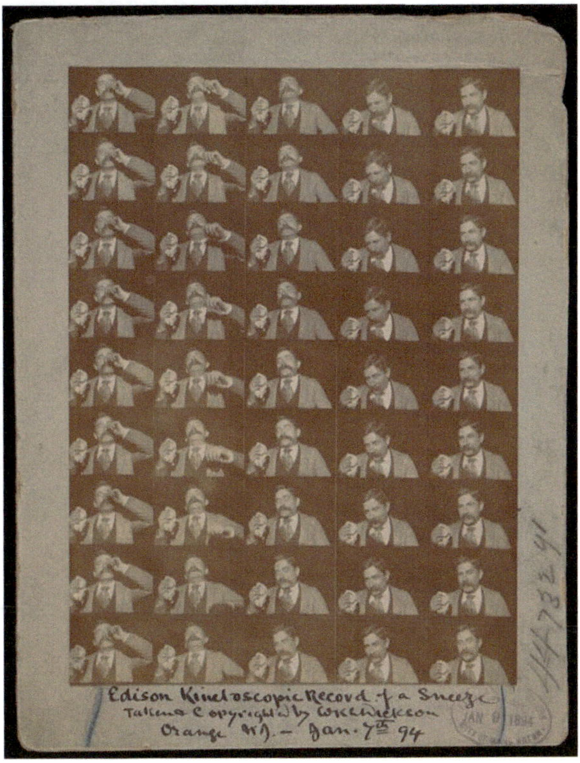

Fig. 7.8 *Fred Ott's Sneeze*. Edison kinetoscopic record of a sneeze/taken and copyrighted by W. K. L. Dickson, Orange, N.J. Courtesy of the Library of Congress

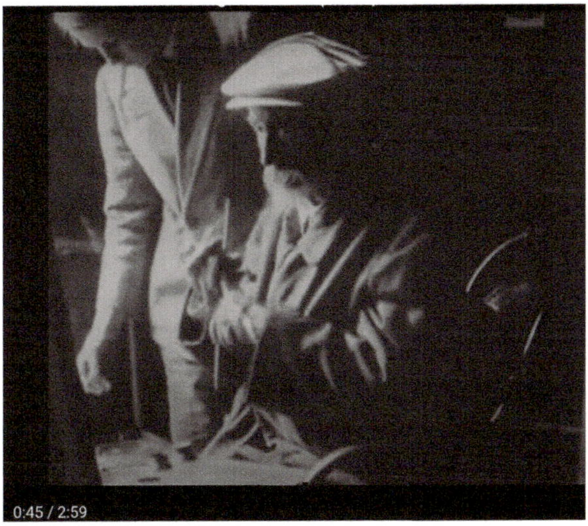

Fig. 7.9 Pierre-Auguste Renoir filmed painting at home (1915). Filmed by Sacha Guitry. Courtesy of JRH Films

[53]. This actuality was shot a few years after his stroke in the spring of 1912, when he was 71. It is unclear if the stroke partially paralyzed his arms and legs or if the paralysis that inhibited his painting was due to his rheumatoid arthritis affecting the

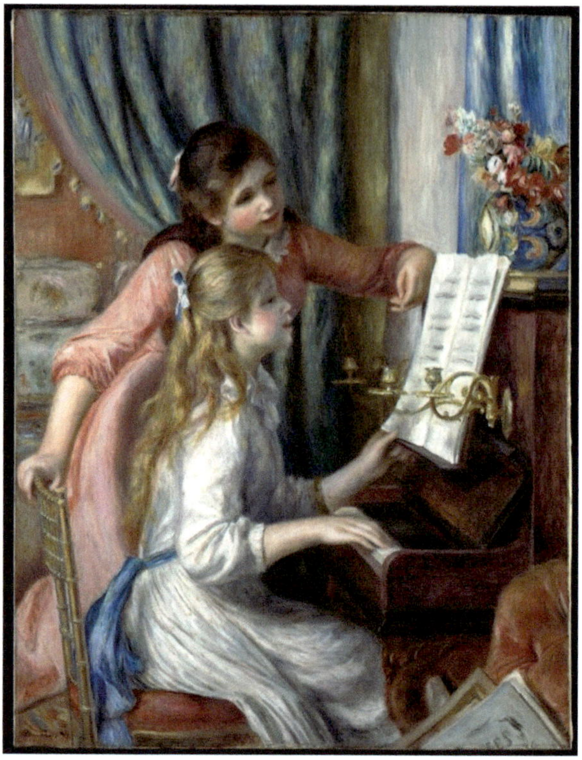

Fig. 7.10 *Two Young Girls at the Piano*. Auguste Renoir. 1892. Oil on canvas. 44 × 34 in. (111.8 × 86.4 cm). Courtesy of the New York Metropolitan Museum of Art. Robert Lehman Collection, 1975

cervical spine. This film also illuminates his life-long health challenges and thus adds to our understanding of how his physical disabilities influenced his painting practice. Because no medical records remain at this point, this short actuality is helpful in providing information about conditions described in his personal letters, and the biographical notes by people who knew him [54].

When Renoir's *Le Déjeuner des canotiers* (*Luncheon of the Boating Party*) was shown in 1882 at the Salon in 1882 three critics identified it as the best painting in the show, including Paul de Charry who described it as "fresh and free without being too bawdy" (p. 413, [55]). Ten years later, in 1892, he painted *Jeunes filles au piano* (Girls at the Piano) (Fig. 7.10). This was about the time he developed rheumatoid arthritis.

This painting, initiated in late 1891 or early 1892, came about after Renoir was invited by the French government to execute a painting for a new museum in Paris, the Musée du Luxembourg, which was to be devoted to the work of living artists. Aware of the intense scrutiny to which his submission would be subjected, Renoir lavished extraordinary care on this project, developing and refining the composition in a series of five canvases. The Lehman painting and the nearly identical version formerly in the collection of Renoir's fellow Impressionist Gustave Caillebotte have long been regarded as the most accomplished variants of this intimate and engaging scene of bourgeois domestic life.

While not a severe problem when executing this work, the arthritis came to limit his mobility for the last 20 years of his life. At the beginning of the disease a pleuritis was reported and later a facial palsy, which was treated with electrotherapy. As a result, he had to give up plein air painting and confine himself to studio work. A photograph taken in 1896, when he was 55, shows swelling of the metacarpophalangeal joints. This typically leads to deformity through the stretching of tendons and ligaments and destruction of joints through the erosion of cartilage and bone, as was the case with Renoir [54]. Despite the expanding loss of sensation and motor function in his hands, feet, and the body Renoir continued to paint.

Five years later, in 1901, at the age of 60, when his youngest son, Claude ("Coco"), was born, Renoir could still use his hands well enough to build a moveable easel that had allowed him to reach around a large canvas despite the limited motion of his right shoulder. No pictures exist of this special easel, and it does not appear that the painter is using it in the film. Nor is he seated in his wheelchair in the film.

It was around 1901 that Renoir began to use one walking stick. In 1904 his wife, Aline, wrote to his dealer Paul Durand-Ruel that it was impossible for him to stand on his legs, but he was getting used to his immobility [56]. In 1908 when it became difficult to walk with the one cane, he began to use two. Deformities of his feet increased over the years, and at some point, the painter was unable to wear shoes. The film does not show his feet, so it provides no information on this score, but he is seated. It has been reported that after the stroke in 1912, his feet had to be wrapped in woolen slippers and he developed fixed flexions of his knees 2 years before his death [54, 57].

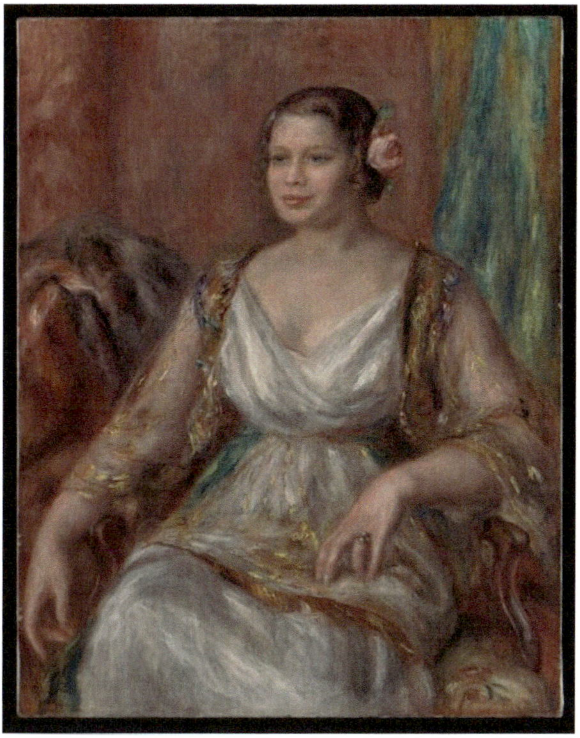

Fig. 7.11 *Tilla Durieux* (Ottilie Godeffroy, 1880–1971). Auguste Renoir. 1914. Oil on canvas. 36 1/4 × 29 in. (92.1 × 73.7 cm). Courtesy of the New York Metropolitan Museum of Art, Bequest of Stephen C. Clark, 1960

The year before this filmed sequence, in July 1914—which was after his stroke and just prior to the outbreak of World War I—Renoir painted the German actress Tilla Durieux (1880–1971) (Fig. 7.11). The painting's classicizing, pyramidal format lends a certain grandeur to the sitter, attired in the costume that the couturier Poiret designed for her role as Eliza Doolittle in George Bernard Shaw's *Pygmalion* in 1913 [58]. When Renoir painted this ambitious portrait, he was so crippled with arthritis that he had to sit in his wheelchair to paint. Whereas it is frequently said that he had to strap the brush to his hand to paint [58], as the film shows (Fig. 7.9), this was not the case. His deformed hands are wrapped, but he still manages to manipulate the brush using them.

The 1915 film footage includes Renoir's youngest son Claude, age 14 at the time, arranging the palette and placing the brush in his father's gnarled hand. His problems grasping a brush due to his distorted fingers [53] are worth noting since holding a pen was harder at this point. This meant that although he could continue to paint, others needed to write letters on his behalf. According to his eldest son Jean:

> His hands were terribly deformed, His rheumatism had made the joints stiff and caused the thumbs to turn inward towards the palms, and his fingers to bend towards the wrists. Visitors who were unprepared for this could not take their eyes off his deformity" (p. 206, [59]).

Although by all accounts the pain he experienced was nearly unbearable, Renoir never stopped painting. Moreover, as his health deteriorated, his popularity as an artist soared. Many of his friends were amazed that in his condition he continued to paint at all. But it seems that painting rejuvenated him,[4] much as Beethoven found composing therapeutic despite his inability to hear his music himself, as discussed in Chap. 4. In December 1912, Durand-Ruel who was 10 years Renoir's senior, marveled at his capacity to keep working:

> Renoir is in the same sad condition, but his strength of character is still amazing. He is unable to walk or even to get up from his armchair. He has to be carried everywhere by two people. What torture! And, despite that, he is still in a good mood and is still happy when he can paint. He has already done several things, and yesterday during the day he painted a whole torso that he began in the morning. Sketchy but superb (p. 277, [56]).

A contrasting actuality, also from 1915, shows Renoir's friend and contemporary, the French Impressionist painter Claude Monet (1840–1926), who was only one year older [60]. Unlike Renoir, Monet was in relativity good health at this point, his cataract surgery took place a few years later [61]. The difference in their mobility is quite evident. In less than 3 minutes, we see Monet's walking about, painting, and in an endearing sequence at the end we see his small dog tagging along as he walks. Shot in his garden at Giverny, this actuality captures the artist working on a water lily-padded pond, a subject for which he is most famous. Although in

[4] Many of Renoir's friends also continued to work despite health problems. For example, Durand's son Joseph related to Renoir in 1914 that Degas could barely see anything and suffered hearing loss as well [56].

black-and-white, those who know his work can easily visualize the colors that are not there [60].

7.5 Film Technique and Syntax

Upon entry into the cultural stream, even unedited early actualities excited people in all walks of life. Vast numbers of films on all topics were quickly available. Innovations such as editing, special effects, variations in shooting distance, camera movement, and so forth followed shortly thereafter. These kinds of variations and others helped develop a language that made cinematic work more accessible, more informative, and more exciting when that was a goal. Even in 1915 we can see this when comparing the 1915 actualities of Renoir and Monet. Whereas the camera is stopped and re-started in the Renoir film, the Monet film was constructed with a mixture of close-ups, long shots and mid-range shots, giving it a livelier dynamic. At this point, just as movement had been an aspiration with the development of still photography, sound became aspirational. The addition of sound in theatres through musical accompaniment and audience participation did, however, make the early films more multisensory.

Adolf Nichtenhauser (1903–1953), a physician, film historian, director, and enthusiast was a key contributor in studying cinematic possibilities. Born in Vienna, Austria, he studied psychology, art history and literature at universities in Berlin and Vienna before he received his medical degree in 1931 from the University of Vienna. Then, while practicing medicine and working as a medical translator, he began to take courses in motion picture technology, fascinated by both the educational and aesthetic possibilities of moving pictures [62]. This phase of his life ended between 1933 and 1937 when the right-wing authoritarian Austrian government banned him from the hospitals where he practiced [62]. This led Nichtenhauser to emigrate to the United States, where he sought to promote health education and quickly became an expert in making medical and health films. He also promoted the creation of film archives and film preservation [62].

The fascinating collection of papers Nichtenhauser amassed became the foundation of an unpublished manuscript on the history of film in medicine [63] and was recently digitalized by the National Library of Medicine (NLM) in the United States. In this work, he encourages effective communication and good practices through building a knowledge of how to use film equipment and how to work with filmic language (its syntax) so as to optimally convey the material [62]. A scathing assessment of a title called "From the Cell" demonstrates his understanding that film has its own grammar, syntax, and logic. His critique of this film notes that the graphics within this film, while excellent, and perhaps useable in another film, were not effective in this one [63].

Like Nichtenhauser, others recognized the need for filmic syntax, particularly in early silent films where behaviors and editing translated the content to aid our brains in constructing the meaning and/or information of the film. Filming from a variety of distances was one early adjustment. Placing the cameras on a moving device like a dolly, was another. Filmmakers also quickly recognized that special effects added information too. Rewinding the camera created the illusion of things going back in time. Stop motion, or stopping the film and removing a person, was a means to advance a story or create a gag. Editing could even place a character in a dream state. An option often useful in educational films and narratives was editing's capacity condense time or expand it.

Because early commercial filmmakers emphasized writing for cinema more than technological innovations, cognitive elements, or the multisensory aspect of the medium, film studies came to concentrate on narrative themes, with a psychological compass. The psychological compass was used primarily as a plot device. As Shimamura points out, psychocinemactics could have been established with the 1916 publication of Hugo Munsterberg's book *The Photoplay: A Psychological Study.* Munsterberg had trained with the physiologist Wilhelm Wundt and in the book he argued for aligning film with attention, memory, and emotion [64]. His approach was largely ignored, however, because it did not align with the Behaviorist fashions of the time [64].

Gestalt psychologists were somewhat more successful than Munsterberg in applying an alternative to behaviorism. Rudolf Arnheim, who studied under the Gestalt psychologist Max Wertheimer, saw value in the use of Gestalt principles to address film in terms of perceptual values such as balance, harmony, and object placement. In his book *Film as Art* [65], Arnheim explained that film as an art form alters our views of the world because it fails to render an exact copy of the real world. In line with Munsterberg's view, he considered dissolves, cross-cues, and point-of-view editing as ways of creating an illusion of space and time that was different from how we actually perceive real space and time [64, 65].

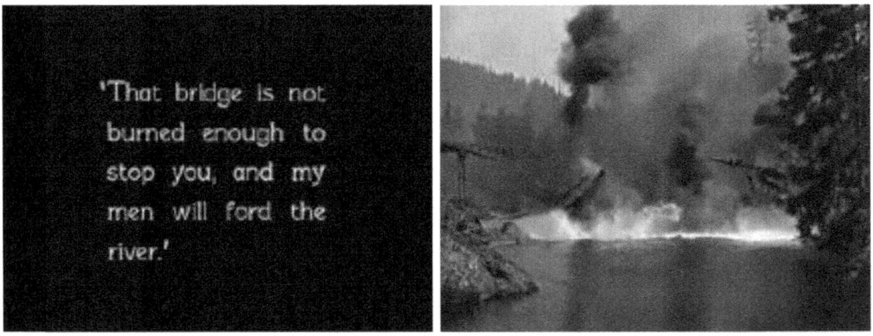

Fig. 7.12 Still shots from *The General*, a 1926 American silent film released by United Artists. The film stars Buster Keaton, who co-directed it with Clyde Bruckman

7.6 Intertitles

The use of intertitles in early cinema is one way silent films communicated before the addition of sound at the end of the 1920s created a second cinematic revolution. Intertitles are textual explanations, edited in between visual sequences to help the viewer follow the film. For example, in Buster Keaton's *The General* [66], a 1926 silent, the intertitle in Fig. 7.12 tells us that the general is directing his forces to cross the bridge, incorrectly assuming it will hold. We see a frame showing these words and then in a subsequent shot we see that the bridge did collapse, despite the general's assurances.

While some silent films, surgeries for example, might use only use textual syntax for opening titles and credits, others inserted intertitles throughout to provide narration or to condense time when presenting an edited case history. As mentioned, one 1923 film produced by Dr. C. C. Wholey, a neurologist at the Western Pennsylvania

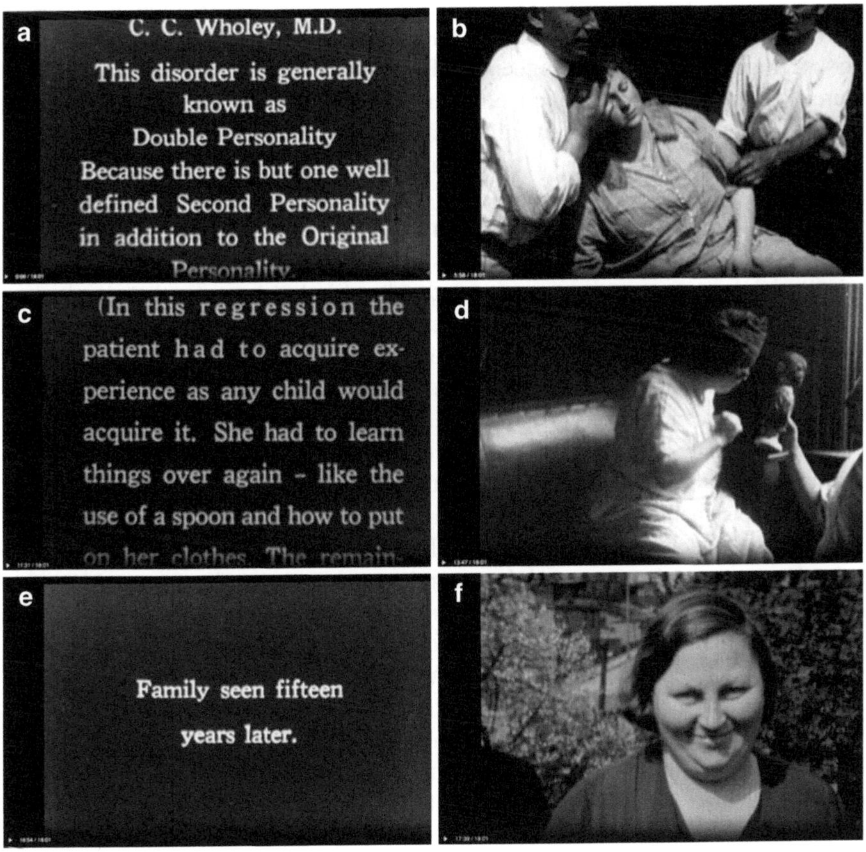

Fig. 7.13 Still shots from a case study of multiple personality, recorded in 1923 by C. C. Wholey and Edward J. Carroll [67]

Hospital and an Instructor of Psychiatry. records an individual with multiple personality disorder, now commonly known as dissociative identity disorder (DID). The patient (Mrs. X) regresses to a childhood state (Susie). She also has another, less well-developed personality (Jack). Presented in under 20 minutes, the case study covers over 15 years and uses intertitles throughout to narrate her story [67].

Figure 7.13, frame "b" (at 3:22 min) shows the patient swooning as she moves into another personality. Frame "d" (at 11:31 min) presents the regressed Susie playing with a doll. Frame "f" (at 16:54 min) advances her case to 15 years later and we see a recovered Mrs. X. Watching the moving and dynamic version presented in the film is much more informative and affective than these still frames accompanied by text. Yet, on the flat page of a book, the text communicates more effectively than the single images about what is going on because the film's life-like movement allows the various personalities to form a full person over the course of its presentation.

Neurological and psychiatric conditions are frequently used as plot devices in commercial films, and many have dramatized DID narratives. One well known adaptation was *The Three Faces of Eve* (1957), a film that presents the case of Christine Costner Sizemore. *The Three Faces of Eve* was based on a book written by her psychiatrists Corbett H. Thigpen and Hervey M. Cleckley [68]. Within the commercial dramatization the affective qualities of the case are intentionally emphasized, a directorial decision that enhances the narrative's impact on the audience.

7.7 Mirror Neurons and Empathy

Directorial decisions and narrative impact go hand-in-hand with studies that look at mirror neurons and empathy. Mirror neurons were first described in a seminal paper in 1992 as a class of monkey premotor cells discharging during both action execution and observation. They were named a few years later [69–72]. A key insight to their discovery came about when the experimenters were training monkeys during a 1996 experiment. The experimenters noticed that the monkeys not only performed particular actions for the experiment, but they also performed the actions when they saw the experimenter demonstrating them outside of the experiment. Seeing that single cells are selective for specific goal-directed actions—whether they are perceived or performed—suggested that neural representations of action could be perceptual, motor, or both [73, 74]. One way to think about this is in communal terms. A mirror system would have no point if you lived on your own. In this respect it is quite unlike having a digestive system. Alternatively, if you lived on your own, a movement system or a visual system still has value.

Furthermore, in terms of film, mirror neurons offer entry into the viewer's experience as they relate to actions, feelings, and conceptualizing another person's point of view. Because it is not normally possible to study single neurons in the human brain, most evidence is indirect. Recent studies have revealed that a variety of cell types distributed among multiple motor, sensory, and emotional brain areas form a "mirror mechanism" that is more complex and flexible than originally thought. This in turn has generated new approaches to social and affective neuroscience [69], a key component of neuroculture. Some have proposed this offers an entry for evaluating affective qualities of film, including empathy, compassion, and expressive translation. The idea is that when watching the screen, we directly understand what another person is experiencing or doing without losing the distinction between the other person and ourselves [74].

Empathy is a word often used to explain this for it is characterized as the ability to identify with or understand another's situation or feelings. The visual arts, film, theater, music, and so forth are known for evoking emotional responses and empathy. Narrative films employ it and indeed are edited to evoke emotion, a point Vittorio Gallese and Michele Guerra stress in *The Empathic Screen* [74]. For example, when a book expands, condenses, or alters time, we use our minds to follow the words. This silent relationship does not include the same measure of stimulation (and simulation) found when the same narrative is reconfigured with music, lighting, editing, costumes, and so forth.

Gallese and Guerra largely look at films in terms of intersubjectivity and what they call *embodied simulation,* "a theory of human perception of and action in the world, including our interactions with human agents" (p. vii, [74]). They ask *why* do films fascinate us and *how* do they engage us [74]. When watching a film, they argue, our motor system activates although we are not moving. Embodied simulation thus speaks to how the motor system relates what we observe back to us, transferring/processing what we have seen into an interior understanding [74]. To summarize, according to Gallese and Guerra, the neural integration of our sensory modalities is the rule rather than the exception with film. Thus, affective neuroscience and affective film help us understand the world from a subjective standpoint because subjectivity is always conditioned by one's relationship with the world.

Gallese and Guerra use words like *cinesthesia* and *cinesthete* to express the cinematic connection with multisensory experience. In doing so they develop a parallel line of thought to synesthesia, a topic discussed in Chap. 12, A similar term, coenesthesia, is defined as the aggregate of impressions arising from organic sensations that forms the basis of one's awareness of body or bodily state, as the feeling of health, vigor, or lethargy.[5] Cinesthetic, of course, also brings kinesthetics to mind (having to do with muscular sensibility and proprioception regulating motor activity) [74].

7.8 Case Studies

Four case studies close this chapter to offer four perspectives on neuroscience and art in film, and to show how broad the reach of film is neuroculturally. The first study on Buster Keaton (1895–1966) is biographical. It looks at how his films

[5] Vivian Sobchack talks about a "cinesthetic subject," combining the words cinema, synesthesia, and coenesthesia [75].

impacted him and his audiences physically and physiologically. The second, Ingmar Bergman's 1966 *Persona*, centers around Bergman's interpretation of mutism, using it as a psychological trope in the story, one that is not intended to represent a neurological condition. In other words, for Bergman, making the condition credible is not a priority. Unlike the pseudo-psychology in *Persona*, the third case examines the strokes of Luchino Visconti (1906–1976) and Federico Fellini (1920–1993), two twentieth century Italian film directors who suffered from right-hemispheric stroke. As a result, both were hemiplegic on the left side. In the final study, I turn to anterograde memory impairment as portrayed in films that allude to the classic case of H.M.

7.8.1 Buster Keaton: "The Boy Who Can't Be Damaged"

Coincidentally, the Lumière brothers premiered their inventive Kinetograph the same year as Buster Keaton (1895–1966) was born. Keaton, one of the finest silent film actors and directors, is best known for his seemingly impossible gags using slow motion and his knowledge of engineering. His engineering prowess was not learned in school, for he had no formal education. Rather, he drew on the knowledge he gained from the entertainers he met in his vaudeville years. Buster joined his parents act around age six. The driving force of the comedic act was Buster's father throwing him about during performances. His apparent invulnerability led them to include "The Little Boy Who Can't Be Damaged" in show billings. As a child, his family was reported for child abuse, but somehow never charged.

During one childhood performance, his father's timing was off, and he misjudged Buster's head causing his kneecap to land on the back of Buster's head. The boy stiffened, fell straight back, hit his head, and was out for 18 hours [76]. Despite these types of incidents, years later he told the *Detroit News* in 1914 that the secret is in landing limp and breaking the fall with a foot or a hand. "It's a knack. I started so young that landing right is second nature with me. Several times I'd have been killed if I hadn't been able to land like a cat." Nonetheless, Keaton claimed that by the end of his life he'd broken virtually every bone in his body [77]. He also claimed that he was able to handle dangerous stunts because he was a first-class athlete.

His concussions, headaches, and other maladies were largely treated as a part of his work product, so we have little information on their long-term impact. Rather, we have anecdotal reports such as this one from his business partner. After asking Buster if he ever got hurt doing these dangerous stunts Keaton replied, pointing to various injuries, and locating them cinematically. The "thing" on the top of his head was assigned to *Seven Chances*. Under his shirt he pointed to a third one from *Battling Butler*. Another injury, on his leg, was from *Sherlock, Jr.* [76].

His surrealistic film *Sherlock Jr.* (1924) centers on a film projectionist who longs to be a detective. After he is framed by a rival who accuses Sherlock, Jr. of stealing his girlfriend's father's pocket watch. Sherlock (Buster) dreams of solving the crime. At the end, when Buster, the projectionist, awakens, he finds his girl has solved the crime and their relationship is saved. When the film was named one of the All-Time 100 Movies in 2005 (Fig. 7.14), a magazine wrote that the film is filled

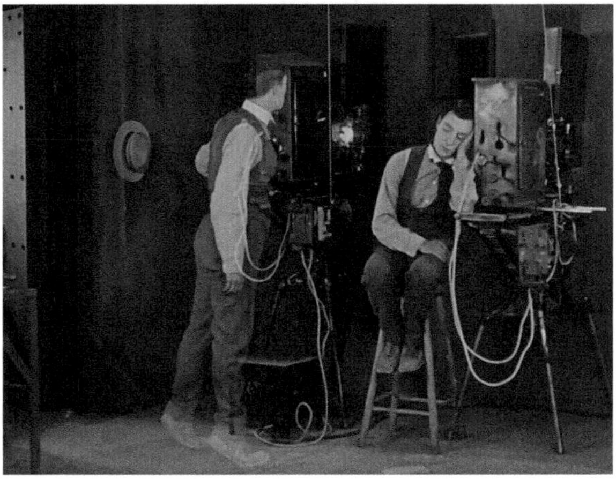

Fig. 7.14 Screen shot from *Sherlock Jr.*, a 1924 American silent comedy film directed by and starring Buster Keaton

with: "simple objects and movement manipulated in casually complex ways to generate a steadily rising gale of laughter.

As is common, this testimonial did not mention Buster's back story or his mishaps. The whole film is only 45 minutes long and Buster's physical feats tempted fate throughout it. During one scene Keaton is riding atop a train and grabs a pipe attached to a water tower to take him back down to the tracks. Instead of the expected smooth descent, the 10-inch pipe suddenly released a flood of water that tore his grip loose and flipped him backward onto the tracks a dozen feet below. Water began gushing out with such force that the back of his neck slammed against the steel rails. He fainted and was brought back with cold water, and the camera kept rolling. For weeks he suffered blinding headaches. Migraines followed, but his incredibly high pain threshold allowed him to continue working [66]. In 1935 an X-ray revealed that at some point he had suffered a major injury. When a doctor pointed to the callus that had grown over a fracture next to the top vertebra and asked if he had broken his neck at some point, Keaton replied never. Later he decided that this incident recorded during *Sherlock, Jr.* was probably responsible.

The most famous stunt in *Three Ages* (1923) came about after the planned gag imploded. Keaton had initially intended to leap from one building onto the roof of another building 12 feet off the ground using a board projecting out from the first. He did not make it. There was a net underneath, but he fell short and smashed into the brick wall before falling onto the off-screen net. He was injured badly enough to postpone shooting for a few days and his crew even convinced him to change the sequence. The camera operators had, of course, continued to film, and they kept the mishap and added to it after Buster recovered. In the finished film he falls through three awnings using a loose downspout that propels him into the firehouse and a slide down a fire pole [76].

The General (1926) is more scripted than most early Keaton films and highlights his ability to innovate and create gags on the fly. It is an historical parody loosely based on an actual United States Civil War incident that occurred 64 years before the film was produced. This derivative mixes comedic gags with a classic narrative that centers around Johnnie Gray's two loves, his locomotive, *The General*, and his sweetheart Annabelle Lee (Marion Mack). The movie begins after the outbreak of the Civil War. Johnnie (Buster Keaton) tries to enlist several times in the Confederate Army and is rejected because officials believe he is more valuable as an engineer than a soldier, although they fail to explain their reasoning to him. Annabelle, her brother, and her father spurn him, incorrectly concluding that Johnnie himself rejected the idea of service.

From this, two lines emerge: how can Johnnie become a soldier and how can he win back Annabelle's affections. Symmetrical train chases, propel the action, accompanied by a series of amusing technologically contrived jokes and his rescue of Annabelle (after she is abducted by Union spies). Johnnie Gray ultimately warns the local Confederate commander of an impending attack and the need to shore up their defenses. At the end, he is rewarded with a commission as a lieutenant. Now a soldier, the once scornful Annabelle comes back to him.

The extraordinary bridge collapse sequence (Fig. 7.12) was filmed with an audience watching on a nearby hill. In other words, this was not a studio production with special effects and dummy actors creating illusions. On a location in Oregon the studio constructed two bridges. One for them to blow up and another that ran parallel to it for filming the bridge as it was destroyed in real time. There was no backup bridge, so the crew and actors had only one opportunity to get it right.

Unseen in the film, citizens from the nearby community watched and felt something dreadful was happening in front of their eyes, despite knowing the event was only enacted for a film. There were explosions, trees falling, injuries. The sparks from the explosion even started a fire in the dry forest. Many assumed the dummy conductor in the train car on the collapsed bridge was an actual person. Nine people were injured during the shooting, and Buster was knocked unconscious for some time as well. After he was aroused, he briefly rested and then went back to work. While there is no way to measure the physiological impact quantitatively or the extent of the injuries to the spectators during this affective shot sequence the reactions speak for themselves and demonstrate the mirror mechanism distributed among visual, motor, sensory, and emotional brain areas is not confined to the screen.

Buster also suffered injuries outside the studio. Hearing loss, sustained during World War I while a soldier in France, continued throughout his life. Sleeping on the ground, where there was always a draft close to the floor, led to a cold. After a month he became almost completely deaf for a time due to being exposed to floor drafts each night. Also, in 1936, Buster Keaton was admitted to the psychiatric ward of a hospital in Los Angeles after suffering from a complete nervous breakdown following a physical breakdown resulting from influenza.

It is tempting to try to link Keaton's later problems in film to the early physical accidents and what is known about his overall health given what we know about his headaches and the many the times he lost consciousness. This is merely a speculative possibility since there are few actual medical details with which to back this up and there were other possibilities as well. In any case, today Buster's blackouts and dizziness would have led to medical interventions [78]. Then, rather than seeking treatment, Buster was inclined to get back to work, after a few hours or a few days.

Another factor that was no doubt influential in how his life unfolded was the switch from the silent pantomime of his early movies to films with sound. It is hard to say how definitively the change impacted him, however, because he also switched studios and lost creative control of his films at this time. Still, it does appear that his subsequent depression and alcoholism are related to having to adapt to studio directed, scripted roles. Buster did not warm to having his stone-faced pantomimed gags transformed into "jokes" that asked him to portray an unhappy guy seeking affection. Asking him to stifle his creative and intuitive style is perhaps another reminder that one's creative impulses are not tied to an "art" center in the brain.

7.8.2 Ingmar Bergman (1918–2007): *Persona* (1966)

Ingmar Bergman's iconic film *Persona* (1966) offers a quite different example. The word *persona,* from which the English word person is derived, means the mask worn by an actor. Thus, the title seems intended to emphasize that to be a person is to wear a mask. Asked what his intention was in creating the film, Bergman responded as one wearing a mask,

> If I've really managed to make a film that has sparked debate, it would be very tactless of me to barge in on that debate and talk about what I really meant by the film. It would be tactless toward the audience because I'm sure they all have their own interpretations, and tactless towards those commenting on it in the media, who might feel hurt if they found they'd misinterpreted the film. Therefore, I prefer not to say anything at all. I played my part in this debate when I made the film [79].

Persona's plot is weak neurologically, and obscure as well. What is strangely paradoxical about Bergman's complex experimental and psychological drama is that while it seems as if it is an effort to clinically probe personal identity in cinematic detail, it also seems it does not want to go in that direction. Thus, the plot's obscurity appears intentional. In *Cinema MD*, Wijdicks correctly concludes that *Persona* is so deliberately confusing that it defies any persuasive psychiatric analysis. Snippets of psychiatry and abnormal psychology, prurient stories, and bewildering dreams do not make a story, even if it does invite a psychological interpretation. This is not totally uncommon in cinema, where filmmakers often use psychopathology in incomprehensible, ambiguous, and oblique ways [80]. Bergman, for example, has used functional mutism twice: in *Persona* and in an earlier film, *The Magician* (1958) [81]. In each case, mutism comes in as a psychological trope to create a story. It is not intended to represent a neurological condition and thus giving the condition credibility on the screen is not emphasized. Like Buster Keaton in *Sherlock, Jr.*, Bergman is reminding us that whatever seems real could also be a fantasy or dream.

Persona is a story about a young nurse named Alma (Bibi Andersson) and her patient, a well-known stage actress Elisabet Vogler (Liv Ullmann), who has suddenly stopped speaking. They move to a cottage, where Alma cares for Elisabet, confides in her, and begins having trouble distinguishing herself from her patient. One of the disputes surrounding the plot centers on whether Elisabet's mute condition was an involuntary mental breakdown or a voluntary moral decision expressing her dissatisfaction with the world. Her psychiatrist, who only appears briefly at the beginning, seems to think it has come about due to a feeling that she can no longer play the roles society wants her to play. Her therapist concludes that since her patient did not commit suicide she could speak again. She not only ridicules Alma's muteness, she also suggests she will eventually renounce her muteness and return to the world:

> Commit suicide? Too nasty? But you can refuse to move or talk, at least you are not lying. You can cut yourself off, close yourself in. Then you needn't play any roles, wear any mask, make false gestures. No one asks whether it is genuine or not, whether you are lying, or telling the truth. I think you should play this part until it is played out, until it is no longer interesting. Then you can drop it, just as you eventually drop all your other roles [82].

In their analysis of *Persona*, Gallese and Guerra focused on the tactile and haptic significance of close-ups of faces, hands, bodies, and out-of-scale objects projected onto the screen. They argue that tactile exchange between the characters in effect "touches" those in the audience. Spectators subsequently embody the experiences portrayed throughout the film through their visual system as well as the motor and tactile system. In other words, looking at the film cinematically rather than narratively, Gallese and Guerra see sensory connections as its outstanding quality. They argue that its sensory elements allow it to connect empathetically with audiences. Therefore, in their view, *Persona's* porous, indeterminate, and opaque narrative plan is less important than the formal elements of the film's construction: these involve how our brains function. These bring the audience into the story, despite how minimal the story line actually is.

Bergman does indeed use many close-ups, doubling, and dramatic effects. In the film's most iconic shot we see Elisabet and Nurse Alma facing each other in the middle of the night in front of what appears to be the mirror of a bathroom cabinet. As the two of them discover (or imagine?) an uncanny resemblance, they move as if to kiss one another and their faces then overlap cinematically. The effect is also quite affective, making us wonder who the face is and what it is looking at so intently as it looks at us and we look at it.

Finally, again, it is hard to underestimate the impact of close-ups in the movie. Part of the film's strangeness is due to how Bergman is emphasizing the similarity of the two women's faces, particularly in the disturbing mirror shot mentioned above, where he combines half of one face with half of the other. In another shot, where Bergman superimposes the two faces, Liv Ullmann later revealed that when she first saw the combination image, she did not see her face, only Bibi's. Bibi's reaction was the same—or perhaps the opposite because, after looking at the composite, Bibi only saw Liv Ullman [83]. Speaking with the film critic Roger Ebert, Andersson reported that she and Ullmann had no idea Bergman was going to do this, and when she first saw the film, she, too, found it disturbing and frightening. Bergman himself said: "The human face is the great subject of the cinema. Everything is there" [84].

7.8.3 Luchino Visconti (1906–1976) and Federico Fellini (1920–1993)

The strokes of Luchino Visconti (1906–1976) and Federico Fellini (1920–1993) offer a biological perspective; the perspective of two directors who both suffered from right-hemispheric stroke. As a result, both were hemiplegic on the left side.

Visconti suffered his first stroke during the editing of *Ludwig* (1972) and showed a tremendous resilience following it. He was working again 2 months later. Although the director never fully recovered, he nonetheless refused to stop working, declaring: "Despite this wheelchair, here I am, ready to direct another movie. Next time, maybe I will do so from a stretcher, but I will never give up" (p. 52, [85]). Eventually he made two more films as an infirm: *Conversation Piece* (1974) and *The Innocent* (1976). Visconti died 4 years later. In the first months of 1976 he was struck by influenza. On March 17, 1976, he listened to the Second Symphony of Brahms in the company of his sister, Uberta. When the music ended, he looked at her and died after saying in Milanese: "That's it. I'm tired" [85].

Fredrico Fellini, by contrast, was unable to direct films after his two strokes. He was only able to comment on his condition through drawings. Fellini suffered his first stroke when he was 73, in August of 1993. This was about 5 months after he received a special Oscar for lifetime achievement in filmmaking at an Academy Awards Ceremony. A CT scan of Fellini one week after the first stroke shows the large lesion encroaching upon the temporoparietal regions of the right hemisphere. Fellini died the following October after a second, massive, stroke.

Despite his inability to continue his professional activities, it appears that Fellini's creative mind was still very much evident as a patient [85]. Usually, patients who show left hemisyndrome and a left inferior quadranopsia suffer from left unilateral neglect and have poor insight into their condition; they behave as if their left visual space simply no longer exists. What is more, they simply do not notice that something is lacking in their experiential field. This was not the case for Fellini. From the beginning, he was not anosognosic of his physical impairment. Rather, he displayed hints of misoplegia, or negative feelings towards his paralyzed limb. He called it "a bloated, damp bunch of asparagus" [85].

Fellini's drawings from this period also offer insight into his neglect profile, which was characterized by severe visual extinction. For example, a left stimulus went unseen when another was presented in the right hemifield at the same time [85]. There was an extrapersonal visual neglect in classic paper-and-pencil tasks like target cancellation, geometric shape copying, and line bisection (all of which did show some improvement after one month). When given such tasks, Fellini would not merely follow the instructions of the neuropsychologist compliantly by bisecting lines or copying complex shapes. Indeed, he would spontaneously add drawings that were not only very creative, but were also evidence that he was aware, at least implicitly, of the errors he made. Neglect was also apparent in his reading: he would often "forget" or "confabulate" the beginning of the words. Humor was perhaps the most intriguing characteristic of Fellini's poststroke drawings, a characteristic present in his films as well. For example, one of the illustrated figure drawings shows that Fellini has inserted a bubble that asks: "Where is the left?" [85].

7.8.4 Anterograde Memory Impairment

Directed by three-time Oscar nominee Christopher Nolan's *Memento* offers a metaphorical neurological perspective on anterograde memory impairment. In this film, we meet Leonard (Guy Pearce) who, following a traumatic accident can no longer form new memories. Nonetheless, he remains determined to piece together the circumstances of his wife's rape and murder. Leonard's closest resemblance to a real patient is to Henry Molaison. At 27, in 1953, Molaison underwent experimental brain surgery to treat epilepsy. This procedure left him with a radical form of amnesia that affected his ability to formulate new long-term memories. Both his short-term memory and his memory of events before 1953 remained intact. In this film, Leonardo, too, can recall details of life before his accident, but is unable to remember what happened 15 minutes ago, or even where he's going, or why.

Christopher Nolan calls the movie a "dis-linear" thriller (a term he created), to explain that the movie's narrative is presented in reverse chronology [86]. It is actually quite linear, but backwards, as is much more apparent on a second viewing. The film itself was based on a short story Christopher's brother Jonathan Nolan later published in *Esquire* in 2001 titled "Memento Mori" [87]. The brothers agreed that the most interesting approach to this concept was the first-person point of view. The question Christopher asked himself when starting was how do you tell a first-person story through the eyes of someone who, when he meets a person, does not know if they have met before or whether that person should be trusted? The answer was to put the audience in that position as well [86].

According to Christopher Nolan, he was attracted to anterior grade memory for his plot because of its metaphorical potential narratively. How can someone who is unable to make new memories know exactly what he is looking for and what he is trying to do? In the film, the concept of revenge drives the story because it raises perplexing questions. Can revenge exist in any real sense in this extreme situation? Or, as the protagonist Leonard asks, "How am I supposed to heal if I can't feel time?" Indeed, if Leonard is unable to form new memories since his brain injury, how does he know he has "this condition"?

The film walks us through Leonardo's contorted view of time through the tattooed information he places onto his body. These markings show some awareness of the passage of time, even if he is unable to say exactly how much time has passed between "now" and the people and events he has recorded. This disparity between his view of his own condition, his "memory," and the cinematic events adds additional complexity to the situation and the issues of memory and identity.

Repetition is also used to translate Leonard's situation. Some of the repetitions that present his pursuit of the killer repeat the same situation differently right after we see them, so what the viewer thought she saw becomes confused and uncertain with the second iteration. The repetitions, particularly their amorphous quality, highlight that the story at times seems clear and at times confused. In some ways this mirrors Bergman intentional obscurity; but here it is also an inflection of Leonard's condition. All in all, we empathize with Leonard's condition but are never certain of who he really is.

A counterpoint film about anterograde memory looks directly at Patient H.M. (Henry Molaison). Crafted by American artist Kelly Tribe (b. 1973), *H.M.* incorporates the physical mechanisms of moving images into a reconstructed portrayal of the patient [88]. As noted above, H.M.'s surgery to treat epilepsy left him unable to formulate new long-term memories, although both his short-term memory and his memory of events before 1953 were left intact. In a format similar to the addition of Oliver Sacks' commentary to the televised version of Michal Nyman's opera on Mr. P.'s case [89], the commentary of H.M. includes observations by Dr. Suzanne Corkin, a neuroscientist who worked closely with Molaison. Found images, text, and animation add another element as does the presentation of the concept. Tribe uses a single film print threaded between two synchronized projectors that goes through a modified looping system. Two parts of the film (20 s apart) are projected simultaneously, creating a dissonance no doubt evocative of what Molaison might have experienced. As the Museum of Modern Art's description of the film notes, the fragility of film—which degrades each time its projected—resonates with the precarious nature of memory [88].

7.9 Conclusion

Many recent books and activities bring film and neuroscience together, and the bibliographic range further underscores how broad this topic is. One author, Eelco F.M. Wijdicks, for example, is oriented toward how cinema depicts major neurological syndromes and clinical treatment of disease [41, 80, 81, 90]. A quite different perspective is found in *Flicker: Your Brain on Movies* [73] by cognitive neuroscientist Jeffrey M. Zacks. He demonstrates how filmmakers have learned to take advantage of the tricks our minds are already playing on us in real life. Like Wijdicks, Zacks' is clearly a lover of films. His interest is in exploring filmmaking practice in terms of perception and memory and emotion. A third volume, *The Empathic Screen: Cinema and Neuroscience* [74] develops a theory of film experience that explores our brain-body interaction when engaging with and watching films. In this book, neurobiologist Vittorio Gallese and film theorist Michele Guerra (now the Mayor of Parma) focus in on the ability of the spectator to orient herself and move in the space-time dimension of film, while remaining immobile in her role of observer [74]. We also see the integration of film and neurology at venues like the annual film festivals presented by the American Academy of Neurology and the American Brain Foundation. These kinds of events often raise awareness about the need for more funding and for research into the prevention, treatment, and cure of brain diseases such as Alzheimer's disease, stroke, brain injury, ALS, epilepsy, Parkinson's disease, and so forth [91].

Finally, one of the fascinating aspects of film is that it is fully neurocultural. It covers all facets of life and thus all topics are explored. Films make us laugh, they make us cry, and they are a research tool as well. I have tried to convey film's reach and related image technologies in this chapter by examining the creative development of technologies that have aided research and in looking at topics that vary significantly (e.g., brain injury, syntax, narrative, biography, and so forth). The next chapter on war and art, shell shock, PTSD, and brain injuries offers a sadder and more focused theme, and it is one that often used film to document the psychological, the physiological, and the pathos of various conditions. It is also one filled with paradoxes for ailments like shell shock and PTSD have stimulated medical advancements as well as artistic statements documenting how trauma impacts both the body and the brain. Whereas illness per se is frequently unexpected and accidental; the biological and psychological disturbances of war are in a category of their own.

References

1. Kevles B. Naked to the bone: medical imaging in the twentieth century. New Brunswick, NJ: Rutgers University Press; 1997.
2. Ione A. Innovation and visualization: trajectories, strategies, and myths. Amsterdam: Rodopi; 2005.
3. Ione A. Art and the brain: plasticity, embodiment, and the unclosed circle. Brill: Leiden; 2016.
4. Azéma M, Rivère F. Animation in Palaeolithic art: a pre-echo of cinema. Antiquity. 2012;86(332):316–24.
5. Lipton L. The cinema in flux: the evolution of motion picture technology from the magic lantern to the digital era, vol. 2021. New York, NY: Springer; 2021.
6. Needham A, Wisher I, Langley A, Matthew A, Little A. Art by firelight? Using experimental and digital techniques to explore Magdalenian engraved plaquette use at Montastruc (France). PLoS One. 2022;17(4):e0266146.
7. Photography History Facts. History of camera obscura—who invented camera obscura: Photography History Facts; 2023. www.photographyhistoryfacts.com. Accessed 25 July 2023.
8. Plato. The collected dialogues of Plato: including the letters (Bollingen Series LXXI). New York, NY: Pantheon Books; 2005.

9. Kosinski DM. The artist and the camera: Degas to Picasso. New Haven, CT: Dallas Museum of Art and Yale University Press; 1999.
10. Gorman MJ. Projecting nature in early-modern Europe. In: Lefèvre W, editor. Inside the camera obscura–Optics and art under the spell of the projected image. Max Planck Institute for the History of Science; 2007. p. 31–50.
11. Gravesande GJ. Physices elementa mathematica, experimentis confirmata: Sive introductio ad philosophiam Newtonianam (3 vols): apud Petrum van der aa; 1720.
12. Morton H. Erecting the inverted image in the magic lantern. J Franklin Inst. 1867;83(6):406–9.
13. Museum DLV. The oldest magic lantern in the world: De Luikerwaal; 2021. https://www.luikerwaal.com. Accessed 27 July 2021.
14. Petroski H. Engineering: next slide. Please Am Sci. 2005;93(5):400–3.
15. Gallerneaux K. The history of the magic lantern. The Henry Ford; 2019:3:19.
16. Barber XT. Phantasmagorical wonders: the magic lantern ghost show in nineteenth-century America. Film Hist. 1989;3(2):73–86.
17. Vogel H. The magic lantern as an aid to instruction in chemical and physical lectures. Dent Regist. 1874;28(4):172–4.
18. Godbey E. Picture me sane: photography and the magic lantern in a nineteenth-century asylum. Am Stud. 2000;41(1):31–69.
19. Edouart A. The magic lantern. New York, NY: The New York Metropolitan Museum, bequest of Mary Martin. 1938; 1835.
20. Curwen J, Nichols CH, Callender JH. Memoir of Thomas S. Kirkbride, M.D., LL.D: E. Cowan & Company, printers; 1885.
21. Verderber S. Innovations in hospital architecture. Abingdon: Taylor & Francis; 2010.
22. Kirkbride TS. On the construction, organization, and general arrangements of hospitals for the insane, vol. 11. Philadelphia, PA: Lindsay & Blakiston; 1854. p. 122.
23. Yanni C. The architecture of madness: insane asylums in the United States. Minneapolis, MN: University of Minnesota Press; 2007.
24. Layne GS. Kirkbride-Langenheim collaboration: early use of photography in psychiatric treatment in Philadelphia. Pa Mag Hist Biogr. 1981;105(2):182–202.
25. Havilla L. Therapeutic architecture. The Pennsylvania Center for the Book. 2010. https://pabook.libraries.psu.edu. Accessed 23 July 2023.
26. Trustees of the University of Pennsylvania. The story of the magic lantern. History of Pennsylvania Hospital [Internet]. 2017.
27. Kirkbride T. Pennsylvania hospital for the insane, Philadelphia. Ann Rep. 1856:17–8.
28. Tomes N. A generous confidence: Thomas Story Kirkbride and the art of asylum-keeping, 1840–1883. Cambridge: Cambridge University Press; 1984.
29. New Pennsylvania Hospital for the Insane building. The Library Company of Philadelphia. 2020. https://www.uphs.upenn.edu/paharc/features/tkirkbride.html. Accessed 27 July 2023.
30. Aubert G. Neurological illustration from photography to cinematography. In: Finger S, Boller F, Tyler KL, editors. Handbook of clinical neurology. 95. 2009/11/07 ed. London: Elsevier; 2010. pp. 289–302.
31. Marvullo J. About photography: on site: writing with light. Archaeology. 1984;37(2):64–8.
32. Darwin C, Ekman P, Prodger P. The expression of the emotions in man and animals. London: Harper Collins; 1872/1998.
33. Prodger P. Darwin's camera: art and photography in the theory of evolution. Oxford/New York, NY: Oxford University Press; 2009.
34. Lorusso L, Venturini S. Cinema and neurology: from history to therapy. In: Colombo B, editor. Brain and art: from aesthetics to therapeutics. Cham: Springer International Publishing; 2020. p. 95–120.
35. Linnett JB. Chronological index of patents applied for and patents granted. Great Britain; 1968.
36. Shimamura AP. Muybridge in motion: travels in art, psychology and neurology. Hist Photogr. 2002;26(4):341–50.
37. Berg AS. Wilson. New York, NY: G.P. Putnam's Sons; 2013.

38. Hendricks G. Eadweard Muybridge: The father of the motion picture. New York, NY: Grossman Publishers; 1975.
39. Muybridge E. Eadweard Muybridge collection, selected items. Philadelphia, PA: University Archives | University of Pennsylvania; 2023.
40. Muybridge E. The human figure in motion; an electro-photographic investigation of consecutive phases of muscular actions. London: Chapman & Hall; 1901.
41. Wijdicks EFM. The legacy of Eugène-Louis Doyen: pioneer in filming craniotomy. Neurocrit Care. 2022;39:724–7.
42. Geesler K. In: Medicine P, editor. Penn Medicine News. Philadelphia, PA: Penn Medicine News; 2021. https://www.pennmedicine.org/news. Accessed 29 March 2023.
43. Barboi AC, Goetz CG, Musetoiu R. The origins of scientific cinematography and early medical applications. Neurology. 2004;62(11):2082–6.
44. Aubert G. Normal and pathological gait as inspiration for the artist. In: Rose FC, editor. Neurology of the arts: painting, music, literature. London: Imperial College Press; 2004. p. 129–40.
45. Lanska DJ. Eugène-Louis Doyen and his Atlas d'Anatomie Topographique (1911): sensationalism and gruesome theater. J Hist Neurosci. 2022;31(2–3):334–50.
46. Danly S, Leibold C. Eakins and the photograph: works by Thomas Eakins and his circle in the collection of the Pennsylvania Academy of the Fine Arts. Washington, DC: Published for the Pennsylvania Academy of the Fine Arts by the Smithsonian Institution Press; 1994.
47. Foster KA, Bockrath M. Thomas Eakins rediscovered: Charles Bregler's Thomas Eakins Collection at the Pennsylvania Academy of the Fine Arts. New Haven, CT: Yale University Press; 1997.
48. Rose B. Picasso, Braque and early film in cubism. New York, NY: PaceWildenstein; 2007.
49. Gunning T. The cinema of attraction[s] early film, its spectator and the avant-garde. In: Strauven W, editor. The cinema of attractions reloaded. Amsterdam: Amsterdam University Press; 2006. p. 381–8.
50. Michaelis AR. Research films in biology, anthropology, psychology, and medicine. New York, NY: Academic Press; 1955.
51. Neupert RJ. French film history, 1895–1946. Madison: The University of Wisconsin Press; 2022.
52. Dickson WKL. Record of a sneeze, Jan. 7, 1894. Washington, DC: Edison Kinetoscopic; 1894.
53. Guitry S. Renoir painting at home. JRH Films; 1915.
54. Boonen A, van de Rest J, Dequeker J, van der Linden S. How Renoir coped with rheumatoid arthritis. BMJ. 1997;315(7123):1704–8.
55. Moffett CS. The new painting, Impressionism, 1874–1886. Geneva/Seattle, WA: University of Washington Press; 1986.
56. White BE. Renoir: an intimate biography. London/New York: Thames & Hudson; 2017.
57. White BE, Renoir A. Renoir: his life, art, and letters. New York, NY: Abrams; 1984.
58. Juan T. Auguste Renoir: Tilla Durieux (Ottilie Godeffroy, 1880–1971). New York, NY: The Metropolitan Museum of Art; 1914.
59. Rishel JJ. Philadelphia Museum of Art: handbook of the collections. Philadelphia, PA: Philadelphia Museum of Art; 1995.
60. Monet C. Claude Monet filmed painting outdoors (1915). JRH Films; 1915.
61. Gruener A. The effect of cataracts and cataract surgery on Claude Monet. Br J Gen Pract. 2015;65(634):254–5.
62. Cantor D. Circulating now. Washington, DC: Historical Collections of the National Library of Medicine; 2022. https://circulatingnow.nlm.nih.gov. Accessed 3 Jan 2023
63. Nichtenhauser A. A history of medical film. Washington, DC: National Library of Medicine Nichtenhauser Papers; 1954.
64. Shimamura AP. Psychocinematics: exploring cognition at the movies. Oxford/New York, NY: Oxford University Press; 2013.
65. Arnheim R. Film als Kunst (Film as Art). München: Hanser; 1933/1975.
66. Keaton B, Bruckman C. The General. United Artists; 1926.

67. Wholey CC, Carroll EJ. Case study of multiple personality. Bethesda, MD: National Library of Medicine; 1923.
68. Thigpen CH, Cleckley HM. The three faces of Eve. New York, NY: McGraw-Hill; 1957.
69. Bonini L, Rotunno C, Arcuri E, Gallese V. Mirror neurons 30 years later: implications and applications. Trends Cogn Sci. 2022;26(9):767–81.
70. di Pellegrino G, Fadiga L, Fogassi L, Gallese V, Rizzolatti G. Understanding motor events: a neurophysiological study. Exp Brain Res. 1992;91(1):176–80.
71. Gallese V, Fadiga L, Fogassi L, Rizzolatti G. Action recognition in the premotor cortex. Brain. 1996;119(2):593–609.
72. Rizzolatti G, Fadiga L, Gallese V, Fogassi L. Premotor cortex and the recognition of motor actions. Cogn Brain Res. 1996;3(2):131–41.
73. Zacks JM. Flicker: your brain on movies. Oxford/New York, NY: Oxford University Press; 2015.
74. Gallese V, Guerra M. The empathic screen: cinema and neuroscience. Oxford: Oxford University Press; 2019.
75. Sobchack V. Carnal thoughts: embodiment and moving image culture. Berkeley, CA: University of California Press; 2004.
76. Gill D, Bronlow K. Buster Keaton: a hard act to follow. Thames Television; 1987.
77. MacLeod D. The sound of Buster Keaton: Buster Books; 1995.
78. Halstead ME, Walter KD, Moffatt K, et al. Sport-related concussion in children and adolescents. Pediatrics. 2018;142(6):e20183074.
79. Bergman I. Ingmar Bergman on the magician. 1966.
80. Wijdicks EFM. Cinema, MD: a history of medicine on screen. New York, NY/Oxford: Oxford University Press; 2019.
81. Wijdicks EFM. Neurocinema—the sequel: a history of neurology on screen. Boca Raton, FL: CRC Press; 2022.
82. Bergman I. Persona. 1966: 84 minutes.
83. Schrader P. Paul Schrader interview: Liv Ullman on Persona. The Criterion Collection; 2014.
84. Ebert R. Persona. 2001. https://www.rogerebert.com/reviews. Accessed 4 May 2023.
85. Dieguez S, Assal G, Bogousslavsky J. Visconti and Fellini: from left social neorealism to right-hemisphere stroke. In: Bogousslavsky J, Hennerici MG, editors. Neurological disorders in famous artists (part 2), vol. 22. Basel: Karger; 2007. p. 44–74.
86. Neff R, Argent D. Interview: remembering where it all began: Christopher Nolan on Memento. Creative Screenwriting; 2015. https://www.creativescreenwriting.com. Accessed 26 July 2023
87. Nolan J. Memento Mori. Esquire. 2001:28.
88. Tribe K. H.M. New York, NY: The Museum of Modern Art; 2009.
89. Rawlence C, Morris M, Nyman M. Michael Nyman—The man who mistook his wife for a hat (1987 TV opera). 1987.
90. Wijdicks EFM. Neurocinema: when film meets neurology. Boca Raton, FL: CRC Press, Taylor & Francis Group; 2015.
91. American Academy of Neurology. Neuro Film Festival. https://www.aan.com/.

War and Art: Brain Injuries, Trauma, and Empathy

8

Released in 1922, the British government's *Report of the War Office Committee of Enquiry into "Shell Shock"* states that over 250,000 British soldiers suffered from shell shock after participating in World War I [1]. Notably, when the archivists put this document online in the twenty-first century, they added notations explaining that this report represents the thinking of its era:

> At the time [shell shock] was believed to result from a physical injury to the nervous system during a heavy bombardment or shell attack, later it became evident that men who had not been exposed directly to such fire were just as traumatized. This was a new illness that had never been seen before on this scale. The condition was poorly understood medically and psychologically. Today, the condition is known as post-traumatic stress disorder and the treatment and attitude to it are very different. The discussion of anxiety within this report does not match current opinion on the condition and may be upsetting for some readers [1].

War injuries such as shell shock, a term coined at the beginning of World War I, are definitively neurocultural. They are integrally interwoven with human behavior for humans initiate and conduct the destructive combat. For this reason, war injuries are in a category of their own and separated here from therapeutics, the subject of the next chapter.

Although the onset of shell shock is associated with World War I, known contemporaneously as the Great War, we can find cases that match shell shock characteristics in earlier literature. In the *Epic of Gilgamesh* (~2100 BCE), for example, we read that after Gilgamesh witnesses the death of his closest friend, Enkidu, he experiences recurrent and intrusive recollections and nightmares related to the event [2]. Recording the Battle of Marathon, in about 440 BCE, the Greek historian Herodotus describes an Athenian, Epizelus, who was suddenly stricken with blindness after seeing his comrade killed in combat. There was no physical wound and the blindness persisted for many years [3, 4]. The Greek physician Hippocrates (~460–370 BCE) described soldiers who experienced frightening battle dreams [4]. We can also

A. Ione, *Neuroscience and Art*, Neurocultural Health and Wellbeing,
https://doi.org/10.1007/978-3-031-62336-3_8

identify treatments for brain injuries that include Neolithic trepanations and the case studies in the *Edwin Smith Papyrus*, an ancient Egyptian medical manual [2, 3].[1]

What changed with World War I was the scale of the brain injuries. Although no firm numbers exist for shell shock cases, the medical and cultural literature in all countries indicates they were pervasive, as was awareness of the invisible aspects of these injuries.

People have since begun to ask if the increase in these types of brain injuries began with World War I, or was it perhaps earlier? Some, for example, Peter Leese [6], see it as a part of a phenomenon that began in the nineteenth century. This agrees with the work of early twentieth century thinkers like the sociologist Georg Simmel who postulated that the industrialization and urbanization that exploded in the nineteenth century had a tremendous physical and psychological impact on human life. Simmel's analyses pointed to a confluence of railway collisions, industrial catastrophes, and traffic accidents, all of which he concluded set off hysterical or neurasthenic symptoms even in those who had survived these calamities otherwise unharmed [7]. Still, by the end of World War I, shell shock, a condition prevalent in so many soldiers who faced combat, demanded attention.

8.1 World War I (The Great War)

Whereas Robson infers that military officers caring for soldiers during the autumn Battle of Loos (1915) were the first to observe what they called "hysterical manifestations" that were later termed shell shock (p. 37, [8]), the term was actually introduced earlier that year by Charles Samuel Myers, a physician and psychiatrist. His paper, "A Contribution to the Study of Shellshock" [9], appeared in *The Lancet* on February 13, 1915. In it, the doctor offers an account of three cases that were admitted into the Duchess of Westminster Hospital, Le Touquet, where he was a captain in the Royal Army Medical Corps. Myers' article describes the intensity of the bombardment and fighting that produced a helplessness appearing variously as panic and being scared, flight, or an inability to reason, sleep, walk, or talk [6, 9]. He concludes:

> Comment on these cases seems superfluous. They appear to constitute a definite class among others arising from the effects of shellshock. The shells in question appear to have burst with considerable noise, scattering much dust, but this was not attended by the production of odor. It is therefore difficult to understand why hearing should be (practically) unaffected, and the dissociated "complex" be confined to the sense of sight, smell, and taste (and to memory). The close relation of these cases to those of "hysteria" appears fairly certain (p. 320, [9]).

Although Myers did not observe significant hearing loss in his initial patients, as the literature expanded, it was soon clear that shell shock extended beyond these first

[1] See my *Art and the Brain* book for more information on the *Edwin Smith Papyrus* and trepanation, a medical procedure used to cure epileptic seizure, migraines, and mental disorders [5].

cases. It caused blindness, deafness, tics, recurring nightmares, and a number of constitutional symptoms. Myers speculated that the shock caused by bursting shells and exploding grenades brought about as yet undetected physical changes (for instance, microscopic lesions) in the brain and spinal cord [9].

The appearance of additional cases did not bring about consensus on causes and solutions; rather, it complicated diagnostics. Due to the lack of apparent physical injury, symptoms like the feelings of helplessness and an uncontrollable shuddering with fear were debated. Obviously, these behaviors were contrary to how people envisioned the bravery and courage of a soldier. Moreover, all countries presumed the soldiers needed to be "fixed" so that they could return to the battlefield.

What a dilemma all faced! Although attributed to brain injury, because physical injuries were often proportionally slight, the condition was hard to locate, as was the extent of the injury. Indications of shell shock were often present even though there was no obvious bodily shock to the spinal cord, the peripheral nerves, or the brain itself. Yet, it was evident that all was not "right" with the afflicted bodies and minds. Some called it "nostalgia" as if to imply the afflicted missed his home. This seemed at odds with cases where one of the characteristics of shell shock was a fixation on a single event, such as remembering a fellow soldier being blown to pieces. Over time, diagnoses were further complicated by evidence that the disability was long-lasting.

Within this puzzle, questions arose that illustrate a tight connection between neurology and psychiatry [10]: Can there be psychological damage without physical cause? Can bodily symptoms be generated and "willed" by the mind? Was it possible to simulate mental disorder if some advantage could be gained from it, like avoiding a return to the front? If this was the case, was it ethical to "heal" soldiers using extraordinary methods like electroshock, particularly if they objected to the treatments, only to send them back into battle?

Some, like Myers, saw shell shock as a somatic condition, a wartime variation of what in 1899 the German neurologist Hermann Oppenheim (1858–1919) had termed *traumatic neurosis* [11]. Oppenheim had hypothesized that traumatic neurosis could be triggered by a sudden physical shock to the central nervous system [12]. He did so within the nineteenth century tradition of psychiatry that decreed all mental or nervous diseases to be the result of physical damage to the brain. He did not attribute symptoms from minor head injuries to neurasthenia or hysteria or traumatic neurosis, for example.

Oppenheim's somatic theory had come under attack during the 1880s and 1890s, particularly by Jean-Martin Charcot (1825–1893). Essentially, Oppenheim and Charcot both favored a similar pathogenetic concept of a not yet understandable but organically based disturbance. Whereas Charcot saw it as dynamic or functional, Oppenheim saw molecular damage [11]. Charcot also believed that he could obtain scientifically verifiable results by applying the psychological method of hypnosis and suggestion. In effect, this indirectly introduced a concept of psychological causation [11].

With World War I, the psychiatric establishment rejected Oppenheim's view. At a 1916 war conference in Munich, they concluded that traumatic neurosis (now war neurosis) was not a physical, undetected, and invisible condition but was imagined and willfully induced by morally inferior minds bent on evading military service [13]. Once shell shock was characterized as a means to avoid service, as was often the case, this further complicated the diagnostic situation. Since a breakdown had life and death implications regardless of whether one remained in an asylum or was returned to the front after release, the question of what constituted a breakdown was front and center.

Among the younger neurologists who grappled with this were Hippolyte Bernheim (1840–1919), Pierre Janet (1859–1947), and Sigmund Freud (1856–1939). They argued that mental and nervous diseases, including hysteria, neurasthenia, and traumatic neurosis, were not necessarily brain diseases but rather disorders of the mind, best treated by mental means: psychotherapy, hypnotism, and psychoanalysis [11]. Strict isolation and extended hot baths were also methods adopted. Another group argued for harsher punishments. The Kaufmann cure, advocated by Fritz Kaufmann, urged the use of electroshock; the assumption behind this was that the painful electric currents would scare patients "into health" and enable them to return to the front [14, 15]. This cure consisted of shocks to soldiers with paralysis. They received the painful electrical impulses to the paralyzed limb until they gave up and began to move it [14].

Similar treatments were adopted in France. As Tatu and his colleagues point out, aggressive therapies were used to promptly return soldiers to battle. One electric method used in 1915–1916 was torpillage, developed by Clovis Vincent (1879–1947). This psychotherapy used faradic and galvanic electric currents to treat soldiers with "intractable" neuroses. The treatment was painful, and soldiers began to refuse it. Some, like Baptiste Deschamps (1881–1953), took their cases to court. After a publicized trial, the method was discontinued [16]:

> Deschamps' trial provoked a heated debate over the question of a patient's right to refuse treatment and demonstrated the polarized character of wartime public discourse. In the press, Deschamps was alternately portrayed as a wounded martyr who had been tortured at the hands of bloodthirsty military authorities, or as a malingering, fearful wimp who refused to do his duty. The debates provoked by the trial demonstrate the persistence of class as a dimension of the war experience and challenge the myth of the *union sacrée*....The trial also represents one of the forums in which the French debated the question of how to define honor and the responsibilities of citizenship in the context of modern, total war. Was honor a tangible or an intangible quality? Could it be defined in medical terms? Was it mandatory? When did a soldier's faltering morale become a medical rather than a legal or a moral problem? (pp. 30–31, [17]).

It was in this context that asylums and rehabilitation centers operated. In England, for example, the soldier-poet Wilfred Owen (1893–1918) was among those who never came home. His poem *Dulce et Decorum est* (see Appendix B) spoke of the honor of fighting for one's country. Drafted in 1917 and published posthumously in 1920, Owen's poem describes the horrific experiences soldiers knew during World War I. The title is a reference to the Roman poet Horace's *Odes*, where he wrote: "Dulce et decorum est pro patria mori" ("It is sweet and fitting to die for one's country") (III.2.13) [6].

Owen saw this as "the old Lie," a lie used to present war as something other than what it is. In *Dulce et Decorum est*, Owen cries out about the horrific experiences' fighting soldiers endured, shouting that the sweetness of war is a mirage. He begins:

> Bent double, like old beggars under sacks,
> Knock-kneed, coughing like hags, we cursed through sludge,
> Till on the haunting flares we turned our backs,
> And towards our distant rest began to trudge.
> Men marched asleep. Many had lost their boots,
> But limped on, blood-shod. All went lame; all blind;
> Drunk with fatigue; deaf even to the hoots
> Of gas-shells dropping softly behind [18].

The draft for the poem came together while Owen was under Dr. William Rivers' care at Craiglockhart Hydropathic Institution, a hospital used to house officers suffering from the symptoms of shell shock. The soldier-poet arrived in May 1917, diagnosed with neurasthenia (or shell shock) and met the soldier-poet Siegfried Sassoon there. Sassoon, who played a crucial role in the dissemination of Owen's poetry following his 1918 death, encouraged Owen to write about the war while they both convalesced and to continue once he was returned to the battlefield. This was a form of therapy that allowed Owen to express the pain he felt and to place it into poems others could read.

Dr. Rivers (1864–1922) was one of four men who met at Cambridge in the 1890s and later studied shell shock. Each man advocated for a different approach, and their contrasting views epitomize how broadly views on mind and brain connections varied as researchers began to unravel the condition. The other three were Grafton Elliot Smith (1871–1937), William McDougall (1871–1938), and Charles Myers (1873–1946) [19].

Myers, as noted, wrote the first paper on shell shock in 1915. He later turned to anthropology. William McDougall, a doctoral student of W. H. R. Rivers, the doctor who cared for Owen and Sassoon, argued that shell shock could be cured through cognitive and affective reintegration. Eventually, as an opponent of behaviorism, with an emphasis on social psychology, he endeavored to forge a grand synthesis of physiology and philosophy of mind [19, 20]. Grafton Elliot Smith was an Australian-British anatomist and Egyptologist. He became a proponent of the hyper-diffusionist view of prehistory [21]. He and Tom Hatherley Pear coauthored *Shellshock and Its Lessons* [22] to explain the condition and its unique set of circumstances. According to these authors, anyone could develop a psychoneurosis if exposed to a sufficiently difficult environment. They endeavored to refute the prewar ideas that mental illness was hereditary and untreatable [19, 22]. Later, Smith traveled the world widely and studied Egyptian mummies as an anatomist, archeologist, and cultural anthropologist.[2]

---

[2] Grafton Elliot Smith is not related to Edwin Smith (1822–1906) who was the American dealer and collector of antiquities who gave his name to the *Edwin Smith Papyrus*, an Ancient Egyptian medical papyrus.

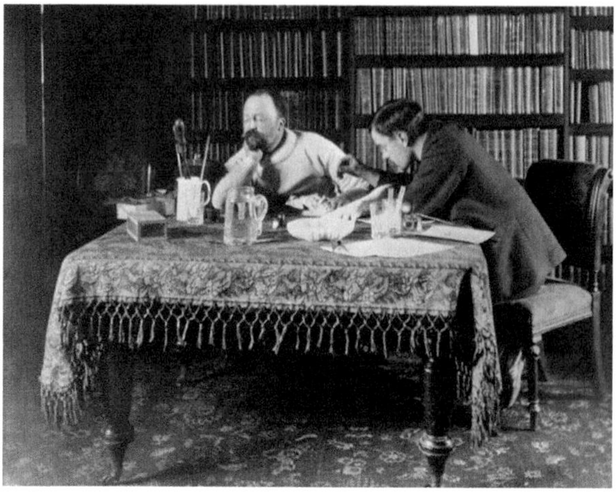

Fig. 8.1 W.H.R. Rivers and Sir Henry Head experimenting on nerve function. Courtesy of the Wellcome Collection

Rivers, a neurologist who advocated for psychological treatments, was a pupil of John Hughlings Jackson (1835–1911). Later, he did outstanding collaborative work with Henry Head (1861–1940) (Fig. 8.1), which has receded into the background [23]. Instead, he is largely remembered today for his friendship with the poets Siegfried Sassoon and Robert Graves, both of whom he met through his work at Craiglockhart Military Hospital [24]. Rivers met Siegfried Sassoon after he was sent to the hospital in response to his "Soldier's Declaration," an anti-war letter. His institutionalization shows how shell shock treatment was interwoven with patriotism. Writing about his experiences later, in his semi-autobiographical novel *Sherston's Progress* [25], Sassoon tells of their first meeting. Rivers verifies that Sassoon's admission to the facility was more political than psychological in his *Conflict and Dream* [26]. In the book, Rivers designated him as "patient B":

> The patient is a healthy-looking man of good physique. There are no physical signs of any disorder of the Nervous System. He discusses his recent actions and their motives in a perfectly intelligent and rational way, and there is no evidence of any excitement or depression. He recognizes that his view of warfare is tinged by his feelings about the death of friends and of the men who were under his command in France. At the present time he lays special stress on the hopelessness of any decision in the War as it is now being conducted, but he left out any reference to this aspect of his opinions in the statement which he sent to his Commanding Officer, and which was read in the House of Commons. His view differs from that of the ordinary pacifist in that he would no longer object to the continuance of the War if he saw any reasonable prospect of a rapid decision [24].

Sassoon eventually returned to the front; he felt guilty for abandoning his comrades. Even as he recognized the contradiction of fighting in a war he could not morally justify, he nonetheless honored his duty. Paradoxically, he was wounded shortly thereafter.

After the war ended, Rivers, who contributed to the British War Commission's report quoted at the beginning of this chapter, summed up his views on shell shock as follows:

> [P]sychoneuroses of warfare are mental, and not physical, it becomes the task of the physician to discover the exact nature of the mental processes involved, and the mechanisms by which these processes are so disordered as to produce the vast diversity of forms in which the morbid state appears… Now that the struggle is over, I believe that we may look to this instability as the source of energy from which we may expect great accomplishments in art and science (p. 1, 158, [27]).

World War II renewed the shell shock debates with cases such as the recovery of Lev Zasetsky described by Russian neuropsychologist A. R. Luria in *The Man with the Shattered World* [28]. Wounded in the head at the battle of Smolensk in 1943, Zasetsky found himself in a frightening world. He could recall his childhood but not his recent past; half his field of vision had been destroyed; he had great difficulty speaking, reading, and writing. Using Zasetsky's own diaries, Luria records Zasetsky's memories in order to reestablish his past and to affirm his existence as an intelligent being.

Once again, in World War II, people raised questions: Was the war responsible? Was it caused by an unwillingness of soldiers to take part in the war? What were the best curative treatments? Were there ethical lines? What if a patient wanted to reject treatment? Again, the military community, the medical community, and the soldiers did not always align. As was the case during World War I and in its aftermath, film once again placed war injuries within the cultural context, as discussed shortly.

8.2 Phantom Limbs

After the Great War, phantom limbs, too, were among the disabilities that came home with returning soldiers. These were definitively physical and they, too, were endemic. Like brain and head injuries, phantom limbs were not new. Indeed, they were documented as far back as the sixteenth century. Then, the French surgeon Ambroise Paré (1510–1590) reported pain occurring in an absent limb, writing: "For the patients, long after the amputation is performed, say they still feel pain in the amputated part" (p. 356, [29]). A self-reported account was presented by the Scottish doctor, William Porterfield (1696–1771). A vision specialist, after his leg was amputated at a young age, he, too, felt sensations in his missing limb [29].

During the American Civil War, Silas Weir Mitchell (1829–1914), a novelist and poet who was also one of the founders of neurology in the United States, similarly described postamputation problems. He used the term "phantom limb" in an 1871 article, followed by a more medical presentation in *Injuries of the Nerves and Their Consequences* published in 1872. This second publication describes several war amputee cases [30]. Mitchell's legacy as a writer and physician places him squarely within the neurocultural landscape.

One returning soldier discussed in Chap. 4 was Paul Wittgenstein (1887–1961). He was shot in the elbow. After his right arm was amputated, he resolved to continue his career as a performer. The musician asked some of the most famous composers of the day to write for him, and this paved the way for later pianists with one arm, or the use of only one arm, to continue their careers as well.

The writer Guillaume Apollinaire (1880–1918) is another case. After he was hit with a shrapnel fragment in the right temporal region, his extroverted personality was replaced with a sad and embittered one. He wrote: "I am not anymore who I was, in any way, and if I listened to myself, I would now become a priest or a religious person" [31].

A third artist who suffered from postamputation complications was Frédéric Sauser, who wrote under the pseudonym Blaise Cendrars (1887–1961). Curiously, before the war, Cendrars had made several allusions to missing hands and amputations, which later would sound like a premonition (in [31]). He was also uniquely positioned to include his physical trauma in his artistic work having studied medicine before turning to writing. Cendrars was shot in the right forearm and lost his limb. Later, in a medical note, he wrote: "A phantom can be seen but doesn't exist, whereas a phantom limb exists but cannot be seen" (p. 151, [32]). Tatu points out that Cendrars, in his own way, also addressed the somewhat fuzzy boundary that exists between psychiatry and neurology [32].

Cendrars suffered from stump pain and phantom limb phenomena for almost half a century following the amputation of his right arm during World War I:

> [M]y severed arm was so painful that I was biting my tongue to stop myself from screaming, and now and then, long shivers shook me for I was cold in the rain, completely naked, lying on my narrow stretcher, immobile, stiff, unable to move at all, restricted as I was, like a mother with her newborn, by the huge bandage, the size of a doll, pressed against my side, this strange thing that I could not move without stirring up a universe of pain, or hold in my good hand without the big white swab becoming soaked with red, feeling an atrocious burning, and realizing that my life was escaping me, was draining away, drop by drop, and I couldn't hold on to it because you can't stop your heart, and my heart was beating regularly, every beat sending out a flow of blood that I could feel, as if I could see it, squirting out of my severed arm—and with these heartbeats, which were morally and physically unbearable, I was able to measure time, which, alone amidst the terror of that awful night, of which I registered every detail, passed relentlessly, as is its true nature of seconds, fractions of seconds, eternity (pp. 358–359, in [29, 33]).

Between 1956 and his death on January 21, 1961, Cendrars suffered two strokes which gradually deprived him of the use of the left side of his body and further compromised his abilities. He became a "handless writer," a situation further complicated by his gradual loss of his ability to express himself orally due to a serious case of dysarthria [32].

8.3 Film: Research and German Expressionism

Films from this era are also illuminating. Several short, silent films in the Pathé catalog, for example, record medical cases of shell shock and shell shock recovery. A British film, probably shot at either Netley Hospital in 1917 or Seale Hayne Military Hospital in 1918, begins in a bare institutional room where we see an emaciated man who shakes, has trouble standing and walking, and is bent into contorted positions. At the end, we see a man leaving the hospital who appears to be the original patient, but he is heavier and smiling. A close-up of the recovered man's fidgeting hand suggests a remnant of his condition remains [34].

Cultural films frequently highlighted the relationship between neurology and psychiatry, and at times, shell shock was used as a narrative device. Robert Wiene's 1920 silent film *The Cabinet of Dr. Caligari* shows that in some quarters psychiatrists were despised after the war [35]. Wiene presents Dr. Caligari's enigmatic character in a plot about an asylum. Whereas we originally assume Caligari is the head psychiatrist, at the end we wonder if he is really one of the patients at the institution. Throughout the film Caligari is also portrayed as a villain who masterminds a string of murders. Cinematically, the plot is buttressed by creative set design, including asymmetrical geometries, and shadows that are used to reinforce ideas about insanity and the idea that all is not right in the world we see [13, 35, 36].

This movie also offers a picture of that time. In the nineteenth century, Germany was the undisputed leader in psychiatry and reformed the asylum as an institution. Schizophrenia and "hereditary degeneration" were common diagnoses [37]. By the time *Caligari* was filmed in 1920, German asylum conditions had deteriorated markedly. They were overcrowded with war patients, housed many underfed patients, and tuberculosis was endemic as well [37]. Indeed, some now debate whether films like *Caligari* influenced the rise of Nazi Germany [13, 35–38], a topic far beyond the scope of this chapter. What is clear is that this film was not alone in asking who the psychiatrists treating patients were [13, 36].

Several important German films on the psychiatry/neurology theme were directed by Fritz Lang (1890–1976) including a series about *Dr. Mabuse*. Born in Vienna, Lang was among the filmmakers who had served in World War I. He volunteered for military service in the Austrian army and fought in Russia and Romania. Wounded four times, he lost sight in his right eye [39, 40]. It was the first of many vision issues he would face in his lifetime. The loss of his eyesight shattered Lang's dreams of becoming an artist. He took up acting, screenwriting, and eventually directing as an alternative [39–41].

His first Mabuse film was a silent, released in 1922. In his 1933 follow-up, *The Testament of Dr. Mabuse* [42], Mabuse is a ghost impersonating a psychiatrist in an asylum. This film was immeasurably important culturally because it was released the year Adolf Hitler came to power in Germany. Hitler's regime quickly banned it as an incitement to public disorder saying it posed a threat to public health and safety [38].

8.4 Biography, Identity, Literature, and Culture

The discussion of Oliver Sacks' life and contributions in Chap. 6 spoke of his emphasis on biography and identity in dealing with patients. In a larger sense, these concepts are equally important here for they help illuminate that phenomena like war affect all people and those in all professions in a quite human way. For example, the year he died, 1915, Alois Alzheimer (1864–1915), now largely remembered for his description of Alzheimer's disease, wrote in *Der Krieg und die Nerven (War and Nerves)*: "We could see ourselves that the war made demands not only on the nerves of the soldiers but also on those who had to stay at home" [43]. Alzheimer may have been speaking of himself. In 1915, when he wrote this, the war was already underway. Alzheimer was also physically deteriorating as a result of a 1912 fall. Although his health problems began before the war, they did coexist with the knowledge that both his son and son-in-law were drafted into the army. Clearly, Alzheimer was among those at home [44].

Dr. William Osler (1849–1919), too, was among those at home. Some have characterized Osler as the greatest doctor in the history of the world [45]. He found the last 2 years of his life clouded by the death of his only son, Lieutenant Revere Osler, an artillery officer, and a youth of great promise. Revere, named after the American Revolutionary hero Paul Revere, was killed in the action at Ypres in 1917 [46].

Rebecca West's first novel, *The Return of the Soldier* (1918) [47, 48], captures the lives of others at home more generically. This book, frequently cited as her finest work, was the only significant work about World War I written by a woman while the outcome of the war remained unknown. In the book, West conveys the intensity of a world struggling to comprehend the inexplicable horrors that ravaged the minds, bodies, and personalities of those who spent time on the front lines.

A slim book, *The Return of the Soldier*, presents the homecoming of a shell shocked soldier, Chris, and how his partial amnesia impacted the lives of three women: his wife, his cousin, and an old innkeeper's daughter he courted many years before the outbreak of the war. Chris' amnesia has left him with no memory of his wife and a longing to reunite with the old sweetheart whom he still recalls. Although she is now married to someone else, she never really forgot him. As the three women help him recover his memory, they struggle with the knowledge that he will need to return to the front once he is cured [47, 48].

We see evidence in the visual arts as well. Otto Dix (1891–1969), who was briefly discussed in Chap. 4 and comes up again in Chap. 11, was among those impacted. Over the course of the war and after it, his painting style gradually simplified and compressed [49]. Enthusiastic at the start of the war, after it, he expressively depicted the events of battle, the fate of his comrades, and the bloody consequences. One 1918 gouache painting, *Soldat mit Tabakspfeife (Soldier with Tobacco Pipe)*, depicts the soldier with a multicolored and fragmented face and a similarly shattered background. His view of World War II, in contrast to his youthful euphoria at the first war's start, was distinctly negative [50], no doubt in part due to the Nazi's condemnation of his work and his placement in a prisoner of war camp after he was captured by French troops. Yet, despite the wars, Dix retained hope.

Dix suffered a right-hemisphere stroke on November 13, 1967. The result was left-sided hemiparesis, left hemineglect, partial hemianopia, and mild spatioconstructive disturbances [51]. His 1969 *Self-Portrait with Marcella*, painted shortly before his death, shows him holding his granddaughter in his arms. By no means pessimistic, this work does reflect a certain physical weakness left over from his stroke, but also portrays a sense of inner freedom and serenity [49]. On July 19, 1969, Dix suffered a second stroke and died on July 25. The last painting standing on his easel was a bouquet of flowers.

By contrast, Walther Benjamin (1892–1940), a German-Jewish philosopher and essayist, committed suicide. Benjamin is best known for his oft-cited essay *The Work of Art in the Age of Mechanical Reproduction* (1935) [52]. Largely unknown while he lived, Benjamin thought of himself as a literary critic. He was rejected by the Army during World War I, and his suicide in 1940 occurred after the Nazis in France confiscated his apartment, complete with his entire library. In "The Storyteller: Observations on the Works of Nikolai Leskov" [53], published shortly before his death, Benjamin writes of the shell shocked soldiers of the earlier war: "Was it not noticeable at the end of the war that the men returned from the battlefield had grown silent—not richer, but poorer in communicable experience?" (p. 84, [53]).

All of these examples speak to how the psychological and biological traumas of war are in a category of their own. Whereas many illnesses and assaults to the body are unexpected and accidental, the injuries of war are a result of human behaviors. Sometimes cultural statements cry out as well.

8.5 Pablo Picasso and *Guernica* (1937)

Like brain and head injuries, artworks documenting war trauma have accompanied human history. One passionate work depicting the victims and the horror of war is Pablo Picasso's 1937 *Guernica*. It is currently housed in the Reina Sofia, Spain's national museum of modern art.

The iconic work was born as a cry against war and over the years has become a symbol for peace. Perhaps more than any other work, it raises the question of the nature of aesthetics in art. Neuroaesthetics is a term frequently employed to correlate art with beauty as if aesthetics' concern is solely the beauty of a work, either within or beyond its cultural context: This raises the question of what we mean by aesthetic. Does aesthetic only encompass the realm of "emotion recollected in tranquility" to borrow the thoughts of Wordsmith Wordsworth [54]?

In *Guernica*, the emotions evoked are horror and pathos. There is not much tranquility in the work. Indeed, the style is deliberately unbeautiful, particularly to the classical eye. Still, few would dispute its primary role as a work of art. Moreover, as noted, art throughout history has had multiple roles, in storytelling, in expressing the magnificence of a ruler, in providing a kind of immortality, in evoking sensory pleasures, in commemorating historical events, and so forth. None of these functions are explicitly beautiful, though beauty may find a place in many of them. More recently, aesthetics has developed a deconstructive function, breaking down the work of art into its component activities, textures, and concepts in a series of revolutionary investigation of the range and scope of the aesthetic domain. *Guernica* is uniquely qualified to probe this schema because Picasso combined formal compositional elements with the artist's emotional reactions to the bombing of Guernica—and he retained all of his sketches.

To reframe a point made at the opening of this chapter about the historical record of shell shock and war injuries, war is also a theme throughout art history. We find evidence on Egyptian steles, in Greek literature such as Homer's *Iliad*, Roman objects, ancient Asian scrolls, and artifacts from cultures throughout the world. All demonstrate the power of visual representation and the arts in general to touch us. Indeed, art is frequently a part of our larger cultural dialogue as it reaches inside us emotionally, perceptually, and empathically.

In terms of Picasso's *Guernica*, historical correlates that come to mind include Peter Paul Rubens' (1577–1640) *The Allegory of Peace and War* (1629) and his *Massacre of the Innocents*. Indeed, *The Allegory of Peace and War* (1629) reflects, and even illustrates, Rubens' strong concern for peace. It was given to King Charles I to help create a peace treaty between London and Madrid. Rubens' strong concern for peace is similarly illustrated in his *Massacre of the Innocents*.

Another figure, and one who surely influenced Picasso's *Guernica*, is Francisco de Goya (1746–1828), another Spaniard. Goya's famous and graphic *The Disasters of War* similarly depicts scenes of horror, brutality, torture, and savagery. The war on his mind was the Peninsular War, which pitted an alliance of Spain, Portugal, and the United Kingdom against France on the Iberian Peninsula during the Napoleonic Wars. Created in the 1810s, but not published until 35 years after Goya's death in 1863, it is likely that Picasso, a Spaniard too, knew these prints. Picasso was likely also inspired by similar Goya works that cry out against the multitude of ways in which war victimizes humanity.

Before *Guernica*, Picasso (1881–1973) was a relatively apolitical artist, being more concerned with the internal logic of his artistic innovations than actions at the forefront of the world community. If *Guernica* marked a major rethinking of how he envisioned art's purpose and his life as an artist, it also retains his concern with a painting's artistic logic. These give us entry into his creative process, or his elaboration of a kernel idea into a final work [55].

The emotional kernel came about after he read the following report of the bombing in the *London Times* on April 26, 1937 [56]:

> Guernica, the most ancient town of the Basques and the center of their cultural tradition, was completely destroyed yesterday afternoon by insurgent air raiders. The bombardment of the open town far behind the lines occupied precisely three hours and a quarter, during which a powerful fleet of aeroplanes…did not cease unloading on the town bombs weighing from 1000-lbs, downward and it is calculated more than 3000 two-pounder aluminum incendiary projectiles. The fighters, meanwhile, plunged low from above the center of the town to machine-gun those of the civilian population who had taken refuge in the fields. The whole of Guernica was soon in flames except the historic Casa de Juntas, with its rich archives of the Basque race, where the ancient Basque Parliament used to sit. The famous oak of Guernica, the dried old stump of 600 years and the young new shoots of this century, was also untouched….. (pp. 18–19, in [56]).

This event coincided with a project Picasso had reluctantly accepted in January of 1937. At that time, the Spanish government in exile, following Francisco Franco's victory in the Spanish Civil War, had asked Picasso to paint a mural for the Spanish Pavilion of the 1937 International World's Fair. He unenthusiastically planned to present a mural of an artist working in his studio, despite the exiled government's hope that he would draw attention to Spain's plight.

After the Guernica atrocity, Picasso decided to respond directly to the bombing. The finished mural depicts Picasso's emotional reaction to the bombing of this Basque city, which, as the *Times* excerpt indicates, was not a "logical" military target. Begun four days after the *Times* report, on May 1, 1937, *Guernica* is among his most famous works. A schematic, almost graphic painting, the mural work includes notations from Picasso's symbolic vocabulary (e.g., his use of the bull) with other forms in an almost diagrammatic manner. It interweaves the narrative of *Guernica* with an almost dreamlike presentation (clearly a nightmare!).

The first sketch already includes the bull in its final location (on the left), the bird in the spot where it will eventually hover over a table, the woman pushing the lamp through the window, and other details. Although quickly contrived, the drawing demonstrates that Picasso was drawn to four basic concepts from the start: the upright bull, the light bearer, the sprawling victim, and the inert base of destroyed bodies. Yet, at the beginning, we find that there is no sun, no flower, and a radically different take on the fallen warrior [56]. Particularly noteworthy is the flower beside the collapsed knee of the horse in the final version. It compels us to ask if it was intended as a symbol of hope and regeneration after destruction and brings to mind the reference to "young new shoots" in the *Times* report.

The sketches also show tears falling from eyes that are easily deciphered. In the final work, however, tear-shaped eyes remain, but formerly falling tears have been solidified into expressive objects [56]. Even with the transformed tears, it appears that (except for the bull) all the figures are crying out in despair. Their emotions seem to mirror a sentiment in the British War Commission report on shell shock quoted at the beginning of this chapter: "that men who had not been exposed directly to such fire were just as traumatized" [1].

Compositionally, the mural's most defining feature is the figures within the central pyramid, with symmetrical correspondence of the flanks [56]. These include the bull at the left, the falling woman at the right, and the culminating feature of the oil lamp. At the tip of the triangle, we find an "eye" of an electric light globe (image of sun/eye) and a woman with the lamp (light holding darkness/bull at bay). In an earlier version, Picasso gave her a blood red tear, but in the end all color was removed from the painting [57] in a remarkable aesthetic appeal to the reversal of the normal expressive role of color in paintings.

Indeed, *Guernica's* monochromatic coloring is among the features that scream out to the viewer, particularly since the bombing did not occur at night and the blood red tear was removed. With the monochromatic tonality, there is no distinction between the complexions of the dead and the living, another deviation from "normalcy" [58]. Rudolf Arnheim postulates that the artist saw the symbolic value in omitting color for it aided him in conveying the depth of tragedy associated with the event [56]. Peter Schjeldahl, by contract, suggests the black, white, and gray tones are indicative of pictorial intelligence on Picasso's part. Schjeldahl's thinking is that color wasn't Picasso's strong point. Limiting his palette to black-and-white was a means to forge his chief handicap into a thematic weapon [58]. In addition, by evoking the look of a newspaper, in Schjeldahl's view, the reportage quality of the painting additionally aids a viewer in comprehending its catastrophic nature.

Larger than anything Picasso had attempted up to this point, measuring 138x308 inches (or a ratio of more than 2:1) and unusually shaped, this work was painted in an amazingly intensive burst of work. Finished in June, the effort required over 200 hours of direct painting and likely another hundred for the sketches that preceded it [59]. His partner Dora Maar recorded his efforts in seven photographic studies of the work in progress. These "status reports" or "state images" allow us to conceptualize how Picasso moved the composition along as he coaxed the elements together [56].

In terms of shell shock, one man on the right is a civilian with arms raised. His eyes are shaped like distorted tears and the image evokes the filmed images of the shell shocked soldiers during the Great War. Another casualty, on the bottom left, screams in agony. It was only in the last stage of the work that Picasso switched the position of this figure's head. Looking upward in the final piece, it was face-down in earlier stages.

The cradle of Basque civilization, the city of Guernica lacked military importance. Picasso's *Guernica* strongly communicates his despair about the cultural loss that came about as a result of the world's first aerial bombardment of a civilian population. Sixteen thousand civilians perished in the bombing. After World War II, German Luftwaffe Chief Hermann Goering acknowledged at the Nuremberg war crime trials in 1945 and 1946 that the bombing was a deliberate act. Guernica was designated as a practice site for the Nazi air force to hone their strategy and skills. It was not until April 26, 1997, that the German government formally apologized to the citizens of the then restored city [57].

8.6 German-American Film Director William Wyler

The idea of the auteur was introduced in 1951 by a group of French critics who had founded *Cahiers du Cinema* (*Notebooks on Cinema*), an increasingly influential magazine in the world of filmmaking. Their view was that a film director is a creative artist with a signature that winds through his or her filmography [60].[3] Discussions about auteurs confirmed and re-evaluated the talents of many Hollywood directors, including two briefly discussed here: William Wyler and John Huston. Both made films informed by their service in World War II.

[3] Auteur theory, introduced in Andrew Sarris' 1962 essay, was a critique of the idea that the director is the author of the films that bear his or her name [60].

German-born director William Wyler (1902–1981) enlisted, and his experience underscores that the war injuries people took home with them extended beyond shell shock. He joined the US Army Air Force at the age of 40 to produce war documentaries, and his health suffered as a result of flying combat missions. While working on *Memphis Belle* (1944) [61], Wyler lost consciousness from lack of oxygen. While working on *Thunderbolt* (1945) [62], he lost his hearing [63–65].

The loss of hearing came about as he recorded the devastation in Corsica. Lying on his stomach as he filmed, the loud engine noise and high-pitched screeching of the wind echoed in his ears. By the time he returned to Allied Headquarters in Grosseto, he had lost the hearing in both ears [66]. As he later recalled it, he couldn't walk straight when he got off the plane, and they thought he was drunk. He couldn't hear at all. Two weeks later, he was sent home by sea to the states [63]. Full hearing never returned.

After several examinations, it was determined that Wyler had suffered irreparable nerve damage [64]. He was told that it was likely that ear plugs or even cotton would have avoided the loss, but he couldn't change the circumstance after the fact. The hearing in his left ear partially returned, and he could make out words when people shouted. His right ear, however, was permanently affected. The Air Force doctors recommended an adenoidectomy, the removal of the lymphatic tissue between the back of the nose and the throat, to improve and preserve the hearing that remained [66].

Depression followed his surgery. All the specialists he visited told him there was no way to fix his hearing and no treatment would restore it [66]. To continue directing, Wyler devised a work around using the sound man's microphones to listen to scenes through a set of headphones plugged into the sound system. In an interview three days before he died of stroke, at the age of 79, Wyler joked that because of this setup he could hear what the actors were saying about him as he directed them [65].

When Wyler returned home in 1945, he wanted to make a film that would combine disability with military experience. He decided to combine them in *The Best Years of Our Lives* (1946) [67], a film that went on to win seven Academy Awards. In the film, he worked with Harold Russell (1914–2002), an American soldier who had lost his hands in a training accident. Wyler asked Russell to play an amputee in a film about three US servicemen readjusting after coming home from World War II. Its narrative centers around the injuries each man sustained and the fact that each served in a different service, had a different rank, and their military ranks did not correspond with their civilian social class backgrounds.

In light of Wyler's hearing problems, it is fascinating to realize how much this movie's sound adds to the emotive quality of the film. Its affective auditory score, composed by Hugo Friedhofer's (1901–1981), stitches the fictional world together. The music serves as an unconscious connective tissue that imparts subjective depth and connects musical passages to convey particular moods to the audience. For example, the minor mode equates with sadness, and each of the main characters is associated with a distinctive musical motif [67].

8.7 American Film Director John Huston

John Huston (1906–1987) was born in the same year as the Italian director Luchino Visconti. Like Visconti, Huston was dying when he directed his last film, *The Dead* [68], a film nominated for several Academy Awards. Written by his son Tony Huston and starring his daughter Anjelica, the film was clearly a family affair. Anjelica later said that her father remained a virtuoso despite his ill health. She then added that she was so stressed during the production that she contracted Epstein-Barr from watching his struggle while also having to play her own part in the film [69].

The Huston family worked with one another in their moviemaking productions. When John Huston directed *Let There Be Light* (1946), a documentary on shell shock victims [70], his father Walter Huston did the voice-over narration. Walter Huston opens with the words: "Every man has a breaking point. And these, in the fulfillment of their duties as soldiers, were forced beyond the limits of human endurance." As he explains, "these are the casualties of spirit" [70].

Researching the project, Huston visited a number of Army hospitals before settling on Mason General Hospital on Long Island, New York, where the officers and doctors were sympathetic and willing. Huston decided to follow one group of patients for 6–8 weeks, from their day of arrival until their discharge [71]. As Huston explained it, the patients arrived "in various conditions of emotional distress. Some had tics; some were paralyzed; one in ten was psychotic. Most of them fell into the general designation of 'anxiety neurosis'" [71]. The purpose of the project was to show prospective employers that they were not insane but were as employable and trustworthy as anyone [71].

Huston's team shot close to 70 hours of film to insure they caught the completely unpredictable exchanges that sometimes occurred during the therapy [71]. The final edited film mixed patient interviews and treatment sequences with narrated montages of patient progress toward recovery. The psychiatric methods are those of the 1940s. For example, one patient with a functional stutter is cured with hypnosis after the administration of sodium amytal to remove any psychological barrier. We also see electroshock treatments given to the severely psychotic, but nothing disturbing appears in the final cut.

On the whole, this hospital treated relatively recent and milder cases for which there was hope of a therapeutic cure. Perhaps this is why the cures look strangely easy. "The things one saw happen were seemingly miraculous," as Huston explained it: "Men who couldn't walk were given back the use of their legs, and men who couldn't talk were given back their voices." The film also explains: "The fact that he can walk now does not mean that his neurosis has been cured. That will require time. But the way has been opened for the therapy to follow" [71].

Uncharacteristically for a documentary of this time, the cameras were acknowledged during the filming. Indeed, the first words of direct sound in the film are: "There is no need to be alarmed by the presence of these cameras," the welcoming officer tells the glum group of arriving men; "they are making a photographic record of your progress at this hospital from the date of admission to the date of discharge" [71]. According to Huston, "As the men began to recover, they accepted the cameras as an integral part of their treatment. The doctors even noticed that the cameras seemed to have a stimulating effect, and that the patients being filmed showed greater progress than those in the other groups" [71].

The finished film included prolonged group therapy sessions where we see black and white soldiers participating. The US military was largely segregated at the time. Integration protocols were not initiated throughout until President Truman's 1948 executive order. As this film shows, some Army hospitals had begun integrating earlier. Huston commented later that he felt that showing black and white men together was one reason the US Army deemed the film unsuitable for public viewing after it was completed in February 1946 [71].

Indeed, the Pentagon immediately shelved the project saying they were concerned about the privacy of the patients. Huston countered that the soldiers has signed release forms and "were enthusiastic about the picture" and "wanted very much for it to be seen" [72]. This was to no avail. He was told these forms had "disappeared" [67].

A few New York critics who managed to see the film objected and this outcry led the Army to commission a remake. Released in 1948 as *Shades of Gray* [73], the second version draws so heavily on the Huston film that it appears likely the military wanted to sanitize the original production. It uses actors to impersonate the returning soldiers and suggests that battle fatigue was a result of the sufferer's upbringing, lack of athletic ability, and inability to stand up to bullying [37]. In this respect, the military's modified version showed that views common in World War I remained.

8.8 Post-traumatic Stress Discover (PTSD)

According to the National Institute of Mental Health, post-traumatic stress disorder (PTSD) develops in people who have experienced a shocking, scary, or dangerous event [74]. Although anyone can develop PTSD, and at any age, likely candidates are combat veterans and people who have experienced or witnessed a physical or sexual assault, abuse, an accident, a disaster, or other serious events [74]. This term is now favored over shell shock, and a range of therapies are used to address it, including psychotherapy, medications, or a combination of psychotherapy and medications, virtual reality (VR) simulations of past events, and so forth.

In the last chapter, the discussion about mirror neurons pointed out that some researchers have shown that the same parts of the brain activate when we are watching and identifying with an action as when we were doing the action ourselves. Based on the hypothesis that mirror neurons form one of the physiological underpinnings of interpersonal attunement and empathy [75], researchers have begun developing computerized tools to activate mirror neurons in PTSD therapeutics and the management of pain, stress, and anxiety. The need for this is clear. Recent military events around the world have reiterated that traumatic stress both directly and indirectly remains an outcome of military engagements.

One preliminary study that postulates the value of VR rehabilitation was by Steuwe et al. [75, 76]. They examined direct eye contact, a common interpersonal experience, in survivors of prolonged child abuse with PTSD. Researchers presented both groups with three-dimensional virtual characters on a computer screen that expressed various expressions. Some expressions were happy, some angry, and others were neutral. Gauging participant responses, they found that irrespective of the displayed emotion, controls exhibited an increased blood oxygenation-level-dependent response during direct vs averted gaze within the dorsomedial prefrontal cortex, left temporoparietal junction, and right temporal pole. Under the same conditions, individuals with PTSD showed increased activation within the superior colliculus (SC)/periaqueductal gray (PAG).

Simulated environments are also used, and these tools build on the kind of neuropsychological interventions that began with World War I and World War II as clinicians tried to develop strategies to improve cognitive, emotional, and psychosocial functioning though inducing brain changes. Contemporary interventions include individual and group cognitive rehabilitation, psychotherapy, psychoeducation, and psychosocial interventions that cover areas in the personal, physical, and social context [77]. In terms of virtual reality simulations, the idea is that immersing the patients into a three-dimensional environment that mimics a traumatic memory helps them recover fragments of memories the mind has blacked out, a common response to trauma [78, 79].

One program at the University of Southern California Institute for Creative Technologies (ICT) was developed with funding by the US Army in 1999. Their work has advanced the development of virtual humans who look, think, and behave like real people, and it is used as a tool to address PTSD [80]. Using interactive tools, including a virtual coach, anonymous participants work through issues and practice with a computer-generated virtual human educational system [80]. The coach is trained to see signs of depression, PTSD, suicide risk, and can simulate what goes wrong when soldiers don't consider variables like cultural sensitivities. This experiential learning to help them work through past events is intended to aid in expressing realistic emotions and communication [80].

VR tools emulate a normal sensory environment that might include recognizable elements like houses, aircrafts, or even military operating theaters [81]. Some have reported positive results from these treatments. One sergeant said it is common for soldiers to push their traumas aside and therefore traditional therapy is not as effective [82]. She then added that virtual therapies put servicemembers into a safe environment, which aids them in reaching the emotional connection necessary to make treatment and recovery possible [82].

In 2018, in an effort to evaluate VR's efficacy, Tieri and his colleagues asked whether virtual reality is effective in rehabilitation or just entertainment for scientists. Their study concluded that within the next 5 years commercial developments would reduce the cost and thus these systems would be easier to use in clinical settings [83]. Now, 5 years later, cognitive and motor functions are being used by physical therapists, cognitive therapists, medical doctors, neuroscientists, computer scientists, and bioengineers, with artists frequently contributing to the design and development of these systems.

8.9 Shell Shock Revisited

As mentioned earlier, during World War I, filmmakers used the war and shell shock as narrative devices to comment on the interplay of injury, neurology, psychiatry, and culture. Contemporary film and literature are still using brain injuries, shell shock, and PTSD as plot lines. For example, in *Disorder* (2105), a film directed by Alice Winocour, we meet an ex-soldier who returned from Afghanistan suffering from post-traumatic stress disorder (PTSD) [84]. *Regeneration*, based on Pat Barker's *Regeneration Trilogy* exposing the causes and effects of shell shock, is another example in this genre [85, 86], and it draws up historical figures discussed at the beginning of this chapter.

In Barker's revisitation of the Craiglockhart Hospital, W.H.R. Rivers is the protagonist, and the stories are those of Owen, Sassoon, Rivers, and Lewis Ralph Yealland. Yealland was a Canadian-born therapist who moved to the United Kingdom to practice medicine during the war. Barker's 1990s cinematic presentation is concerned with the social issues surrounding shell shock treatment and military justice. She sees the issues in terms of contemporary culture, human rights issues, the nature of war, class, and PTSD.

Barker presents shell shock treatments in terms of identity, a variable Oliver Sacks emphasized as important in his own work with patients, as discussed in Chap. 6. Whereas Sacks' concern was seeing patients as individuals rather than simply in terms of their pathologies, Barker's focus is on how modern war reshaped lives within the larger cultural milieu [6, 85]. She asks, first, what does it show us about human nature and identity? Second, what guidelines should treatments follow?

Among the most thought-provoking sequences she presents are those that show and comment on differing therapeutic approaches. Having heard of the success Lewis Yealland (1884–1954) was having in curing his shell shock patients, Rivers goes to London to observe a treatment. Cinematically, the portrait is scathing. In the Yealland sequence, we see him using electroshock on a patient suffering from hysterical mutism who is told "you will speak before you leave, there is no alternative" [86]. Rivers, who was adamant about treating soldiers as ill rather than malingers who refused to fight, was strongly and adversely affected. Rivers' reaction to Yealland's success showed that physicians addressed the need to heal with a range of prescriptions. To Rivers, Yealland wasn't curing the patient as if they were a human. Rather, Rivers saw someone addressing the cure as if the person was a fighting unit being repaired so that he could return to the front lines [86]. Yealland's book *Hysterical Disorders of Warfare*, published in 1918, details 44 cases and thus offers a comprehensive overview of his clinical practice [87],

Rivers, unlike Yealland, had a sense of guilt about returning people to the front. *Regeneration* portrays him struggling with his role as a healer who returned his patients, people he came to known and grew to care for as individuals, to the killing fields. At one point, another doctor at the institution suggests Rivers was even taking on shell shock symptoms from the daily stress. Indeed, his lifelong stuttering problems were more pronounced, and he was having bad dreams. One rather obscure reference to his past comes up when he mentions a nightmare about his collaborative work with Henry Head (1861–1940), without providing any informative detail.

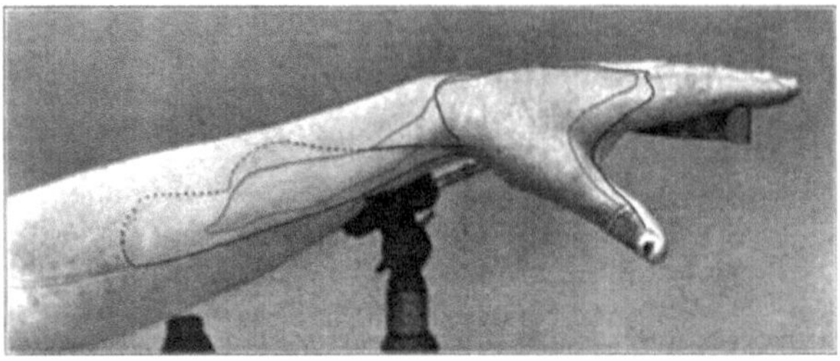

Fig. 8.2 *A human experiment in nerve division.* W.H.R. Rivers and Henry Head. 1908. Brain 1908: 31; 323–450. To show the extent of the loss of sensation 50 days after the operation (June 14, 1903). The dark black line encloses the loss to cutaneous tactile stimuli; the lighter line encloses the cutaneous analgesia. Wherever these lines are broken, the border was an indefinite one. Courtesy of Oxford University Press

He says in a dream he experienced as a nightmare, he saw himself serving as the patient rather than Head in an experiment they began on April 25, 1903. At that time, Rivers severed the radial nerve in Head's left arm (Fig. 8.2). A portion of the skin at first, on the forearm, and hand were numbed, and the restoration of feeling was periodically chartered over the next 5 years. Although these findings are no longer considered fully explanatory of the regeneration process, they are still held to be an excellent illustration of what occurs after such a wound [23, 88].

Finally, a truly informative element of *Regeneration* [86] is how it chronicles the ways in which the war upsets the balance of both doctors and patients, much as we saw earlier in the *Dr. Caligari* discussion. In this later film, however, it is clear that the injuries of the war touched the nerves of all involved. This is particularly evident in Rivers, the doctor whose job it is to uphold the values of the state and collective society. As noted, he expresses increasing doubts about his work aimed at healing the patient's mind so that he might return these people to killing and being killed as soldiers [6, 85]. As also pointed out earlier, the paradox was also evident in the lives of the soldier-poets Owen and Sassoon who objected to the way the war was conducted and yet believed they should be in the field, fighting with their comrades.

8.10 Conclusion

Barker's trilogy and the various examples covered in this chapter underscore that wounds of war bring up human issues for healers, those who suffer from combat, and those in the culture at large. As noted in this chapter's opening, the activities of war are interwoven with all aspects of human life. Before turning to the next chapter on therapeutics, neuropsychology, and art therapy, I want to point out that war has a unique position neuroculturally. Indeed, one paradox of war is that the events stimulate medical advancements and a range of artistic reactions, as Picasso's *Guernica* shows. Picasso's *Guernica* also underscores that even those uninjured and promoting peace feel war's trauma, impact, and destruction. Humanity's urge to heal, to prevent illness, and to alleviate pain and injury is the subject of the next chapter.

References

1. War Office Committee of Enquiry. Report of the War Office Committee of Enquiry into "Shell shock" Catalogue ref: WO 32/4748. The National Archive; 1922.
2. Sandars NK. The epic of Gilgamesh. Harmondsworth, Middlesex: Penguin Books; 1960. (2100 BCE/1960)
3. Herodotus. The histories, revised (Penguin Classics). Marincola JM, editor. London: Penguin Classics; c. 450–420 BCE/2003.
4. Warner CH, Castro CA. Veteran and military mental health: a clinical manual. Cham: Springer International Publishing; 2023.
5. Ione A. Art and the brain: plasticity, embodiment, and the unclosed circle. Leiden: Brill; 2016.
6. Leese P. Shell shock: traumatic neurosis and the British soldiers of the First World War. New York: Palgrave; 2002.
7. Simmel G. The metropolis and mental life. In: Wolff K, editor. The sociology of Georg Simmel. New York: Free Press; 1950.
8. Robson S. The First World War. London: Taylor & Francis; 2014.
9. Myers CS. A contribution to the study of shell shock. Lancet. 1915;185(4772):316–20.
10. Bogousslavsky J. Hysteria: the rise of an enigma. Basel and New York: Karger; 2014.
11. Holdorff B. The fight for 'traumatic neurosis', 1889–1916: Hermann Oppenheim and his opponents in Berlin. Hist Psychiatry. 2011;22(88 Pt 4):465–76.
12. Oppenheim H. Die traumatischen Neurosen. Beobachtungen: Hirschwald; 1892.
13. Kaes A. Shell shock cinema: Weimar culture and the wounds of war. Princeton: Princeton University Press; 2009.
14. Baloh RW, Bartholomew RE. Psychogenic illness on the battlefield: from the Civil War to Gulf War syndrome. In: Havana syndrome: mass psychogenic illness and the real story behind the embassy mystery and hysteria. Cham: Springer International Publishing; 2020. p. 59–70.
15. Scull A. Hysteria: the disturbing history. Oxford: Oxford University Press; 2011.
16. Tatu L, Bogousslavsky J, Moulin T, Chopard J-L. The "torpillage" neurologists of World War I. Electric therapy to send hysterics back to the front. Neurology. 2010;75(3):279–83.
17. Roudebush M. A patient fights back: neurology in the court of public opinion in France during the First World War. J Contem Hist. 2000;35(1):29–38.
18. Owen W. Poems. London: Chatto & Windus; 1921.
19. Shephard B. Headhunters: the search for a science of the mind. London: The Bodley Head; 2014.
20. McDougall W. An introduction to social psychology. London: Methuen & co.; 1908.
21. Smith GE. The migrations of early culture: a study of the significance of the geographical distribution of the practice of mummification as evidence of the migrations of peoples and the spread of certain customs and beliefs. London and New York: University Press; Longmans, Green & Company; 1915.
22. Smith GE, Pear TH. Shell shock and its lessons. Manchester: University Press; 1918.
23. Rivers WHR, Head H. A human experiment in nerve division. Brain. 1908;31(3):323–450.
24. Wilson JM, Dr W, H. R. Rivers: Siegfried Sassoon and Robert Graves' 'fathering friend'. Brain. 2017;140(12):3378–83.
25. Sassoon S. Sherston's progress. Garden City, NY: Doubleday, Doran; 1936.
26. Rivers WHR. Conflict and dream. London and New York: Harcourt, Brace & Company; 1923.
27. Rivers WHR. Instinct and the unconscious: a contribution to a biological theory of the psycho-neuroses. Cambridge: University Press; 1920.
28. Luria AR. The man with a shattered world; the history of a brain wound. New York: Basic Books; 1972.
29. Tatu L, Bogousslavsky J, Boller F. Phantoms in artists: the lost limbs of Blaise Cendrars, Arthur Rimbaud, and Paul Wittgenstein. J Hist Neurosci. 2014;23(4):355–66.
30. Mitchell SW. Injuries of nerves and their consequences. Philadelphia: J.B. Lippincott; 1872.
31. Vitturi BK, Sanvito WL. The neurology behind three wounded French artists during the great world war. Arq Neuropsiquiatr. 2020;78(6):380–2.

32. Tatu L. The missing hands of Blaise Cendrars. Front Neurol Neurosci. 2010;27:143–59.
33. Cendrars B, Loignon S. J'ai saigné. Paris: Hatier; 1938. /2012
34. British Pathé. Wonderful shell shock recovery. London: British Pathé; 1914.
35. Wiene R. The Cabinet of Dr. Caligari (Das Cabinet des Dr. Caligari). Decla-Bioscop Studio; 1920.
36. Kracauer S. From Caligari to Hitler, a psychological history of the German film. Princeton: Princeton University Press; 1947.
37. Wijdicks EF. Cinema, MD: a history of medicine on screen. Oxford: Oxford University Press; 2019.
38. Kracauer S. Bollingen Foundation Collection (Library of Congress). Theory of film: the redemption of physical reality. New York: Oxford University Press; 1960.
39. Lang F, Grant BK. Fritz Lang: interviews. Jackson: University Press of Mississippi; 2003.
40. Stevens G, American Film Institute. Conversations with the great moviemakers of Hollywood's golden age at the American Film Institute. New York: Alfred A. Knopf; 2006.
41. Llewellyn J, Southey J, Thompson S. "Weimar cinema," Alpha History; 2014. http://alphahistory.com/weimarrepublic/weimar-cinema/. Accessed 3 July 2023.
42. Lang F. The testament of Dr. Mabuse. Germany: Nero-Film; 1933.
43. Alzheimer A. Der Krieg und die Nerven. Breslau: Preuß & Jünger; 1915.
44. Goedert M, Ghetti B. Alois Alzheimer: his life and times. Brain Pathol. 2007;17(1):57–62.
45. Bliss M. William Osler: a life in medicine. Oxford and New York: Oxford University Press; 1999.
46. Garrison FH. Sir William Osler (1849–1919). Science. 1920;51(1307):55–8.
47. West R. The return of the soldier. London: Nisbet; 1918.
48. Klinkenborg V. Introduction: the return of the soldier (Rebecca West). New York: Modern Library; 2004.
49. Gallery T. Otto Dix, 1891–1969. London: Tate Gallery; 1992.
50. Karcher E. Otto Dix, 1891–1969. Köln: Taschen; 2010.
51. Bäzner H, Hennerici MG. Painting after right-hemisphere stroke—case studies of professional artists. Front Neurol Neurosci. 2007;22:1–13.
52. Benjamin W. The work of art in the age of mechanical reproduction. In: Arendt H, editor. Illuminations. New York: Schocken Books; 1999. p. 217–52.
53. Benjamin W. The storyteller: observations on the works of Nikolai Leskov. In: Eiland H, Jennings MW, editors. Selected writings: 1935–1938, vol. 3. Cambridge: Belknap Press; 1996. p. 143–166.
54. Wordsworth W, Coleridge ST. Lyrical Ballads. London: Penguin Books Limited; 2017.
55. Weisberg RW. On structure in the creative process: a quantitative case-study of the creation of Picasso's Guernica. Empirical Studies Arts. 2004;22(1):23–54.
56. Arnheim R. Genesis of a painting: Picasso's Guernica. Berkeley: University of California Press; 1962.
57. Harris JC. Picasso's Guernica. Arch Gen Psychiatry. 2010;67(9):878.
58. Schjeldahl P. Spanish lessons. The New Yorker; 2006. p. 90–1.
59. Martin R. Picasso's war: the destruction of Guernica and the masterpiece that changed the world. New York: Dutton; 2002.
60. Sarris A. Notes on the Auteur theory in Auteurs and authorship: a film reader. Hoboken: John Wiley and Sons; 2008. p. 29–35.
61. Wyler W. The Memphis Belle. United States: Paramount Pictures; 1944.
62. Wyler W. Thunderbolt. United States: Monogram Pictures; 1947.
63. Herman J. A talent for trouble: the life of Hollywood's most acclaimed director, William Wyler. New York: G.P. Putnam's Sons; 1995.
64. Miller G. William Wyler: the life and films of Hollywood's most celebrated director. Lexington: University Press of Kentucky; 2013.
65. Slesin A. Directed by William Wyler. United States: American Masters; 1981.
66. Macor A. Making the best years of our lives: the Hollywood classic that inspired a nation. Austin: University of Texas Press; 2022.

67. Wyler W. Best years of our lives. US: RKO Radio Pictures; 1946.
68. Huston J. The dead. Vestron Pictures; 1987.
69. King S. Anjelica Huston's magical movie life, from 'Prizzi's Honor' to 'John Wick'. Los Angeles Times. May 17, 2019.
70. Huston J. Let there be light. US Army; 1945/1981.
71. Simmon S. Let there be light (1946) and its restoration. Film Preservation. 2012. https://www.filmpreservation.org. Accessed 3 July 2023.
72. Lowe B. Let there be light. Washington DC: Library of Congress; 2012. https://www.loc.gov/. Accessed 24 July 2023
73. US Army. Shades of Gray. US Army; 1948.
74. National Institute of Mental Health. Post-traumatic stress disorder (PTSD). Washington, DC: National Institute of Mental Health; 2023.
75. Walker A. The role of mirror neurons in relational dysfunction in posttraumatic stress disorder. PhD Dissertation. Pepperdine University; 2020.
76. Steuwe C, Daniels JK, Frewen PA, Densmore M, Pannasch S, Beblo T, et al. Effect of direct eye contact in PTSD related to interpersonal trauma: an fMRI study of activation of an innate alarm system. Soc Cogn Affect Neurosci. 2014;9(1):88–97.
77. Yi A, Belkonen S. Neuropsychological rehabilitation. In: Kreutzer JS, DeLuca J, Caplan B, editors. Encyclopedia of clinical neuropsychology. New York: Springer New York; 2011. p. 1766–7.
78. Johnson DA, Rose FD, Rushton S, Pentland B, Attree EA. Virtual reality: a new prosthesis for brain injury rehabilitation. Scottish Med J. 1998;43(3):81–3.
79. Park MJ, Kim DJ, Lee U, Na EJ, Jeon HJ. A literature overview of virtual reality (VR) in treatment of psychiatric disorders: recent advances and limitations. Front Psych. 2019:10.
80. Institute for Creative Technologies (ICT). Military & Veterans Initiatives: Institute for Creative Technologies (ICT). Los Angeles: University of Southern California; 1999.
81. Schouten KA, Van Hooren S, Knipscheer JW, Kleber RJ, Hutschemaekers GJM. Trauma-focused art therapy in the treatment of posttraumatic stress disorder: a pilot study. J Trauma Dissociation. 2019;20(1):114–30.
82. Collins TG. Virtual reality therapy may help PTSD patients. Military News. 2012; https://www.military.com/daily-news. Accessed August 2, 2023
83. Tieri G, Morone G, Paolucci S, Iosa M. Virtual reality in cognitive and motor rehabilitation: facts, fiction and fallacies. Expert Rev Med Devices. 2018;15(2):107–17.
84. Winocour A. Disorder. Darius Films: Dharamsala; 2015.
85. Barker P. Regeneration Trilogy. Harmondsworth: Penguin; 1998.
86. MacKinnon G. Regeneration. United Kingdom: Artificial Eye; 1997.
87. Yealland LR. Hysterical disorders of warfare. London: Macmillan; 1918.
88. Compston A. A human experiment in nerve division by W. H .R. Rivers MD FRS, Fellow of St John's College, Cambridge and Henry Head MD FRS, Physician to the London Hospital, Brain 1908: 31; 323–450. Brain 2009;132(11):2903–2905.

9 Neurocultural Therapeutics and Applications

Because art and artists intersect with neuroscientific therapeutics in more than one way, how we examine art in terms of neuroscience makes a difference. If an individual is already acquainted with an art form or the arts, additional factors are a part of the equation. So, just as art therapy and an art practice are not the same thing, physiological and psychological issues are not necessarily in alignment when examining neuroscience and art in tandem. For example, as discussed in Chap. 4, composing music was psychologically therapeutic for Beethoven after his loss of hearing, but this creative activity was unable to restore his hearing. The color-blind artist Jonathan O. similarly found his art therapeutic although, like Beethoven, this creative activity did not restore his color vision or synesthesia. The composer Ravel, who suffered aphasia, apraxia, agraphia, and amusia, found that music only added to his despair: As noted, "music [was] trapped forever in his aphasic mind, Maurice Ravel, frustrated beyond words, could only cry" (p. 62, [1]).

There are also artists such Yayoi Kusama (b. 1929) who works out her mental problems through her art. She calls her work art-medicine. Starting in the 1970s, she walked to her nearby studio every day from the Tokyo mental hospital where she lives. As she explains it: "I fight pain, anxiety, and fear every day, and the only method I have found that relieved my illness is to keep creating art…I followed the thread of art and somehow discovered a path that would allow me to live" [2].

A fourth example is the novelist Howard Engel. His case follows the opening section of this chapter on theoretical approaches and strategies. Engel, like Jonathan O. and Beethoven, found a way to continue working professionally after he acquired *alexia sine agraphia.* In other words, after a small stroke, he lost his ability to read but retained his ability to write. His remedial therapy, coupled with a large dose of determination, helped him develop strategies that made it possible for him to write several books after he left the hospital. Even though he never fully regained his ability to read, he continued to practice his art, writing, professionally. His therapy was not "art therapy" but a series of adaptations that allowed him to resume his career path.

A. Ione, *Neuroscience and Art*, Neurocultural Health and Wellbeing,
https://doi.org/10.1007/978-3-031-62336-3_9

9.1 Therapeutic Approaches and Strategies

The word therapeutics means to cure or treat medically and is derived from the Greek *therapeutikos* [3]. In a broad sense, therapeutics covers serving and caring for a patient in a comprehensive manner, preventing disease and managing specific problems. Prevention includes exercise, diet, and mental factors. Management can include treatments such as drugs to relieve pain or infection, surgery, and counseling or psychotherapy to relieve emotional distress [4].

As noted in Chaps. 1 and 2, all cultures have interwoven therapeutics, neurology and psychology with their views on health, the mind, and the body. Conceptually, neurology and psychology trace their roots back to Thomas Willis' (1621–1675) *Cerebri Anatome*. In his work, he made a distinction between neurology and psychology, using neurology to characterize his doctrine of the nerves, and psychology to speak of the soul [4].

With the ascendency of Darwinian theories and nineteenth century experimental studies, the relationship between the nervous system and the soul was redefined. Neurology came to refer to the branch of medicine that deals with the nervous system and its disorders. Psychology became the term used for the study of behaviors of the mind, rather than the soul [5]. Art was often the odd man out here because the tendency had been to equate art with the soul and emotions, as if cognition and the mind were outside of art's purview [6, 7]. With the soul removed from the picture in the nineteenth century, there was less clarity as to where art as well as the characteristics coupled with the soul belonged in the scheme of things [6, 7].

In reviewing this, it is helpful to remember that in the nineteenth century, with the emergence of Darwin's evolutionary theory, discussions on how to perceive the human brain in biological terms were intense [7]. These were paralleled by an experimental neuroscience that had added the use of electricity to its toolbox. At that time, observation, microscopes, and dissection were available to analyze details. Researchers like Paul Broca (1824–1880) began to establish localized functions within the brain, and significant emphasis turned to unraveling the way that the nervous system automatically governs the body's functions. Discoveries on how messages are sent between neurons and synapses soon followed. Nonetheless and despite extraordinary advances, questions about the relationships among the mind, the body, the brain, human behavior, and culture persisted, as the controversies in dealing with shell shock, discussed in the last chapter, demonstrated.

Art, or the arts, too, has a long, complicated history. It, too, is dynamic definitionally. We do find healing within the various categories of art. Music, in particular, is frequently mentioned in texts in terms of its healing properties. For example, in the Old Testament (1 Samuel 23), we learn: "And it came to pass, when the evil spirit from God was upon Saul, that David took an harp, and played with his hand: so Saul was refreshed, and was well, and the evil spirit departed from him" [8] (Fig. 9.1).

The term art therapy was coined by the artist Adrian Hill (1895–1977) in 1942 to link art and psychotherapy [9]. Hill, while serving as a member of a Scouting and Sniping Section in World War I, was the first artist commissioned by the Imperial War Museum to record the conflict on the Western Front [10]. After the war, he

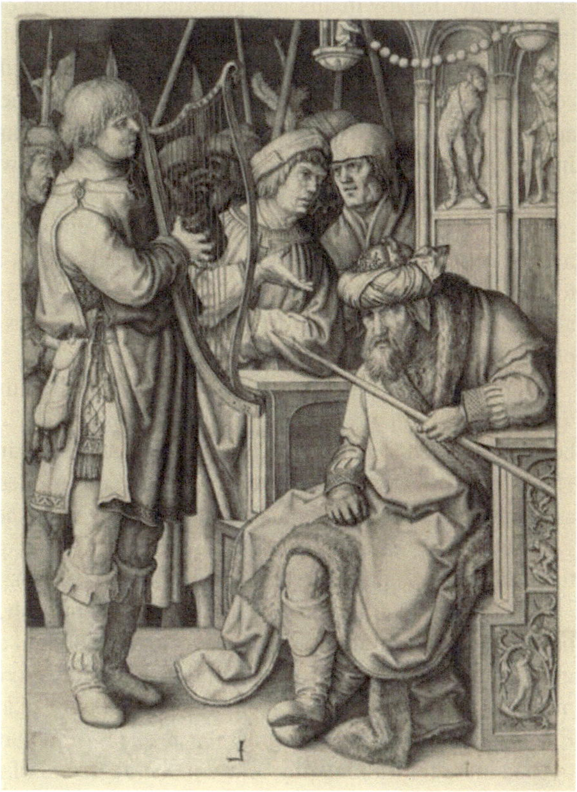

Fig. 9.1 David Playing the Harp Before Saul, ca. 1508. Artist: Lucas van Leyden (Netherlandish, Leiden ca. 1494–1533 Leiden). Engraving; first state. Sheet: 10 × 7 1/4 in. (25.4 × 18.4 cm). Courtesy of the New York Metropolitan Museum of Art, Rogers Fund, 1918

studied at the Royal College of Art, became a professional artist, and was admitted to the King Edward VII Sanatorium in 1938 with tuberculosis. During his convalesce, he passed his time drawing.

The next year, in 1939, the institution invited him to teach art to other patients as a part of their occupational therapy program. This non-pharmacological or alternative form of therapy was made available both to those who had returned from war and to civilian patients [11]. Shortly thereafter, Hill began to frame art therapy as a process to enhance well-being, as outlined in his 1945 book *Art Versus Illness* [9].

Occupational therapy is a cousin to art therapy, which no doubt explains why Hills' program was joined with it. Both build on the idea of engaging the body and the mind. Occupational therapy is a concept that goes back to ancient times. One can think of Asclepiades (124–40 BCE), who prescribed creative and enjoyable methods of treatment, such as massages, sea baths, and hydrotherapy [12]. A later physician associated with this methodology is Philippe Pinel (1745–1826). At Bicêtre Hospital in France, he prescribed participation in exercise, dance, and crafts for asylum inmates as an alternative to the use of metal chains and restraints.

World War I elevated the status of occupational therapy. At this point, it was seen as a way to aid those disabled from the war, a need that increased with World War II

[12]. Today, therapy in this area provides both a range of services including enhancing attention span and improving memory, problem-solving, and self-esteem [13].

As an approach to crafting mind/body harmony, occupational therapy began conceptually with the entry of "moral" treatment in the nineteenth century, a theoretical approach discussed in Chap. 7 in relation to Dr. Thomas Kirkbride's work at the Institute of the Pennsylvania Hospital. The approach was also linked with the Arts and Crafts Movement in England, which was started by artists, architects, and critics such as John Ruskin (1819–1900) and William Morris (1834–1896). Their movement was a part of the impulse for social reform that developed as a reaction to the problems caused by the Industrial Revolution [13]. As such, their critique was along the lines of sociologist Georg Simmel's analyses of the stresses urbanization placed on the mind [14].

In nineteenth century America, we see occupational therapy taking root in institutions like St. Elizabeths Hospital in Washington, DC, which opened in 1855, as a hospital devoted to the development of advancements in mental health care [15]. In 1892, St. Elizabeths began to transform its focus to medical and scientific concerns as "long-held notions that environment was the main cause of mental illness slowly gave way to a greater reliance on neurology, anatomy, physiology, organic chemistry, bacteriology, and general medicine" (p. 121, [15]).

After World War I, when the massive numbers of shell shock casualties returned home, occupational therapy was elevated and adopted at many medical and mental institutions. The idea was that giving people things to do or make would aid them mentally and in moving their minds to a "better" place. This was the case at St. Elizabeths as well. In 1922, under the auspices of the Veterans Bureau, a school was opened there for training physicians in the treatment of mental and nervous diseases [15]. The facility's best-known practitioner was probably Walter Jackson Freeman II (1895–1972), the first neurologist in Washington, DC. As a laboratory director, he performed transorbital lobotomies to treat epilepsy and some psychiatric conditions [15].

Another practitioner was Dr. John E. Lind (1888–1944) who worked with the criminally insane. At St. Elizabeths, he advocated for occupational therapy in treatments. Figure 9.2 displays a project by one of his patients entitled "dimensional divertissement." While Lind frequently offered expert testimony in criminal courts [16], his own murder involved him with criminal court in a quite unlikely way. Dr. Lind sometimes treated private patients, including the wife of criminal lawyer, Robert I. Miller. He helped with her emotional problems [17]. In 1944, Miller was accused of shooting Dr. Lind twice at close range, after learning of an affair with his wife. Miller was acquitted in less than an hour after pleading self-defense [17, 18].

Contemporary occupational therapy practitioners address the physical, cognitive, psychosocial, sensory, and performance in a variety of contexts to support engagement in everyday life activities that affect health, well-being, and quality of life [13]. Arts and crafts are frequently a part of its approach. The methodology is now used with a range of disorders such as stroke, cancer, traumatic brain injury, and polytrauma from vehicle crashes or explosive devices [13].

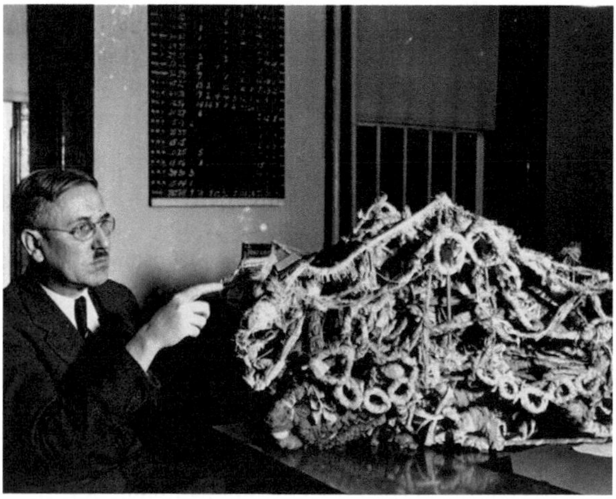

Fig. 9.2 Art therapy in St. Elizabeths Hospital. Dr. John Lind displays the handiwork of one of his patients entitled "dimensional divertissement." Courtesy of the National Library of Medicine

Neuropsychology, a relatively new branch of psychology, is another component of this therapeutic tree. The word entered the lexicon at about the same time as occupational therapy. The term was first used by Sir William Osler, in 1913. He mentioned "neuropsychology" in the context of students being able to take specialized courses dealing with mental disorders. He did not define his new word and presented it only in passing in an address at Johns Hopkins Hospital for a training at the Phipps Psychiatric Clinic [19].

Kurt Goldstein (1878–1965), the German-born neuropsychiatrist, used the term decades later in his 1934 book *The Organism: A Holistic Approach to Biology Derived from Pathological Data in Man* [5, 20]. It is Karl Lashley (1890–1958), a famous American experimentalist, who is generally credited with introducing the term neuropsychology. He used it in 1936 in the context of brain damage and behavior during a presentation before the Boston Society of Psychiatry and Neurology that appeared in print in 1937 [5].

All of these men viewed mental disorders as brain diseases and saw neuropsychology as the study of higher functions following brain injuries or diseases [5]. Efforts to treat mental illness were woven into their studies, as were questions about what constitutes mental illness. For example, as discussed in Chaps. 1 and 2, the long meandering history of defining mental illness even includes designating physiological abnormalities as mental disorders. Within this, we can identify instances where ideas lingered. One example is found in Philippe Pinel's (1745–1826) work. Despite his history as a mental health advocate in the eighteenth century, he used imbecility interchangeably with "idiotism" to indicate both an intellectual and an emotional disorder he ascribed to the deaf [21].

Chapter 2 noted that some experimentalists like Johannes Müller (1805–1858), Hermann von Helmholtz (1821–1894), and Wilhelm Wundt (1832–1920) were influential in combining psychology with physiology. Müller, for example, said, "Nobody can be a psychologist, unless he first becomes a physiologist" (p. 46,

[22]). Helmholtz, while not a psychologist, influenced psychology as well as neuroscience through his studies of sensory physiology. His student, Wilhelm Wundt, is often considered the founder of modern psychology. Wundt established his experimental laboratory at Leipzig University in 1879, about the same time that the philosopher/psychologist William James set up his laboratory in America.

William James (1842–1910) (Fig. 9.3), who had trained as a physician, is linked to philosophy, occupational therapy, and brain plasticity. In *William James, MD: Philosopher, Psychologist, Physician*, Sutton argues that he abandoned his plans to be a doctor due to his ill health, and his early medical studies were the beginning of his lifelong occupation with questions about the essential nature of health, healing, and invalidism, including their implications for society [23]. His views on brain plasticity were developed as a result of his conclusion that work habits offer a mechanism for changing one's brain. He introduced this concept in his books *The Principles of Psychology* and *Habit* [24, 25], or well before the idea of neuroplasticity was validated scientifically. For example, in *Habit*, he says, "the brain-matter is

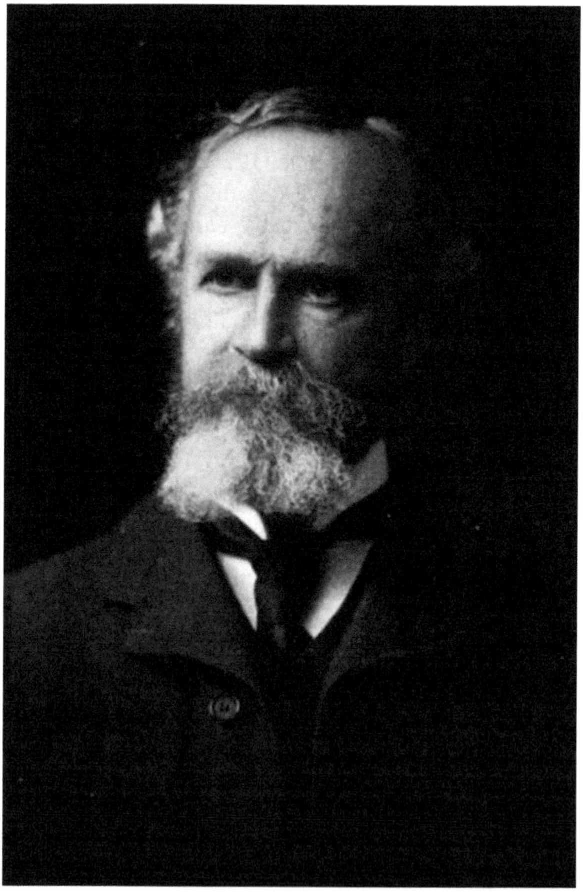

Fig. 9.3 William James (1842–1910). Photographer: William Notman. Courtesy of the National Library of Medicine

plastic" and "it is only a question of time when the exact mechanical explanation of the case shall be found out" (p. 11, [25]). James was referring to hypothetical changes in nervous activity that underlie the formation of habits, an interesting precursor to our use of the term today.

Both William James and the American educator John Dewey (1859–1952) worked with Swiss-American Adolf Meyer (1866–1950), the first psychiatrist-in-chief of the Johns Hopkins Hospital. Meyer's approach integrated occupational therapy and employed techniques developed by both Carl Jung and Sigmund Freud to discern subconscious factors [26]. Trained in neurology before coming to America and influenced by Kraepelin and Freudian ideas, Meyer had an impact on formulating the central tenets of occupational therapy and clinical psychiatry in the United States.

Meyer defined scientific medicine as the union of pathology and therapeutics. According to Meyer, practical medicine was not a basic laboratory science and was rather a modality in pursuit of new therapeutic applications [26]. As he puts it in his *Philosophy of Occupation Therapy* [27], occupational therapy helps to bring a patient in contact with reality [27]. He also relates that he recognized this connection when observing patients participating in activities in an asylum.

Neurologist Jean-Martin Charcot (1825–1893), sometimes characterized as the founder of modern neurology [28], is another figure who bridges neurology and psychiatry/psychology. He was also "a gifted painter who used his artistic abilities and strong visual memory to make associations about patterns of disease in the field of medicine and anatomy" (p. 46, [28]). As indicated in the last chapter, Charcot worked on hysteria, a condition used in shell shocked diagnostics and one that underscored neurological and psychiatric connections. Shell shock included both psychiatric and electrotherapeutic treatments, which in effect affirmed mind, body, and brain relationships. As noted in the last two chapters, disagreements among healers and patients raised questions about treatments within the neurocultural terrain and thus broadly impacted the culture at large. Essentially, while neuropsychological rehabilitation strategies used on brain-injured soldiers were designed to improve cognitive, emotional, and psychosocial functioning through inducing changes in the brain, in practice, they varied widely, as did patient responses to them [29].

Charcot's student Sigmund Freud (1856–1939) and Freud's colleague Carl Gustav Jung (1875–1961) are also interwoven into this history, for both Freud and Jung have had a great influence on how "art therapy" is now defined culturally. The work of Josef Breuer (1842–1925) is indirectly included here because he influenced both Freud and Jung. Despite being known as a neurophysiologist, Breuer stimulated Freud's interest in the mind through their collaboration on the case of Anna O (Bertha Pappenheim, 1860–1936) [30]. Anna's case was published in their *Studies on Hysteria* [30, 31].

Freud trained as a neurologist and is known for his psychiatric writings on the mind. Since his views did not fit well with neurological thinking at that time, he founded a new field, psychoanalysis, to promote his psychodynamic approach to mental illness [32]. In terms of art, Freudian views of the unconscious were interwoven with interpretations about subjectivity, as were his analyses of artists such as

Michelangelo and Leonardo da Vinci [33–36]. Freud's approach largely overlooks relational aspects of art that speak to how the making and viewing of art involves many systems (motor, somatosensory, visual, etc.). Also, as neuroscientist Eric Kandel points out in *The Age of Insight: The Quest to Understand the Unconscious in Art, Mind, and Brain* [37], when Freud attempted to link art and science in his writings, he focused on the psychology of the artist, paying little attention to related studies on the psychology of perception or critiques prominent in arts literature. Curiously, although Freud studied with Charcot, who had trained as an artist and integrated art into his medical practice, their inclinations toward art were as unlike as their legacies.

Freud's influence is quite evident culturally, particularly in film narrative and art interpretation. In terms of film, this was despite Freud's reservations [38]. Indeed, Freud did not see film as an appropriate medium to communicate psychoanalytic ideas, and he declined many requests to consult, including one in 1925 from MGM's Samuel Goldwyn. Goldwyn had offered him $100,000 to discuss a silent film about love stories. Without Freud's blessing, the Austrian film director G.W. Pabst, in 1926, went ahead with a production of a Freudian-based story. For this silent German drama, *Secrets of a Soul*, Pabst consulted with Freud's close associates Karl Abraham and Hanns Sachs. The film features a professor who consults a renowned psychoanalyst after experiencing nightmares that have resulted in paranoia and a fear that he will murder his wife [38].

John Huston's 1962 *Freud* also shows Freud's cultural presence, and it serves as a reminder that embellishment often accompanies efforts to present scientific figures and therapies through entertainment. Huston's impetus to film *Freud* grew from his experience with *Let There Be Light* [39], discussed in the last chapter. The director was so impressed after seeing the successful psychotherapeutic treatments with the shell shocked soldiers that he set his sights on doing another film that looked at Freud himself [38].

Freud mixes known facts with a fictional narrative. On the one hand, we see efforts to achieve historical grounding in the recreation of the 1887 André Brouillet painting *A Clinical Lesson at the Salpêtrière* (Fig. 9.4), a painting depicting an "hysteria" patient being presented to postgraduate students by Jean-Martin Charcot. On the other hand, some scenes in the film, such as Freud's massaging the back of his patient Dora [40] and the script's melding of patient biographies, take a great deal of creative license, accentuating how easily film narratives stray from reality [38].

For example, Huston's use of the iconic Charcot painting further reinforces the link between the two, even as it takes creative license since Freud was not depicted in the painting. Freud did, however, have the work over his couch in his office. The etching included here, created by A. Lurat in 1888, after the 1887 Brouillet painting, was popular because it was suitable for neurologists such as Sigmund Freud to hang in their consulting rooms [41].

As mentioned, Carl Jung also worked outside of the neuropsychological paradigm. He proposed an analytic psychology that emphasized achieving harmony [30], seeing art as a tool that could aid in alleviating feelings of trauma, fear, or anxiety and also aid in healing [42]. Whereas these kinds of words seem applicable to

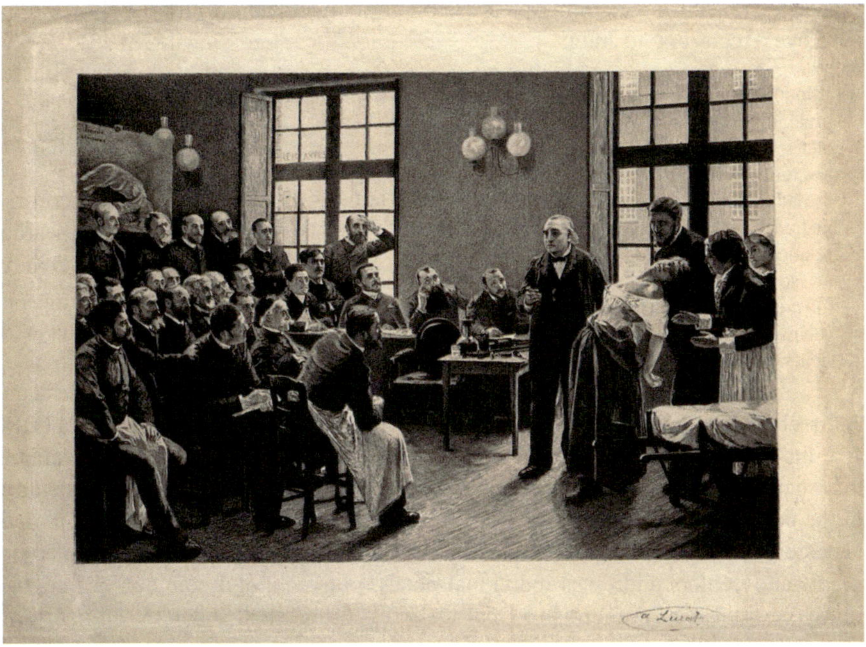

Fig. 9.4 Jean-Martin Charcot demonstrating hysteria in a hypnotized patient at the Salpêtrière. Etching by A. Lurat, 1888, after P.A.A. Brouillet, 1887. Courtesy of the Wellcome Collection

cases like Beethoven's urge to continue composing, Jung's theoretical premises do not mesh well with cases like Beethoven's where physiological degeneration is so prominent. Concentrating on archetypes, individual differences, myth, cross-cultural thinking, and related subjects, Jung's complex system was quite subjective and foundationally intwined with his own art, which had helped him personally when he was dealing with his own depression [43]. His art output had deepened his interest in symbols, religion, the occult, and mythology, which influenced his views [30, 43]. Clearly, like Freud, Jung's path was removed from the neuroscientific tenets of the time, and, like Freud, his work, too, is frequently integrated with contemporary patient art therapies [44]. Suffice it to say that both Freud and Jung are now characterized in terms of neuroticism:

> Neuroticism is a key construct in psychoanalytical theories where distress is triggered by unconscious defense mechanisms for unresolved conflicts (Sigmund Freud) or conflict between two psychic contents, one of which must be unconscious (Carl Jung). Neuroticism is also one of five domains of the NEO-Personality Inventory, which is comprised of six facets: anxiety, depression, angry hostility, self-consciousness, impulsivity, and vulnerability (pp. 1774–1775, [45]).

Paradoxically, many art therapy programs profess to Jungian thinking even as the expressive impulse behind them does not align with Jung's own theoretical views on subjective abstraction. Jung's views are more in line with the idea that art is primarily subjective and with Plato's vision of prisoners who watch the shadows in the Cave. According to Jung:

> [Modern art] … is really performing a work of psychological education on the public by breaking down and destroying their previous aesthetic views of what is beautiful in form and meaningful in content. The pleasingness of the artistic product is replaced by chill abstractions of the most subjective nature which brusquely slam the door on the naive and romantic delight in the senses and their obligatory love for the object. This tells us, in plain and universal language, that the prophetic spirit of art has turned away from the old object relationship and towards the – for the time being – dark chaos of subjectiveness. Certainly art, so far as we can judge of it, has not yet discovered in this darkness what it is that holds all men together and could give expression to their psychic wholeness. Since reflection seems to be needed for this purpose, it may be that such discoveries are reserved for other fields of endeavor (p. 122, [46]).

As is obvious, these kinds of diverging theories do not lend themselves to a one-size-fits-all approach to therapeutics in terms of neuroscience and art. Therefore, this chapter will leave theory behind at this point. In the next section, a case study of the novelist Howard Engel focuses on how identifying what an individual's case needs enhances the therapeutic experience. Following this, a brief survey of programs and centers addresses individual needs, outreach, and community integration. The final section returns to Deaf Culture. Just as shell shock raised question about the role of the patient in the implementation of therapies, cochlear implants have fostered similar debates.

9.2 Novelist Howard Engel: The Writer Who Could Not Read

The Canadian novelist Howard Engel (1931–2019) unknowingly suffered a small stroke in 2001. At the hospital, Engel learned he had a condition called *alexia sine agraphia. Alexia* or word-blindness is a purely visual problem. *Sine agraphia* presents a dissociation or disconnection syndrome [47] and indicates that the condition did not affect his ability to write.

Pure *alexia sine agraphia* was first described by Jules Dejerine to the Biological Society in Paris on February 27, 1892 [48, 49]. His patient was a highly intelligent 68-year-old man (Monsieur C) who was in excellent health until his cerebrovascular accident on October 25, 1887. After a walk 2 days earlier, on October 23, he had experienced some right-sided numbness of his arm and leg. He was able to read the advertisements in shop windows at that time. The next day, he began to feel numbness in his arm and leg. It persisted on the following day, when Monsieur C found himself completely unable to read, even though he had no difficulty writing, communicating, or understanding spoken language. Thinking that his problem was visual in nature, he visited an ophthalmologist (Landolt), who referred him to Dejerine [48].

In Monsieur C's case, and Howard Engel's as well as, the tension among seeing, reading, and writing is present and produces a word-blindness that is the result of the visual cortex being cut off from the language area of the same brain. *Aphasia*, an inability to comprehend or formulate language because of damage to specific brain regions or a disturbance in language, was not a problem.

In Engel's case, the various tests at the beginning showed his small stroke had only affected a limited area of the occipital cortex, the visual part of the brain, on the left side. Aside from the *alexia*, there were initial difficulties in recognizing shapes and colors produced by quadrantanopia. These diminished relatively quickly, but the *alexia* did not [47, 50, 51]. Early on, there was also a blind spot that hid the right-hand side of a line from him, and he needed to shift his eyes to see it. The good news was that by the time of his recovery, words no longer appeared as if written in some exotic language. Still, he was unable to hold a complicated structure in his head long enough to get it down on paper. While he could easily see letters, he could neither read nor interpret them. He also had problems spelling because he could not perceive words internally or in his mind's eye [47, 50, 51].

Another problem was that even as he began to develop therapeutic strategies, Engel found that he sometimes would read the first letter of the next word instead of the word he was looking at directly. This means that "director general" might come out as "girector general." He was also distorting familiar objects. The moon, for instance, regardless of phase, looked flat on top.

The upshot was that after the stroke there were only minimal problems with his eyes and nothing wrong with his intelligence. He could *see* letters but was unable to interpret them. Essentially, what he had lost was the higher-level ability to interpret what he was seeing and a quarter of his vision in the upper right-hand side of the visual screen. This defect, however, wasn't usually noticeable as he moved about because his brain filled in the gaps [52].

A few things are worth noting here. First, unlike Mr. I., the color-blind artist, whose total way of life felt out of kilter due to his finding his world turned into a monochromatic gray, Howard Engel didn't notice his *alexia* at all most of the time. The sky looked blue, the sun shone as usual, and the world hadn't suddenly become an alien space. His *alexia* only existed to him when he tried to read, and reading was his passion from a young age—and it was connected with his profession as a writer [52]. His urge to read was evident even while in the hospital just after the stroke. He kept trying to read almost from the start. Through his efforts, he began to slowly decipher words letter by letter, learning to make groups to help him along.

Second, his personality remained largely unchanged. At first, he reacted with his trademark sense of humor. He quipped that he could tell his publisher that he could still write but didn't do rewrites as a matter of principle [52]. Even still, his humor couldn't mask his sense that his career as a writer was over. How could he write novels if he had lost his ability to read English, the language he used to write, read, and edit his books? [52].

Third, as is frequently the case with artists, he vowed to find a way to continue creating. At the hospital, he learned to write things down in what he called his "memory book": appointment times, phone numbers, directions, and so forth. Engel had always kept journals. This therapeutic and constant companion in effect returned a piece of his pre-stroke identity to him. His first poststroke book, *Memory Book*, used this therapeutic appendage as the handle for its plot.

A fourth and quite important factor in renewing his creative work was having therapists at the hospital who were sensitive to his needs. He said they had dropped the colored stones that gave him a path to follow out of the forest. For example, at the beginning of his rehabilitation, one therapist commented on his willingness to embrace strategies, accept his deficits, and go about accomplishing his goals in a different way. Another stone was the way the people in the hospital emphasized that even if he couldn't read he could still write. He felt that their emphasizing what he could do prepared him and helped him believe working again would come in time [52].

It was the memory book that fully brought his therapy and aspirations together. An important healing mechanism, it also showed him that his therapists conceptually understood his need to write. Seeing Engel as an individual, they helped him envision and develop a path forward. One told him later that she began to work with him to explore how he could best decode written material because his descriptions were much more detailed and elaborated than those of other patients. Another prodded him to think about writing by asking him what he might try to do in a new work of fiction. A third taught him how to deal with words of different lengths using what he called the "Kim's Game" technique:

> In Rudyard Kipling's novel *Kim*, the boy hero is introduced to the game of recognizing unrelated objects exposed for a limited time on a tray. When the tray has been covered, the boy is asked to recall in detail the objects he has seen. All of this was to train him for the work of a spy for the British Raj (p. 75, [52]).

Another aid in his recovery was his awareness of others. Writers are observers and at the hospital he observed. As he did so, he learned about and from those helping him, the others on his ward, and the visitors as well. As a result, as time progressed, his distress over the loss of his ability to read was also tempered by his awareness that many people in the hospital had far more debilitating conditions [52]. This, too, helped open him to learning new strategies. In his memoir, he repeatedly refers to how often he would list all the things he could still do as if in articulating them he was assuring himself that he had tools he could use to achieve his aspirations.

Engel also used other senses and other people as aids in completing tasks. He learned he could transfer words from his visual memory to his motor memory. In other words, the problem was that he could not internally "see" words as words in his mind, but he found could write them once he translated them using other sensory systems. Sometimes after encountering a word, he would trace the letters again and again, using repetition to form the motor memory of what motions created the letters that spelled the word. He might write it in the air or trace the words along the roof of his mouth with the tip of his tongue to tell himself what the words he was looking at were. This helped him increase his reading speed. Tracing letters on the roof of his mouth meant a book he might have completed in an evening before the stroke might now take a month or two. When writing a book, having his editor read the entire book to him was key to fixing its overall structure [52].

Generally, with practice, those with *alexia sine agraphia* can learn to read most letters out loud. It is also possible to spell them aloud and recognize them auditorily. During his rehabilitation, Engel used this actively too. He realized that when he looked at a word, a couple of letters would sometimes jump out. With the help of therapists, Engel learned to puzzle out words on signs and newspaper headlines. Sometimes, if a few letters were clear, he could infer a word or puzzle it out much like a child would. Still, as Engel explained it, sometimes, he felt as if he had been told that his right leg had to be amputated, but he could keep the shoe and sock [52].

His therapy also led him to Oliver Sacks. Engel and one of his therapists often compared his problem to case studies Dr. Sacks had published such as *The Man Who Mistook His Wife for a Hat* and *Awakenings* and Sacks' own struggles to reconnect with his leg after a climbing accident as recorded in *A Leg to Stand On*. Deciding that "The Writer Who Could Not Read" sounded like an Oliver Sacks' case study, Engel sent a letter to the doctor detailing his situation. A polite replay followed, and then a second letter asked if he might quote Engel's letter in an article for *The New Yorker* magazine. Engel had received so many rejections from that magazine over the years that he couldn't believe that his name was even going to be mentioned in it. This gave him even more resolve to write again and a hope that he could publish again himself [52].

A few things to keep in mind here are, first, a*lexia* is an acquired deficit that speaks to the ability to interpret written language. Impairments that prevent the normal acquisition of reading skills, such as developmental dyslexia or simply dyslexia, are congenital or acquired early in life. Sometimes, similar strategies are used for both. Therefore, their foundations differ significantly, as the next chapter points out when discussing Laura Jane Stevenson, but there are also commonalities.

Second, Engel's *alexia sine agraphia* is one of several types of *alexia*. In cases with *agraphia*, the ability to read is lost as well as the ability to write. Because Engel retained the ability to write and comprehend written material, he had more tools to work with in returning to his career as a writer.

Third, Engels ultimately wrote three books following his 2001 stroke: *Memory Book* [53] his memoir *The Man Who Forgot How to Read* [52], and *East of Suez* [54]. In his first poststroke book, *Memory Book*, he explains that this fictional work was in effect a recovery story and that his therapeutic memory book served as the kernel for it [47]. It is particularly insightful as an autopathography, for his writing about his own illness even in a fictionalized manner allows us to see it from the inside. He uses the following analogy to try to capture how difficult the project was from his perspective:

> In the mid-1960s there was a film called *Cat Ballou* starring a young Jane Fonda. In it the veteran star Lee Marvin plays the part of a dissolute former gunfighter named Kid Shaleen, a washed-up, drunken wreck, scarcely able to stand. Looking him in the face, an old timer says to him, "Your eyes look *terrible!*" Kid Shaleen replies, "You should see them from this side" (p. 237, [53]).

Engel continues: "In writing this book, after my illness of a couple of years ago, I've tried to show what the changes in perceptions and cognitions are like as seen from 'this side'" (p. 237) [53]). In the winter of 2007, Engel wrote his memoir, *The Man Who Forgot How to Read* [52], an incredible book that tells the story of his life and details his return to writing from the inside. When he began *Memory Book*, he had no idea if he could pull it off. With his memoir, he had the assurance of one who had done it before. In effect, the memoir was an extended acknowledgment of the people who aided him find a road back to writing [52].

In 2007, Engel was no longer seeing therapists, and when he went to the hospital, it was by invitation, to speak to people there about his case and how he was able to cope. He still had trouble with his *alexia*, but there was much improvement [52]. The next year, 2008, he published another fictional work, *East of Suez*. On July 16, 2019, Engel died in Toronto at the age of 88. The cause was pneumonia, after a second stroke.

9.3 Related Alexia Cases

The German writer Gert Hofmann (1931–1993) demonstrates that Howard Engel's case is not a singular one. An itinerant professor of German literature, Hofmann published his first novel at the age of 48 in 1979. According to his son, his stroke at the age of 57 came about because he pushed himself to write nearly a book a year to make up for lost time having started to publish later in life. Like Engel, Hofmann struggled with *alexia*. He dictated his books to his wife after his stroke. She read the drafts back to him to correct them and he would offer his corrections aloud. The three books he wrote after his stroke were *The Film Explainer*, *Luck*, and *Lichtenberg and the Little Flower Girl* [55, 56]. The main character in the final book, *Lichtenberg*, is always on the lookout for the secrets of language, sometimes just for "words, words." This adds a particularly poignant element to the work and Hofmann's *alexia* [55].

Another related case is Ian McDonald's. With his 2006 article "Musical *alexia* with recovery: a personal account," he entered the literature of doctors who describe a personal neurological event. He writes:

> I describe the experience of an acute loss of ability to read music and play the piano accurately and expressively following an embolic infarct of the right angular and supramarginal gyri in a setting of chronic migraine. Other parietal deficits included a small visual field defect, visual hallucinations, prosopagnosia, topographical disorientation, disturbance of perception of velocity of moving objects and dyscalculia. Recovery began within a month of the ictus after instituting a regular practice routine. The ability to read and play polyphony recovered before the ability to read homophonic music. A substantial degree of recovery of musical function occurred within 6 months and of the other parietal deficits over a year. Failure to maintain regular practice led to marked though recoverable deterioration. An increased frequency of migraine persisted for some 18 months. [57]

McDonald began studying piano at seven and it was a chief recreation throughout his life. He had experienced migraines since adolescence, which had rarely interfered with his work. An academic neurologist with no special skills in cognitive neurology, he wrote up his experience after seeing that the literature on acquired musical *alexia* is small, and in 2006, the literature did not include patient accounts of the disorder [57]. It is worth noting that this doctor ignored his symptoms at first:

> [D]espite my neurological background, I did not at first identify these as specific deficits attributable to cerebral dysfunction: in short, I did not realize that I had had a stroke… In retrospect, I realized that I had been experiencing a related [topographical disorientation] deficit from early in 2005 [57].

As the case study notes, his embolic stroke resulted in musical *alexia* and included problems with calculation (*dyscalculia*), face blindness (*prosopagnosia*), and topographic disorientation. He also mentions that his musical *alexia* recovery fluctuated from day to day and correlated with his ability to practice. After he recovered from the stroke, he surmised that many of the difficulties he had in reading music and playing it on the piano, and in writing it down, derived from a visuospatial deficit [57]. What remained unclear to him when he wrote his description was the short-lived impairment of the ability to instinctively recognize the emotional content of a passage and to convey it while playing. He also could not account for the defect in rhythmic awareness he experienced. There were no follow-up articles.

9.4 Therapeutic Outreach: Centers and Programs

A short chapter is unable to cover the myriad of ways in which art is now used to help the elderly, disabled, and brain-damaged individuals. Among the countless measures of response are community centers and programs that aid in identifying individual needs and offer outpatient services while serving as growth centers.

One ongoing and established program is Creative Growth, a nonprofit art center based in Oakland, California. Begun in 1974, and therefore about 50 years old as I write, this organization provides a community, a supportive studio environment, and offers gallery representation to artists with developmental disabilities. Their outreach also advanced the inclusion of artists within the larger art community. The focus on disability rights in the United States at the end of the twentieth century led to landmark legislative acts, including the Rehabilitation Act of 1973. These laws provided federal protection for people with physical and developmental disabilities. Creative Growth was one of the programs developed at that time. Founded by Florence Ludins-Katz, an artist and an educator, and Elias Katz, a clinical psychologist, the Center's mission is predicated on the belief that art is fundamental to human expression and that all people are entitled to its tools of communication.

Creative Growth now works with more than 150 artists with developmental, mental, and physical disabilities. Their artists use a wide array of media, and many have found mainstream success [58]. Judith Scott (Chap. 3), an artist with Down syndrome, worked at the Creative Growth Center. Dan Miller, an autistic artist, who exhibited with Judith Scott at the Venice Biennale in 2017, was another success story. Miller works with paper, ink, pencil, and paint to make visible "the cacophony that is daily life" [58]. Another autistic artist who works there is William Scott, who has no relation to Judith. William Scott uses meticulous detailing of a futuristic urban utopia in his drawings and paintings. These works communicate his desire for an optimistic future [58]. He, too, has exhibited throughout the world [59].

Another therapeutic group is Creative Forces, a military healing network based in Washington, DC, that offers an updated approach for veterans with post-traumatic stress injuries. Funded by the US National Endowment of the Arts, in partnership with the US Departments of Defense and Veterans Affairs, Creative Forces helps those who served in the military unlock the trauma and nightmares that are residues of their service [60]. Here, too, creative arts therapies are a central component of the overall clinical approach. This network provides both community arts engagement and resources to further aid their target populations [60].

Physiotherapy is of value for people with some neurodegenerative disorders like Parkinson's disease (PD). Dance for PD is one organization that reaches out to this cohort. Classes are offered for people with motor and nonmotor symptoms such as bradykinesia, resting tremor, postural instability, and cognitive and neuropsychiatric manifestations. Small studies have shown that dance therapy improves mind-body intervention and improves the symptoms and outcomes in patients with PD, especially motor symptoms. In other words, this population benefits from the kind of complex motor activity dance offers, as well as pharmacological treatment. These therapies show positive effects on balance, functional mobility, and cognition [61, 62].

As the number of Alzheimer's sufferers grows, many have developed and/or turned to programs to help. The impetus behind several Alzheimer's program stemmed from the success of art therapy in programs with cancer, post-traumatic stress disorder, and the elderly in general. Yet, art therapy's efficacy in diseases that directly affect the nervous system is less clear at this point [63]. One study that looked at the effect of visual art training on people with dementia utilized a randomized control trial design to investigate the effects of an 8-week visual art training program on cognition. Their research emphasized overall cognition, delayed recall, and working memory, which show deficits in people with dementia. According to their results, there were no group differences in overall cognition or working memory [64]. Delayed recall yielded an undetermined difference, based on post-test/pre-test difference scores. In this study, groups were comparable at baseline on all measures [64].

Another related study by Miller et al. describes the clinical, neuropsychological, and imaging features of five patients with frontotemporal dementia (FTD) who acquired new artistic skills in the early stages of dementia. Four of the five patients had the temporal variant of FTD in which anterior temporal lobes are involved but the dorsolateral frontal cortex was spared. Their visual skills were spared but their language and social skills were devastated. The investigators concluded that patients with the temporal lobe variant of FTD may offer a window into creativity and that the loss of function in the anterior temporal lobes may lead to the "facilitation" of artistic skills.

Community programs do suggest there is value even when there isn't solid quantitative data to back up the program. For example, one program supported by the Alzheimer's Association is "Memories in the Making" [65]. A study on its efficacy evaluated 76 patients with middle- to late-stage dementia. They concluded that the quantitative evidence for well-being outside of the art class is ambiguous, but anecdotal observations indicate that the program is beneficial for some participants, if only fleetingly [65]. Another analysis of this program concluded that the participants in the program showed more interest in art than activities centered around current events or crafts [66]. Another program, "Meet Me at MoMA," was initiated by the New York Museum of Modern Art (MoMA). It couples 90-minute sessions of group discussions led by curators with small patient groups of about seven to eight people. Qualitative evaluations affirm that these programs seem to improve patient moods [63].

Additionally, the Hilgos Foundation program is dedicated to the memory of Hilda Gorenstein, a lifelong painter who worked to the end despite her Alzheimer's. Hilgos was her chosen name as an artist and her painting career spanned 75 years. When she died at age 93, she was known for her beautiful marine paintings, which are now in collections all over the world. During the last 3 years of her life, Hilgos struggled with profound memory loss and yet retained her artistic sense, as shown in the award-winning documentary and the book *I Remember Better When I Paint* [67].

Both offer a positive perspective on the impact of art and other creative therapies on people with Alzheimer's and how these approaches can change the way we look at the disease. The film includes outreach to caregivers and the general population through interviews with renowned neurologists who explain how creative activities engage areas of the brain that are not damaged by the disease and thus reawaken a sense of personality, identity, and dignity.

A sidenote here is that practicing artists who develop Alzheimer's, e.g., William Utermohlen and Willem De Kooning, have presented creative results when they painted despite their Alzheimer's. In these cases, the art itself presents another perspective on information gleaned from disease progression diagnosed through clinical evaluation [68–71]. In de Kooning's case:

> [D]espite de Kooning's major cognitive deterioration, his artistic style evolved to more pure abstract expressionism in the context of his neurological disorder and reached the ultimate abstract expressionist level. The public and critic perception of his work remained unchanged. There is no mathematical measure which would allow defining the rationale of persistence of de Kooning's artistic skills in the midst of dementia. In addition to this, de Kooning's techniques did not change and the fractal complexity decreased. Overall, his artistic style remained unchanged within the theme of abstraction and expression despite a major change in mentation (p. 175, [72]).

9.5 Deaf Culture and Therapeutic Controversies

As noted earlier, controversies surrounding shell shock therapies demonstrated that not all therapies are comparable nor are they greeted comparably. The Deaf Community's response to cochlear implants offers another perspective on how and why an individual or a group might object to a particular therapy. Briefly, many have detailed the continuing debates about whether deaf people should be educated in their own unique, rich culture-based visual language, Sign, or forced into "oralism," a term used when verbal speech is given preference over signing [73–77]. Here, suffice it to say, a main factor in these debates is that Sign is not just a linguistic language; it is also a social tool, a topic discussed in Chap. 2. It gives users a language that is easy for them to use in interacting with others who know it. Using their own language also accentuates the capability of our cognitive system to adapt and coherently use, integrate, and share a variety of sensory inputs [73, 76–78]. In other words, each of us has our own sensory spectrum, and those born totally deaf offer a cognitively different profile from those who becomes deaf. Moreover, many of the congenital deaf are resistant to therapies they are told will aid them "hear" sound, even as they do not oppose enhanced sensory awareness unilaterally.

Chapters 3 and 4 discussed the considerable difference between those born profoundly deaf and those who have a later loss of their hearing, noting that those in the former cohort have never heard natural sounds or spoken language. Integrated into the earlier discussion was commentary on the band Beethoven's Nightmare to point out that they, as deaf musicians, in effect sense music and that sensory awareness remains a part of the Deaf lifestyle. Beethoven's Nightmare's concerts also show that deaf people enjoy attending shows and other musical events even without neurotypical hearing. Indeed, musical productions are an important part of Deaf Culture, where audience members attending events are provided with balloons, which they hold on their fingertips to "feel" the musical vibrations.

One of the challenges the profoundly deaf face is that the deaf and the hearing interact throughout their lives and, to date, no therapies have successfully and fully treated congenital deafness or age-related hearing loss, although there are relatively recent auditory-based therapies or auditory-verbal therapies (AVP) that provide a type of early intervention therapy for young children who are profoundly deaf and those who have difficulty hearing. These include those who use hearing technology such as hearing aids or cochlear implants, a small, complex electronic device surgically implanted underneath the skin behind the patient's ear.

Some in the Deaf Community see cochlear implants, which do not give a user neurotypical hearing, as a technology that exposes the urge of hearing people to cure them [79–81] for there are key differences between the sounds a cochlear implant delivers and normally heard sounds. With the implant, a patient receives auditory signals from the environment that are transmitted to the patient by electronically bypassing nonfunctional parts of the ear and directly stimulating the auditory nerve. Unlike a hearing aid, it does not merely amplify sound nor does it restore normal hearing. Rather, the additional input provided by the implant often improves sound detection and may increase speech understanding. Therein lies the point of controversy.

Cochlear implants, surgically implanted neuroprosthetics, are intended to provide a person who has moderate-to-profound sensorineural hearing loss with enhanced sound perception. Although improved since first introduced in the 1970s and 1980s, they remain a substitute for hearing, not a means of restoration [79–81]. The controversy stems from the fact that implantation requires both a surgical procedure and significant therapy afterward. Moreover, not everyone derives similar benefit from the device. Even adults who have acquired hearing loss acknowledge that the technology has limitations. For example, Oliver Sacks reported on a patient who came to see him after having musical hallucinations that she associated with her increasing hearing loss. She called with an update after she had cochlear implant surgery. According to her, it was the best decision she ever made. She could hear herself speak again. She could even once again distinguish voices. The downside, she relayed, was that she could no longer enjoy music. It sounded crude, and with the relative insensitivity of her implant, she could hardly detect the tonal intervals that are the building blocks of music. On top of her auditory loss of music, her musical hallucinations remained [82].

The fiercest debates surrounding the cochlear implants concern their use with children. Those who encourage the therapy argue that it aids a child in understanding sound and in developing the use of speech to some degree. Advocates of this view argue that implants allow integration comparable to that of a child who isn't hearing impaired and helps a child function in making friends and eventually working within the mainstream economic system. The claim is that this will let them eventually live independently.

The counterargument is that the typical deaf child, particularly one prelingually deaf, will probably not be able to speak intelligently even after receiving the implant and even with speech training afterward. Therefore, they argue, it is misleading to say that, as an adult, one with an implant will move around in the hearing world seamlessly. In part, this is because the sounds heard through an implant, as mentioned, are different from sounds normally heard. Users who have heard both describe them as artificial or robot-like. As the Sacks case showed, even many who respond enthusiastically express dissatisfaction with their perception of nonspeech sounds like music [83].

Moreover, as discussed in Chap. 2, quantitative research has shown deaf people use their brains differently than hearing people [84–87]. One example discussed there is that the auditory cortex—the sound-related part of the brain—is used to help process peripheral vision and motion. Researchers have also found that their deaf participants have more activity from the touch and sight stimuli. One study of congenitally deaf adult humans has also examined the neuroanatomical plasticity resulting from altered sensory experience. As these researchers pointed out, attributing the source of the brain's structural variance in the deaf is complicated by the fact that deaf individuals also differ in their language experiences (e.g., sign vs spoken), which likely influence brain anatomy independently.

In terms of art, as John Carlin case study (Chap. 3) and the Beethoven's Nightmare case study (Chap. 4) point out, deaf artists who work with their deafness rather than trying to fix it can follow their passions. In the case of Beethoven's Nightmare, we grasp this through the group's self-written songs that reflect the emotions of deaf people and relay how the band members experience living in a hearing world [88].

9.6 Conclusion

This chapter examined how neuroscience and art are interwoven with therapeutics. The initial theoretical section introduced approaches in terms of neurology, psychology, neuropsychology, psychiatry, occupational therapy, and art therapy. An array of case studies followed. The next chapter begins Part III, which focuses on practices. The thrust of the closing three chapters is twofold. They both encompass how individuals learn about and/or translate specific neuroatypical conditions and also cover how novices, specialists, and the atypical differ. It begins with education and translation, a topic intrinsically related to therapeutics. When a problem is genetic and/or congenital, the kinds of educational, translational, and therapeutic responses include helping those at a young age adapt and grow. Sometimes, congenital disabilities are not diagnosed until later in life and how individuals adapted to and were educated with an undiagnosed condition offers valuable insight into condition nuances and spectrums. With acquired conditions, education and translation take another form. Later in life, adaptation must also address the sense of loss that frequently comes with this type of biological trauma. This, too, is discussed in the next chapter.

References

1. Seddon AM. Music inexpressible: the tragedy of Maurice Ravel. Med Probl Perform Art. 1995;10(2):62–5.
2. Swanson C. The art of the flame-out: Yayoi Kusama. New York: New York Magazine; 2012.
3. Online Etymology Dictionary; 2014. https://www.etymonline.com. Accessed 12 Aug 2023.
4. Rakel RE. Therapeutics. Encyclopedia Britannica. https://www.britannica.com/science/therapeutics. Accessed 12 Aug 2023.
5. Finger S. History of neuropsychology. In: Zaidel DW, editor. Neuropsychology. San Diego: Academic Press; 1994. p. 1–28.
6. Ione A. Innovation and visualization: trajectories, strategies, and myths. Amsterdam: Rodopi; 2005.
7. Ione A. Art and the brain: plasticity, embodiment, and the unclosed circle. Leiden: Brill; 2016.
8. Carroll R, Prickett S. The Bible: authorized King James version. Oxford: Oxford University Press; 2008.
9. Hill AKG. Art versus illness. A story of art therapy. London: G. Allen & Unwin; 1945.
10. Gough P. A terrible beauty: British artists in the First World War. Bristol: Sansom and Company; 2010.
11. Hogan S. Healing arts: the history of art therapy. Philadelphia: Jessica Kingsley; 2001.

12. Carroll MA, Lawson K. The intermingled history of occupational therapy and anatomical education: a retrospective exploration. Anat Sci Educ. 2014;7(6):494–500.
13. Andersen LT, Reed KL. The history of occupational therapy: the first century. Thorofare, NJ: Slack Incorporated; 2017.
14. Simmel G. The metropolis and mental life. In: Wolff K, editor. The sociology of Georg Simmel. New York: Free Press; 1950.
15. Otto TJ. St. Elizabeths Hospital: a history. Washington, DC: US General Services Administration (GSA); 2013.
16. Lind JE. The cross-examination of the Alienist. Crim L Criminol. 1922;13(2):228–34.
17. Bly D. Dr. John Edward Lind (1888–1944). WikiTree; 2023.
18. New York Times. Jury Frees Miller of Dr. Lind Killing. New York Times. June 1; 1944.
19. Osler W. Specialism in the general hospital. Baltimore: Johns Hopkins Hospital; 1913.
20. Goldstein K. Der Aufbau des Organismus: Einführung in die Biologie unter besonderer Berücksichtigung der Erfahrungen am kranken Menschen (The organism: a holistic approach to biology derived from pathological data in man). Netherlands: Springer; 1934.
21. Pinel P. Traité médico-philosophique sur l'aliénation mentale, ou la manie. Paris: Brosson; 1800.
22. Bringmann WG, Lück HE. A pictorial history of psychology. Chicago: Quintessence; 1997.
23. Sutton EK. William James, MD: Philosopher, psychologist, physician. Chicago: The University of Chicago Press; 2023.
24. James W. The principles of psychology. New York: Henry Holt; 1890.
25. James W. Habit. Baltimore: Henry Holt; 1890.
26. Lamb SD. Pathologist of the mind: Adolf Meyer and the origins of American psychiatry. Baltimore: Johns Hopkins University Press; 2014.
27. Meyer A. The philosophy of occupation therapy. Reprinted from the archives of occupational therapy, Volume 1, pp. 1–10, 1922. Am J Occup Ther 1922/1977;31(10):639–42.
28. Kumar DR, Aslinia F, Yale SH, Mazza JJ. Jean-Martin Charcot: the father of neurology. Clin Med Res. 2011;9(1):46–9.
29. Yi A, Belkonen S. Neuropsychological rehabilitation. In: Kreutzer JS, DeLuca J, Caplan B, editors. Encyclopedia of clinical neuropsychology. New York: Springer New York; 2011. p. 1766–7.
30. Wade NJ. Art and neurologists. In: Colombo B, editor. Brain and art: from aesthetics to therapeutics. Cham: Springer International Publishing; 2020. p. 53–66.
31. Breuer J, Freud S. Sigmund Freud collection (library of congress). In: Studies on hysteria. New York: Basic Books; 1957.
32. Leigh H. Neuropsychiatry. In: Kreutzer JS, DeLuca J, Caplan B, editors. Encyclopedia of clinical neuropsychology. New York: Springer New York; 2011. p. 1759–61.
33. Freud S. Leonardo da Vinci: psychosexual study of an infantile reminiscence. London: K. Paul, Trench, Trubner; 1910/1932
34. Freud S. The Moses of Michelangelo (Eine Kindheitserinnerung des Leonardo da Vinci); 1910/1969, vol. 10. Bildende Kunst und Literatur, pp. 87–160.
35. Schapiro M. Leonardo and Freud: an art-historical study. J History Ideas. 1956;17(2):147–78.
36. Stites R. A criticism of Freud's Leonardo. College Art J. 1948;7(4):257–67.
37. Kandel ER. The age of insight: the quest to understand the unconscious in art, mind, and brain: from Vienna 1900 to the present. New York: Random House; 2012.
38. Demons of the Mind. Freud on film: psychoanalysis and cinema. London: Science Museum Group; 2023.
39. Huston J. Let there be light. US Army; 1945/1981.
40. Wijdicks EFM. Neurocinema—the sequel: a history of neurology on screen. Boca Raton: CRC Press; 2022.
41. Wellcome Collection. Catalog entry: Jean-Martin Charcot demonstrating hysteria in a hypnotised patient at the Salpêtrière. In: Trust W, editor. Etching by A. Lurat, 1888, after P.A.A. Brouillet, 1887. London: Wellcome Trust; n.d.
42. Malchiodi CA. The art therapy sourcebook. New York: McGraw-Hill; 2007.

43. Jung CG. Memories, dreams, reflections. London: Collins; 1967.
44. Galbiati E. The significance of art therapy. In: Colombo B, editor. Brain and art: from aesthetics to therapeutics. Cham: Springer International Publishing; 2020. p. 169–82.
45. Loftis C. Neuroticism. In: Kreutzer JS, DeLuca J, Caplan B, editors. Encyclopedia of clinical neuropsychology. New York, NY: Springer New York; 2011. p. 1774–5.
46. Jung CG. The undiscovered self. Boston: Little, Brown and Co; 1958.
47. Sacks O. Afterword: memory book by Howard Engel. New York: Carroll & Graf; 2006. p. 238–48.
48. Bub DN, Arguin M, Lecours AR. Jules Dejerine and his interpretation of pure alexia. Brain Lang. 1993;45(4):531–59.
49. Dejerine J. Contribution à l'étude anatomo-pathologique et clinique des différentes variétés de cécité verbale. Paris: G. Masson; 1892.
50. Sacks O. Afterword to The man who forgot how to read. New York: Thomas Dunne Books/St. Martin's Press; 2008. p. 149–57.
51. Sacks O. The mind's eye: a man of letters. New York: Alfred A. Knopf; 2010. p. 53–81.
52. Engel H. The man who forgot how to read. New York: Thomas Dunne Books/St. Martin's Press; 2008.
53. Engel H. Memory book. New York: Carroll & Graf; 2006.
54. Engel H. East of Suez: a Benny Cooperman mystery. Toronto: Penguin Canada; 2008.
55. Hofmann G. Lichtenberg and the little flower girl. New York: New Directions Books; 2004.
56. Hofmann M. Afterword: Lichtenberg and the little flower girl. New York: New Directions Books; 2004. p. 245.
57. McDonald I. Musical alexia with recovery: a personal account. Brain. 2006;129(Pt 10):2554–61.
58. Wagner L. Art 21: Creative Growth Art Center in San Francisco Bay Area. In: Ravich N, editor. Public Broadcasting Company film; 2018.
59. Creative Growth Art Center. Oakland: Creative Growth; 2023. https://creativegrowth.org. Accessed 15 Sept 2023.
60. Creative Forces. National Endowment for the Arts: Creative Forces. Washington, DC: National Endowment for the Arts; 2023. https://www.nih.gov. Accessed 3 Sept 2023
61. Hasan SM, Alshafie S, Hasabo EA, Mm S, Elnaiem W, Qasem A, et al. Efficacy of dance for Parkinson's disease: a pooled analysis of 372 patients. J Neurol. 2022;269(3):1195–208.
62. Michels K, Dubaz O, Hornthal E, Bega D. "Dance Therapy" as a psychotherapeutic movement intervention in Parkinson's disease. Complement Ther Med. 2018;40:248–52.
63. Chancellor B, Duncan A, Chatterjee A. Art therapy for Alzheimer's disease and other dementias. J Alzheimers Dis. 2014;39(1):1–11.
64. Johnson KG, D'Souza AA, Wiseheart M. Art training in dementia: a randomized controlled trial. Front Psychol. 2020;11:585508.
65. Gross SM, Danilova D, Vandehey MA, Diekhoff GM. Creativity and dementia: does artistic activity affect well-being beyond the art class? Dementia (London). 2015;14(1):27–46.
66. Kinney JM, Rentz CA. Observed well-being among individuals with dementia: memories in the Making©, an art program, versus other structured activity. Am J Alzheimer's Dis Other Dementias®. 2005;20(4):220–7.
67. Huebner BG, Ellena E. I remember better when I paint. French Connection Films; Hilgos Foundation; 2009.
68. Filippi M, Piramide N, Elisabetta S, Agosta F. Neurodegenerative diseases and changes in artistic expression. In: Colombo B, editor. Brain and art: from aesthetics to therapeutics. Cham: Springer International Publishing; 2020. p. 27–39.
69. Bougousslavsky J, Hennerici MG, editors. Neurological disorders in famous artists: Part 2. Basel: Karger; 2007.
70. Espinel CH. de Kooning's late colours and forms: dementia, creativity, and the healing power of art. Lancet. 1996;347(9008):1096–8.
71. Chatterjee A. The neuropsychology of visual artistic production. Neuropsychologia. 2004;42(11):1568–83.

72. Piechowski-Jozwiak B, Bogousslavsky J. Abstract expressionists and brain disease. Front Neurol Neurosci. 2018;43:8–22.
73. Lane HL. When the mind hears: a history of the deaf. New York: Random House; 1984.
74. Lane HL. The mask of benevolence: disabling the deaf community. New York: Vintage Books; 1993.
75. Lane HL, Philip F. The deaf experience: classics in language and education. Washington, DC: Gallaudet University Press; 2006.
76. Davis LJ. Enforcing normalcy: disability, deafness, and the body. Brooklyn: Verso Books; 1995.
77. Sacks O. Seeing voices: a journey into the world of the deaf. Berkeley: University of California Press; 1989.
78. Pavani F, Bottari D. Frontiers in neuroscience: visual abilities in individuals with profound deafness a critical review. In: Murray MM, Wallace MT, editors. The neural bases of multisensory processes. Boca Raton: CRC Press/Taylor & Francis; 2012. p. 423–48.
79. Blume S. The artificial ear: cochlear implants and the culture of deafness. New Brunswick: Rutgers University Press; 2009.
80. Mauldin L. Made to hear: cochlear implants and raising deaf children. Minneapolis: University of Minnesota Press; 2016.
81. National Institute on Deafness and Other Communication Disorders. Cochlear implants (NIH Publication No. 00-4798). Bethesda: NIH; 2016.
82. Sacks O. Musicophilia: tales of music and the brain. New York: Alfred A. Knopf; 2010.
83. Moir Z, Overy K. The impact of cochlear implants on musical experience. In: Bates V, Bleakley A, Goodman S, editors. Medicine, health and the arts: approaches to the medical humanities. London and New York: Routledge; 2014. p. 246–63.
84. Graham S. Brain scans show deaf subjects 'hear' vibrations (87th Scientific Assembly and Annual Meeting of the Radiological Society of North America). Scientific Am. 2001;28:661–95. https://www.scientificamerican.com/. Accessed 15 Aug 2023
85. UW Medicine. Deaf people sense vibrations in auditory cortex. UW Medicine Dean's Report 2002 (July 1, 2001–June 30, 2002). https://depts.washington.edu/. Accessed 18 Aug 2023.
86. Neary W. Brains of deaf people rewire to 'hear' music. UW News; 2001. https://www.washington.edu/news/. Accessed 18 Aug 2023.
87. Olulade OA, Koo DS, Lasasso CJ, Eden GF. Neuroanatomical profiles of deafness in the context of native language experience. J Neurosci. 2014;34(16):5613–20.
88. Lawrence Livermore National Laboratory. "Beethoven's Nightmare"—a dream turned reality for Longo. Livermore, CA: LLNL News; 2007. https://www.llnl.gov/. Accessed 1 Sept 2023

Part III

The Neurocultural in Practice

10 Translation and Education

As noted in the last chapter, of his three poststroke books [1–3], Howard Engel's memoir is the clearest in offering a view of what writing with *alexia sine agraphia* is like from the inside, from the patient's perspective. As he explains, when he found himself unable to read, all of the faculties he retained helped him convince himself that he could return to his career as a writer. In his memoir [2], Engel tells the reader how his perception and cognition worked poststroke. One intriguing feature of his recovery that he shares with readers is that there were times he didn't notice his *alexia* at all. The sky looked blue, the sun shone as usual, and the world hadn't suddenly become an alien place.

Jean-Dominique Bauby's book *The Diving Bell and the Butterfly* [4] is another autopathography that translates a patient's viewpoint on his condition. Bauby was totally paralyzed after a cerebrovascular seizure left him with locked-in syndrome. Unlike Engel, Bauby was only able to blink his left eyelid to communicate with others. Therefore, he was constantly aware of how radically his life was transformed by his stroke.

What should we make of how dramatically neurological differences and neurological disturbances impact the ways in which we act, think, live, learn, and feel? This is a question that many have wrestled with throughout time. Bauby's inside look opens this chapter about translation and education because of how eloquently he translates his experience of a pathology and because his work intersects with many locked-in syndrome descriptions in art and literature. After examining translation, this chapter turns to education on its own terms.

A. Ione, *Neuroscience and Art*, Neurocultural Health and Wellbeing, https://doi.org/10.1007/978-3-031-62336-3_10

10.1 Jean-Dominique Bauby: The Diving Bell and the Butterfly

A well-known Parisian journalist and a 43-year-old editor of *Elle* France, Jean-Dominique Bauby (Jean-Do) experienced his stroke on December 8, 1995. It was a gray, muted day. He drove to his office with the Beatles song *A Day in the Life* playing on the car radio. His plan was to take his son to the theater after work. There were some rough spots during the day, including a 45-minute phone call that left him feeling "like a trampled doormat." He also felt tired. The thought of canceling his outing with his son crossed his mind, but he decided to fight the sense of utter exhaustion.

While driving, after picking up his son, everything turned blurry. Bauby also started to feel as if he were functioning in slow motion. Near the home of his sister-in-law, a nurse, he started sweating excessively. Half-conscious, he stopped the car, collapsed in the rear seat, and sent his son to get her. By the time they got him to a clinic, 10 miles away, he was unable to talk, and his head was wobbling on his neck. His last thought before slipping into a coma was that they would have to cancel the play.

Twenty days later Bauby awoke, paralyzed. His first thought upon regaining awareness was that he felt as if there was something like a giant invisible diving bell holding his body prisoner. This was how he came up with the first part of the title of his book, *The Diving Bell and the Butterfly*. As he put it, the book describes his life "since taking up residence in my diving bell" (p. 77, [4]. At this point, he also learned about his stroke:

> Up until then, I had never even heard of the brainstem. I've since learned that it is an essential component of our internal computer, the inseparable link between the brain and the spinal cord. I was brutally introduced to this vital piece of anatomy when a cerebrovascular accident took my brainstem out of action. In the past, it was known as a "massive stroke," and you simply died. But improved resuscitation techniques have now prolonged and refined the agony. You survive, but you survive with what is so aptly known as "locked-in syndrome." Paralyzed from head to toe, the patient, his mind intact, is imprisoned inside his own body, unable to speak or move. In my case, blinking my left eyelid is my only means of communication. (pp. 3–4, [4]).

In his book, Bauby also tells us that despite feeling enclosed in a diving bell, his mind, his memory, and his imagination were free to take flight. He uses these faculties to roam freely like a butterfly:

> Through the frayed curtain at my window, a warm glow announces the break of day....something like a giant invisible diving bell holds my whole body prisoner.... and my mind takes flight like a butterfly. There is so much to do. You can wander off in space or in time, set out for Tierra del Fuego or for King Midas's court (pp. 3, 5 [4]).

Part of the book's mesmerizing quality comes from Bauby's ability to bring the reader into both his inner and outer experiences. He explains being reduced to silence, only able to talk to others through blinking his one eye. He explains how people treat him, the incessant hospital sounds, and how sensations like smell remained tantalizing within his environment, although he was unable to eat or even swallow. He explains how his imagination became a fertile companion. In other words, as each chapter unfolds, it is clear that even as Bauby feels totally boxed in and misses his old life, he has still managed to retain his sense of humor, his intelligence, and his joy in smelling French fries cooking when wheeled to the beach. He also talks about how lonely and quiet the hospital becomes on Sunday; how angry he gets at those on the staff who turn the television on and off without any sense of his desires, about his 93-year-old father who is similarly locked in his apartment because he is no longer able to negotiate the stairs in his building; and how his joy is mixed with sadness when his children visit.

He also tells the reader that his case was not a classic locked-up syndrome case because he is able to swivel his head, to turn it 90°. This gave Bauby a much appreciated, expanded field of vision out of his one eye. He also retained feeling on a small area of his face, so he could manage a half-smile.

The most important addition to his new life was learning to use his alphabet/letters, a communication tool that allowed him to translate his thoughts and feelings blink by blink. His speech therapist designed his alphabet chart with each letter on it placed according to the frequency of its use in the French language. Anyone wanting to talk to him could go through the letters, and his one eye would blink when they reached the letter he needed. He learned to talk to visitors through this blink-by-blink system, and he wrote his memoir this way as well. Claude Mendibil was the amanuensis hired to transcribe it. Letter by letter, blink by blink, words emerged followed by paragraphs and then chapters. At the end, *The Diving Bell and the Butterfly* presents the psychological torment of being trapped inside his body and where his imagination takes him.

Throughout his book we learn how Bauby felt as he discovered his new life, and how he recalls life before the diving bell enclosed him. Sometimes, for example, when he comments on two medical visits to Paris, he brings the two together. On arriving he says: "My old life still burns within me, but more and more of it is reduced to the ashes of memory," and when leaving he notes: "Nothing was missing [in Paris], except me, I was elsewhere" (pp. 77, 79, [4]). We learn that when he was well enough to use a wheelchair, he realized that he had graduated from a patient whose prognosis was uncertain to an official quadriplegic. It was while in the wheelchair that he saw his reflection in a glass window and discovered that he looked like "a man who seemed to have emerged from a vat of formaldehyde" (p. 25, [4]).

He also talks about the hospital staff. The majority would not dream of leaving the room without making sure everything was fine, but there is also a minority who just don't want to be bothered. The ophthalmologist who sewed his right eyelid shut was in the latter category. This doctor used a needle and thread, "just as if he were darning a sock" (p. 53, [4]), and Bauby was shocked that the doctor did the procedure without offering any explanation as to why he was sewing his eye closed. The patient eventually learned that the lid was no longer fulfilling its function as a protective cover and was sutured shut to prevent corneal ulceration [4].

Bauby's sense of humor is a part of the book's poetry, and it serves to humanize some of the pathos. At one point, Bauby mentions that when he started a diet a week before his stroke he never thought he'd lose 66 pounds in 20 weeks. His humor also comes through when he explains that a witticism doesn't work when he tries to tell a joke to a visitor letter by letter or blink by blink. We also see his sense of humor when he talks about *The Count of Monte Cristo.*

Before his stroke, Bauby had signed a contract to write a book, and his plan was to do an updated version of Alexandre Dumas' 1844 book *The Count of Monte Cristo* [5]. Monsieur Noirtier De Villefort, a character in that book, is considered one of the first literary descriptions of locked-in syndrome [5, 6]. When outlining this piece of history, Bauby jokes about learning the hard way that one should not tamper with masterpieces. After the stroke, Bauby wrote *The Diving Bell and the Butterfly* instead. He died at the age of 44, two days after the book's French publication. He ends his book pensively, and in doing so retains the spirit of the man we have encountered throughout the book: "Does the cosmos contain keys for opening up my diving bell? A subway line with no terminus? A currency strong enough to buy my freedom back? We must keep looking. I'll be off now" (p. 131–2, [4]).

10.2 Locked-in Syndrome

When Bauby published *The Diving Bell and the Butterfly* instead of a revision of *The Count of Monte Cristo,* he added his lived experience to the locked-in syndrome literature. Dumas' fictional description of Monsieur Noirtier De Villefort's locked-in condition [5] nonetheless seems very similar to Bauby's text:

> Sight and hearing were the only senses remaining, and they, like two solitary sparks, remained to animate the miserable body which seemed fit for nothing but the grave; it was only, however, by means of one of these senses that he could reveal the thoughts and feelings that still occupied his mind, and the look by which he gave expression to his inner life was like the distant gleam of a candle which a traveler sees by night across some desert place, and knows that a living being dwells beyond the silence and obscurity (Chap. 58, [5]).

The condition locked-in syndrome was first described and presented in 1875 by Camille Darolles, an intern supervised by François Damaschino, at a monthly meeting of the Société Anatomique de Paris chaired by Jean-Martin Charcot in 1875 [7, 8]. The name was introduced by Plum and Posner in their 1966 book *The Diagnosis of Stupor and Coma.* These authors acknowledged that descriptions dated back to the nineteenth century and reference Dumas' character specifically when describing a patient with "brainstem lesions sparing consciousness." They note:

> Like M. Noirtier de Villefort in Dumas' *The Count of Monte Cristo*, this patient was awake but had lost all power to communicate except with his eyes — "a corpse with living eyes." Although both akinetic and mute, he did not suffer from the syndrome of akinetic mutism in which patients are unconscious and make no appropriate responses. Great care must be exercised in distinguishing such locked-in patients with "a mind…encumbered by a body over which it has lost the power of compelling obedience" from typical akinetic mutism in which self-aware consciousness is lost. The implications for humane patient care are profound (p. 160, [9]).

Georges Simenon's 1963 novel *The Patient* (*Les Anneaux de Bicêtre*) [10, 11] was published at about the time Plum and Posner introduced the term locked-in syndrome. In this fictional account, we also find a renowned journalist, René Maugras, who like Bauby has a stroke and wakes up both paralyzed and unable to speak. There is no indication that Bauby knew this work, and the patient in this fictional account regained some mobility. Both accounts do include sections about the physiotherapy, the wheelchair, memories of childhood, thoughts on food, the desire to regain speech, and so forth.

Thérèse Raquin written in 1868 by Émile Zola potentially describes another case of locked-in syndrome, although it is not as clear-cut. Zola's *Madame Raquin* [12, 13] is saddened by her son's death, and after a series of medical events, she becomes completely paralyzed except for her eyes [6]:

> Thérèse and Laurent rose from their chairs, terrified at this stroke, which had contorted the old mercer in less than five seconds. When she became rigid, and fixed her supplicating eyes on them, they pressed her with questions in order to ascertain the cause of her suffering. Unable to reply, she continued gazing at them in profound anguish. They then understood that they had nothing but a corpse before them, a corpse half alive that could see and hear, but could not speak to them…At the bottom of their hearts, they cared little for the suffering of the paralyzed woman [6].

Haan proposes that literary examples stem from the way locked-in syndrome serves as a strong metaphor for human existence [12]. Perhaps this also accounts for the many derivative works. Bauby's book was followed by the Jean-Jacques Beineix documentary *Assigné à residence* [14] and Julian Schnabel's award-winning commercial film *The Diving Bell and the Butterfly* written by Ronald Harwood [15]. Schnabel, although an American, chose to make the film in French because it was the language Bauby used to write his book [16]. Released in 2007, Schnabel's award-winning film is particularly noteworthy in terms of neuroscience and art because the film director is also a professional painter. His inspiration to cinematically develop the film was his own dying father. Schnabel received Harwood's script while trying to come to terms with his father's death.

Schnabel decided to do everything in the camera when translating the book to film. For example, when the patient's visual field is presented to the viewer, his blinks were accomplished by having the cinematographer move large objects in front of the camera's lens [16]. Camera manipulation also presents both his double vision and his blurry vision. The visuals additionally add affective texture to the book by demonstrating how the therapist and visitors use the alphabet to communicate with the patient. As the book relates and the movie expresses through behaviors we can perceive, some friends were eager to "talk" and embraced this tool. Others find this kind of exchange trying, confusing, and their impatience left them unable to find a comfort zone.

Other films have presented locked-in syndrome as well. There are at least three English film versions of *The Count of Monte Cristo,* and the overall number is more than twice this when other languages and television adaptations are added into the mix. *Thérèse Raquin*, too, has spawned many variants in several languages.

10.3 Painting: A Translation Tool After Brain Injury

Although Bauby translated his condition into words, it is more common for painting to serve as a medium of translation after a neurological incident because visual art syntax easily allows for ambiguity. For example, physician Claudia Osborn turned to painting after she suffered a traumatic brain injury (TBI) after falling off her bike. The week before her accident, she had taken up painting and enjoyed it. After the accident, she had trouble concentrating, and painting was the only post-injury activity that sustained her interest.

According to her account, before her injury, she was no more likely to paint than to sing *Carmen* at the Met. After her accident, painting was one of her few self-initiated activities. She enjoyed it even if it was not a part of the sophisticated rehabilitation program in which she participated [17]. It appears that it just happened:

> I got out my acrylics and put on Strauss' moving "Death and Transfiguration." Using my left hand, I painted a picture of my brain at work — cobalt-blue "thinking" overlaid with intense scarlet "feelings" and many brilliant yellow disconnections. I think the brain in the picture was more vigorous than the one in my head. I also illustrated my imprisoned language in a second picture. These acts of creating images pierced small holes in my sense of isolation. The joy I once had in spoken language, the release in confiding and sharing, the pleasure in intellectual exchanges with others, might now have two other expressions, however inchoate and primitive. If I could not speak what I felt, I would draw and write it (p. 127, [18]).

Other individuals who have similarly found that visual art provided a means to communicate after a brain injury are Tommy McHugh and Jon Sarkin. Both became prolific, full-time artists. While some have characterized their painting as acquired savant syndrome [19], I think this is a misnomer and fails to give adequate acknowledgement to the evidence that both men say that they took up painting to communicate their symptoms to themselves and to others. The extraordinary abilities of those with congenital savant syndrome most often surface during childhood and in relation to some underlying developmental disability [19].

Tommy McHugh's artistic turn began in 2005 [19]. His skills emerged following a sudden-onset brain injury without any associated dementia or impaired verbal abilities. At 51 and a right-handed man, McHugh had worked in construction. He was admitted to the hospital in January 2001 with a sudden onset of severe headache. McHugh suffered a stroke on both sides of his brain that resulted in two burst blood vessels. This sent him into a coma for a week. After emerging, he showed some left hemispatial neglect, which is suggestive of dysfunctional right frontal/parietal connection. He scored within or above the average range on tests of letter, category, and design fluency; however, his scores dropped dramatically on a more demanding version where he was required to alternate between categories. Generally, he showed greatest difficulties in tasks requiring cognitive switching [19]. In addition, his personality changed, and he became extremely chatty, in a way that triggered a stream of associations that he found difficult to inhibit: "the context of his artwork, the stream of images and thoughts that are triggered in this way may well be a source of his creativity, at the same time that the process of expression through the artwork may help him to make sense of it" (p. 398, [19]).

McHugh was never involved with or interested in the creative arts previously. He explained his art as a way of watching his own brain. When he emerged from his coma, he didn't know who he was or what was happening. He recalls that his memory was spotty, and the fragments others gave him helped little. He couldn't put the pieces together. He stopped taking the pills he was given for his intense headaches, believing they didn't help. In McHugh's case, we can correlate the emergence of his new and extraordinary ability with a specific brain injury and his desire to find a tool to help him explain his transformed brain to himself and to explain his pain to others.

Jon Sarkin (b. 1953) is often coupled with Tommy McHugh because he, too, acquired his passion for making art. Sarkin, a self-taught contemporary American artist, worked as a chiropractor before becoming an artist. When he was 35, a well-known vascular neurosurgeon did an operation intended to cure his tinnitus and vertigo [20]. A severe postsurgery hemorrhage led to a coma and necessitated an immediate craniotomy, removal of the hematoma, and partial resection of his cerebellum [21]. He awoke from the surgery deaf in one ear, his vision splintered, and his balance permanently skewed. Told his brain had been permanently changed through the surgery, Sarkin needed to discover how best to move on [22].

Evidence showing that his brain and personality had changed accompanied his slow recovery. Sarkin explained to an interviewer that once he returned home he began to draw a cactus over and over and over again. When asked why, his reply was that deserts are inhospitable places and cacti thrive in the desert. He saw this plant as a metaphor for his life [23], and the way it gave him strength recalls how Howard Engel observed that despite all his problems he had abilities he could still draw upon after his stroke.

Sarkin had returned home disorganized, manic, and driven to philosophical ramblings and strange poetry. He began to create intricate, humorous paintings. His newly discovered obsession with color was paired with a very powerful urge to paint that he could not resist. This is why some have characterized him as a savant [19]. Before his stroke, Sarkin had sometimes produced very focused sketches. His new works, reengineered through his trauma, display distorted cartoon faces with symbols that drew on comics and popular culture. What was most noticeable was that the images kept emerging [22].

10.4 Translation and Adaptation: Stereopsis and Strabismus

Another individual who has used art to show others what his neurological problem looked like from the inside is James Robinson, who has strabismus. For a student project at Duke University, he produced a 12-minute video on his condition; later, he developed a series of videos on disability for the *New York Times* [24, 25]. Robinson has a condition called anomalous retinal correspondence (ARC), which means for him each eye works individually. Normally, our eyes have retinal correspondence. Most of us see binocularly because our eyes fuse the images from each eye in the occipital cortex of the brain.

Growing up, he had surgery twice and tried everything from patches to 3D glasses, all to no avail. Robinson has learned to adapt to how he sees but, as he puts it, whereas he doesn't have a problem with the way he sees, he has a problem with the way he is seen. This is because, for example, when he is looking at a person it appears to them that only one of his eyes is looking at them. This makes people uncomfortable for the few minutes they engage. The many short encounters in which people demonstrated their discomfort with his eyes has made him uncomfortable for his entire life.

Robinson's video is titled "How Life Looks Through My 'Whale Eyes.'" Using graphics and his family, the filmmaker walks the viewer through his world using optical tests and visual effects to distinguish normal vision from his own. He calls his condition whale eyes because we can only look into one whale eye at a time. At the end of the video, he calls for more acceptance of people who are neuroatypical saying: "I don't have a problem with the way that I see, my only problem is with the way that I'm seen" [25].

Technically, strabismus is characterized as the misalignment of one eye in relation to the other, resulting in failure of the two eyes to simultaneously focus on the same image and the loss of binocular vision [26]. It is estimated that it affects about 2–4% of the population. In the United States, this is about 13 million people [26, 27]. Concomitant strabismus, which may be inherited, is the dominant form. It occurs when the angle of deviation of the eyes remains constant, independent of the direction of gaze, and includes the common forms of strabismus (e.g., esotropia, exotropia, hyperopia, microstrabismus, and monofixation syndrome) [26]. Incomitant strabismus accounts for about 5% of strabismus cases. These share a common presentation as congenital, nonprogressive restrictive ophthalmoplegias. This form includes various forms of Duane retraction syndrome (DRS), horizontal gaze palsy, and congenital fibrosis of the extraocular muscles (CFEOM) [26].

Strabismus can occur from a childhood strabismus that was not previously treated or has recurred or progressed. In other words, sometimes, characterization is tricky [28]. What we do know is that with this condition the eyes do not move as a pair. As Robinson's film shows, they may alternate when looking at an object, or they may drift. Strabismus may be present occasionally or constantly. Often, if present during a large part of childhood, amblyopia may result. This is also known as lazy eyes and includes the loss of depth perception. People with amblyopia also have problems with binocular vision, and squinters with a lazy eye tend to use just one eye, which flattens what they see. Therefore, there is limited stereoscopic depth perception, and they usually have difficulty seeing three-dimensional images.

10.5 Historical Strabismus

Strabismus in effect simulates an exercise often done in the art classroom when learning to draw representationally. Instructors often tell art students to close one eye to better see how to translate the three-dimensional form one sees into a two-dimensional image on a flat surface, such as a painting, because the perspective and shadows appear more like shapes. In other words, when the stereopsis is removed by shutting one eye, the artist is looking monocularly, and the perception of depth is removed as well. This exercise shows it is easier to perceive how lines and shapes are translated from three dimensions to two when an individual uses one eye rather than fusing the images seen with both eyes.

The first author to describe eccentric fixation in a patient with strabismic amblyopia was the German eye surgeon Albrecht von Graefe (1828–1870) in 1854 [29]. His achievements included finding ways to use German physiologist Hermann Helmholtz's ophthalmoscope (a perforated mirror used to inspect the interior of the eye) as he developed several effective surgical remedies for eye disorders.[1] A consummate clinician [30], he contributed importantly to our understanding of eyeball motion and the integration of the brain and the eye in the formation and location of images. He additionally described in intricate detail the physiology of the extraocular muscles [30–32].

Historical paintings offer depictions of strabismus even before this neurological condition was scientifically described. Raphael's paintings of Tommaso Inghirami (also known as Phaedra) illustrate this condition. In both paintings, he appears lost in thought and gazing up. Scholars assume this angle was chosen to diminish the impact of the sitter's lazy right eye [31].

Another compelling example is the artist Giovanni Francesco Barbieri (Fig. 10.1), known as "il Guercino" or the squinter [33]. Scholtz and his team point out that he squinted because one eye was amblyopic (i.e., lazy). They assume it was strabismus, and we see this in a contemporaneous portrait of him with misaligned eyes by Ottavio Leoni (Fig. 9.1). The Scholtz team further relates the legend that his cross-eyed appearance was somehow due to an inattentive nursemaid who left him alone while he was sleeping. A loud noise woke and frightened him causing one of his eyes to fix at an angle in its socket. As they point out, it is more likely that his strabismus was congenital and, as is sometimes the case in newborns, was initially unnoticed [33].

Between 1621 and 1623, Barbieri lived in Rome, where he played an important role in the evolution of Roman High Baroque art. One commission in Rome included decorating the sixteenth century *Casino dell'Aurora*. Built in 1570, *Casino dell'Aurora* takes its name from the Sala dell'Aurora, the ceiling painted in 1621 by Giovanni Francesco Barbieri. Agostino Tassi (1580–1644) assisted him and painted the architectural details. The villa also includes ceiling paintings by Caravaggio, Domenichino, and many others. On Guercino's ceiling, we see *Aurora*, goddess of the dawn, riding through the brightening sky in a horse-drawn chariot. He painted the fresco so that it appears as though there is no ceiling; the viewer perceives

[1] In 1857, Albrecht von Graefe introduced iridectomy (surgical removal of part of the iris) for the alleviation of glaucoma, a disease resulting in opacity of the lens. In 1860, he showed that blindness and visual defects connected with cerebral disorders are often traceable to optic neuritis, or inflammation of the optic nerve. He identified central retinal artery embolic occlusion, papilloedema, and optic neuritis. Von Gräfe also developed, in 1867, a surgical treatment for cataract by extraction of the lens [30].

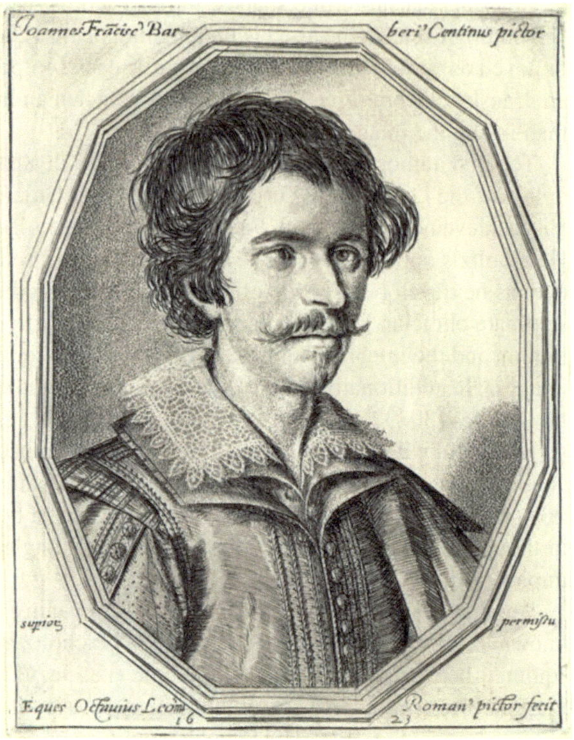

Fig. 10.1 Giovanni Francesco Barbieri, called Guercino. 1623. Artist: Ottavio Leoni. Engraving. National Gallery of Victoria, Melbourne. Everard Studley Miller Bequest, 1959. Courtesy of the NGV Collection Online through the generous support of the Joe White Bequest

Aurora's chariot moving directly over the building. One tantalizing feature of Guercino's story is that he was known as the "Rembrandt of the South" because he so skillfully painted with light and shadow. This comparison is intriguing because researchers have claimed that Rembrandt, too, had strabismus, in his case exotropia.

A study by Margaret S. Livingstone and Bevil Conway [34], two neuroscientists, hypothesized that Rembrandt's eyes failed to align correctly through an evaluation of his self-portraits and etchings. All the art they examined showed the same outward displacement of the left eye, which appears on the right in the etchings because the prints are reversed from the plates. One alluring quality of this reversal is that it indicates Rembrandt (Fig. 10.2) was not consciously depicting the eye displacement but, rather, depicting what he saw in the mirror with great accuracy. Thus, we can admire his talents even as we speculate on what it means that he did not see in a normal fashion (i.e., binocularly) because his brain automatically switched to one eye for such visual tasks.

The upshot is that although strabismus does not add to an artist's talent, artists may have still used their own stereo blindness as an asset [35]. Thus, it is not a handicap for an artist, and it may even be an asset that depth perception is reduced [36]. Moreover, a survey Conway and Livingstone conducted showed that artists are more apt to have strabismus than nonartists [37]. It may also be connected to the

Fig. 10.2 Rembrandt: Self-Portrait. Artist: Rembrandt (Rembrandt van Rijn) (Dutch, Leiden 1606–1669 Amsterdam). Date: 1660. Medium: Oil on canvas. Dimensions: 31 5/8 × 26 1/2 in. (80.3 × 67.3 cm). Courtesy of the New York Metropolitan Museum of Art, Bequest of Benjamin Altman, 1913

evidence that people with poor binocular vergence are often dyslexic as well. With dyslexia or poor reading skills, individuals are likely encouraged to pursue activities such as art because it uses a different type of language and syntax [35].

People experience strabismus in many ways, and there are thus several possibilities in Rembrandt's case. It is possible that he had the condition as a child. If it were ignored, it would likely have led to a loss of his binocular vision as he reinforced a monocular way of looking. This would have helped him in creating two-dimensional art. It is possible that Rembrandt worked with the fluctuations of exotropia. This would mean he had more control of it, but this seems less likely given the consistency of the condition as represented in his work. In either case, this eye problem could have aided him in translating the three-dimensional world onto two-dimensional surfaces. This would also mean that Rembrandt's physiognomic deftness is of a quite different quality than someone working without the condition.

In response to the Livingstone and Conway study, Michael F. Marmor and Saad Shaikh pointed out five reasons to display caution when evaluating their conclusions: First, there are no controls in the study, such as a survey of other artists with seemingly normal vision. Second, many artists take liberties with facial and ocular characteristics (e.g., El Greco and Modigliani). Third, the range measured by Livingstone and Conway is in the range of 5°, a small amount by clinical standards. Normal persons can in fact spread their eyes to this degree without losing stereopsis.

Fourth, as an artist fixates on one eye to paint it and then shifts his gaze to the other eye, the shift in the angle of view is likely to cause more white of the eye to be seen on each nasal side [38].

Another artist associated with strabismus is Leonardo da Vinci. Tyler, who studied six images (two sculptures, two oil paintings, and two drawings), concluded Leonardo had intermittent exotropia with the resulting ability to switch to monocular vision. This would perhaps explain his great facility for depicting the three-dimensional solidity of faces and objects in the world and the distant depth recession of mountainous scenes. In exotropia, the divergent eye alignment is typically manifested as an outward shift in the locations of the pupils within the eyelid aperture. The condition was assessed by fitting circles and ellipses to the pupils, irises, and eyelid apertures images that are identified as portraits of Leonardo da Vinci and measuring their relative positions [39]. Shakarchi and Guyton have questioned these studies [40].

10.6 Translation and Education: Academic Projects

Like James Robinson, a number of disabled artists have used academic projects as a launching pad for understanding their own worlds and showing their worlds to others or designing tools to aid their artistic pursuits. Visually challenged artist Elizabeth Goldring founded The Vision Group at the Center for Advanced Visual Studies (CAVS) at the Massachusetts Institute of Technology (MIT) with the goal of studying perceptual and artistic awareness in order to design machines that extend sentient, creative capabilities for visually challenged persons like herself [41]. Laura Jane Stevenson used her Master's thesis to describe the cognitive benefits of the dyslexic processing style and to reveal how public education often fails dyslexic students [42]. Laura Cechanowicz, after a diagnosis of rheumatoid arthritis, designed a dissertation to come to terms with this invisible disability [43].

Turning first to CAVS, a transdisciplinary team of artists, scientists, engineers, physicians, and students designed the Seeing Machine Camera (SMC). SMC is a device that opened a path for artistic expression for those with decreased vision and allows sight-impaired people to live with an increased independence [41]. Unlike magnifying glasses and video magnifiers, the SMC helps those with advanced visual impairments caused by retinal diseases like retinitis pigmentosa, macular degeneration, and glaucoma that distort views, blacken out areas, and blur vision.

Laura Jane Stevenson's project was about the learning disability dyslexia. It encompassed her experience and the condition in general. Dyslexia is a learning disability that is neurobiological in origin. Dyslexia had no relationship to intelligence, hearing, or vision, and there is no cure. It is characterized by difficulties with accurate and/or fluent word recognition and by poor spelling and decoding abilities [44]. Therapies often help with reading, and as noted, many turn to other forms of communication like art. Due to their reading problems, many dyslexics think they are "dumb" while in school. Indeed, many turn to art because it uses a visual syntax that bypasses the reading problems. Studies on art students have confirmed a high incidence of dyslexia among artists that seems to confirm the anecdotal reports [44–47].

Like autism, discussed in Chap. 3, the condition has often gone undiagnosed. As Sally Shaywitz has pointed out, we now know that dyslexia is not uncommon; it affects one out of every five men and women. It affects both spoken language and the ability to read; it is also a condition that does not go away, although it is possible to use remedial programs to aid learning [48].

The term "dyslexia" was coined by Rudolf Berlin, a German ophthalmologist and professor in Stuttgart in the 1880s, after observing that some of his patients had problems reading printed words and showed no sign of visual problems. Several publications late in the twentieth century brought this learning disability to the public's attention [49, 50].

Since there was so little awareness of the condition until recently, it is not surprising that Stephenson's case was missed for the cause of dyslexia is not known and there is no physical evidence of it (e.g., such as damaged tissue) that highlights it in the brain. Functional MRI studies do suggest that the brains of people with dyslexia work differently than the brains of other people. Essentially, when people read, three areas in the brain, one in the front and two in the back, are generally activated. In people with dyslexia, there are significant differences in Broca's area and the left inferior temporal region for nonword reading and a phonological transformation task [51].[2]

Stevenson, who now works as a full-stack engineer, crafted her thesis around her dyslexia and the condition more generally [42, 52]. In the thesis, she examines the cognitive benefits of the dyslexic processing style and looks at the ways in which the public school system fails dyslexic students [42]. The research combined qualitative interviews with twenty dyslexics with quantitative findings in neuroscience and educational psychology. Titled *Neither Contradictory Nor Coincidental*, the text of her research also served as the foundation for the artwork that accompanied it. The artwork was comprised of a mounted, interactive video installation that aimed to raise awareness about the dyslexia issues covered in her research. Much like "H.M." by Kelly Tribe, discussed in Chap. 7, and the Michael Nyman television presentation of Oliver Sacks' *The Man Who Mistook His Wife for a Hat*, discussed in Chap. 6. Stevenson mixed an array of media, textual elements, and commentary. Stevenson also included interviews as audio material and relevant information to create a three-dimensional environment that endeavors to highlight the dyslexic mind's strength in processing three-dimensional material [42].

One interesting outcome in creative terms is that the work was not a "designed" project, but one that evolved as she worked with the information. Programming was a key component in devising the installation, and it, in turn, led her to engineering after her graduation [52]. Thus, not only was the finished product vastly different

[2] People with phonological dyslexia have trouble isolating individual phonemes, which leads to difficulty sounding out and pronouncing words.

than what she envisioned at the start; her career path changed as well. Writing about dyslexia in her thesis, she notes:

> [T]he piece is intended to provide viewers with a well-rounded understanding of dyslexia and demonstrated why disadvantaging dyslexic students is a disservice to society. By extension, it encourages the audience to look at all non-neurotypical minds through this framework, questioning what strengths are found in cognitive abnormalities, rather than viewing these abnormalities as something to be corrected (p. 2, [42]).

Laura Cechanowicz's diagnosis of rheumatoid arthritis led her to similarly design a dissertation to reckon with her invisible disability [43]. She created a fictional scenario that centers around the idea that everybody will develop an invisible disability in their lifetime and within 30 years. Her idea was to give them time to think about this so that they don't feel desperate or have too many problems. The idea is that given the information people will redesign society spatially, culturally, and with the idea of making a better living situation for everybody so that they are comfortable living with their disabilities.

Although not factual, Cechanowicz's impetus was to produce real-world solutions. Therefore, she included interviews with the disabled and those suffering with mental illness, aging, etc. [43]. She has explained it in terms of her own situation, which was the impulse behind the work:

> I have constant pain, which is—without taking pain medicine—usually a five to an eight out of ten, all the time. Luckily I have pain medicine and I found a wonderful doctor who really asked me how bad it was and said "Ok, let's see if we can get that down to four to six" and that's what we did, we got it down to four to six and, for me, it is a really big difference in functionality. My pain is also very unpredictable, I could be walking around fine for a while and then, suddenly, my knee feels like it is breaking from the inside, so decision making is highly altered for people with chronic pain. You just don't know when it's going to happen, how long it is going to last or if it will prevent you from working [43].

Cechanowicz's hope is to find ways to create technology that will change the world and to build programs for elementary schools. These range from social technology to innovative ways to tell stories and use design [43]. She characterizes this mixture of design and social thinking in terms of "technologies of the imagination" [43] and calls the work a worldbuilding project. She is creating a design philosophy for envisioning solutions to real-world problems that incorporates holistic real-world research in crafting fictional worlds with transmedia outputs [53]. The goal is to offer hope for a better future.

10.7 Temple Grandin Revisited

In Chap. 3, Temple Grandin's biography was briefly introduced. Her outreach over the years shows the degree to which she wanted to give back to others in comparable situations, and her emphasis on education and translation has coincided with her other accomplishments.

It was in the mid-1980s that Grandin first spoke in public about autism, at the request of Ruth C. Sullivan, one of the founders of the Autism Society of America

(ASA). While waiting for a transfer flight to the annual National Society for Autistic Meeting (NSAC), now the Autism Society of America (ASA), Sullivan saw Grandin listening to an autism discussion among conference goers while awaiting their transfer flight. It wasn't until later in the week that Sullivan realized that this high-functioning individual, Grandin, was someone with autism. This led Sullivan to ask if she would share her experiences at a future meeting. As Sullivan recalls it:

> The audience couldn't get enough of her. Here, for the first time, was someone who could tell us from her own experience, what it was like to be extremely sound sensitive ("like being tied to the rail and the train's coming"). She was asked many questions: "Why does my son do so much spinning?" "Why does he hold his hands to his ears?" "Why doesn't he look at me?" She spoke from her own experience, and her insight was impressive. There were tears in more than one set of eyes that day (pp. xviii–xix, [54]).

Grandin, who has a PhD in Animal Science, quickly became a much sought-after speaker in the autism community. After Oliver Sacks published her story, her name was more broadly associated with the condition. Sullivan's forward to Grandin's book *The Way I See It: A Personal Look at Autism and Asperger's* talks about Grandin's influence on not just educating the public about the condition but also serving as a translator [54]. When she first spoke to Sullivan's group, few people had heard of autism, and even fewer could communicate well enough to tell others about it from the inside. As Temple Grandin has put it:

> I didn't become social overnight. There wasn't a point where some magic switch turned on in my brain and the social stuff made sense after that. I'm the person I am today because of all the experiences I've had, and the opportunities those experiences offered me to learn, little by little. It wasn't easy; sometimes it was really difficult. I've made a lot of mistakes, but I just kept going until I got it right. And, I'm still learning today! That's what I want other people on the spectrum to learn: You just can't give up. You have to keep trying (p. xxi, [54]).

Grandin has also said that she believes in early intervention with little kids to get them talking if possible and that intervention is harder for older children. Asked if she would have achieved as much professionally if she were not autistic, Grandin didn't think so:

> [T]here was a motivation that I had that a nonautistic person doesn't have. And I had a visualization skill that goes beyond what most people have. When I designed a piece of equipment, I could actually test-run it in my head like these virtual-reality computer programs. And I didn't even know that other people couldn't do this. That's the positive side of autism. Now obviously, if you have a child who's still nonverbal and not toilet trained, that's the negative side of autism. It's a continuum…. [55].

10.8 Conclusion

As this chapter has detailed, many artists have found ways to translate their own conditions and/or those of others into a work others can perceive. This educates others about what conditions look like from the inside. The range of examples is far more than a short chapter can present. The next two chapters, on face blindness (*prosopagnosia*) and synesthesia, introduce more examples in this category. In closing out this chapter, a few additional vignettes will serve to look back at examples earlier in this book and broaden the spread.

Educational films and documentaries are standards in presenting reenactments and archival material. The film *Iris*, discussed in Chap. 5, presented her Alzheimer's disease through a script based on books by her husband. *Still*, a 2023 film release, takes a different tact in showing actor Michael J. Fox's Parkinson's disease. Michael J. Fox (b.1961) is a Canadian-American activist and retired actor. Fox started displaying symptoms of early-onset Parkinson's disease in 1991 while shooting the role of a flamboyant plastic surgeon Ben Stone in *Doc Hollywood.* He was diagnosed shortly thereafter and waited until 1998 to reveal the diagnosis. Though his initial symptoms were only a twitching little finger and a sore shoulder, he was told that within a few years he would not be able to work. Shortly thereafter, he became an advocate for finding a cure and in 2020 founded The Michael J. Fox Foundation to help fund research. When reflecting on a return to acting, he once jokingly said "I can play anybody, as long as the character has Parkinson's disease" [56].

Sequences throughout the movie *Still* stress his need of dopamine and his initial reactions to learning he had a progressive, degenerative, and incurable condition. One was drinking too much. Another was trying to hide it from the world. He also found ways to take his pills when needed on sets while filming before he had gone public. As he explained, when his dopamine is low, he struggles to smile, show expressions, and takes on a blank countenance. Low dopamine levels also cause the slowed movements and tremor symptoms of Parkinson's disease. The film both explains and visually communicates his condition.

Finally, the educational projects presented here serve as a validator of art education. In the United States, debates about art education and developmental education tend to center around whether art is a luxury. This misses many other sides of art, particularly its power as a form of communication. According to the "art is a luxury" view, basic education should center around the meat and potatoes such as reading, writing, and science. Yet, as the studies detailed in this chapter indicate, art has a role in health and well-being, regardless of whether one is neurotypical, is neuroatypical, or is coping with the onset of an unexpected health challenge. As discussed throughout this book, all arts have their own syntax, and working with a syntax that allows one to draw on one's strengths offers a mechanism for creating meaning and translating who one is. Similarly, those with brain injuries benefit in discovering alternative means of expression as they develop new forms for translation and communication. In addition, those with alternative goals or neuroatypical profiles can find expression through art. Thus, art's complexity adds to interpersonal exchanges, and we can point to studies that have shown the arts boost academic performance. Finally, engagement fosters empathy, aids in social and emotional learning, and helps in understanding others [57–59].

References

1. Engel H. Memory book. New York: Carroll & Graf; 2006.
2. Engel H. The man who forgot how to read. New York: Thomas Dunne Books/St. Martin's Press; 2008.
3. Engel H. East of Suez: a Benny Cooperman mystery. Toronto: Penguin Canada; 2008.
4. Bauby J-D. The diving bell and the butterfly. New York: Alfred A. Knopf; 1997.
5. Dumas [père] A. The Count of Monte Cristo: The Project Gutenberg eBook of the Count of Monte Cristo; 1844/1998.
6. Zola E. Thérèse Raquin. The Project Gutenberg eBook; 1868/2006.
7. Darolles M. Ramollissement des protubérances: thrombose du tronc basilaire. Prog Méd (Paris). 1875;3:629.
8. Teive HAG, Ferreira MG, Coutinho L, Camargo CHF, Munhoz RP, Walusinski O. The locked-in syndrome: the early French descriptions. Rev Neurol (Paris). 2022;178(10):996–8.
9. Plum F, Posner JB. The diagnosis of stupor and coma. Oxford: Oxford University Press; 1982.
10. Simenon G. The patient. London: H. Hamilton; 1963.
11. Simenon G. Les anneaux de Bicêtre; roman. Paris: Presses de la Cité; 1963.
12. Haan J. Locked-in: the syndrome as depicted in literature. Prog Brain Res. 2013;206:19–34.
13. Pearce JMS. The locked in syndrome. BMJ. 1987;294(6566):198–9.
14. Beineix J-J. Assigné à résidence. France: Bibliothèque publique d'information; 1997.
15. Schnabel J. The diving bell and the butterfly. English language. United States: Miramax Films; 2007.
16. Schnabel J. Charlie Rose interviews Julian Schnabel. In: Rose C, editor. Charlie Rose conversations: https://charlierose.com; 2008. Accessed 3 Aug 2023.
17. Kaplan F. Art, science and art therapy: repainting the picture. New York: Jessica Kingsley Publishers; 2000.
18. Osborn CL. Over my head: a doctor's own story of head injury from the inside looking out. Kansas City: Andrews McMeel Publishing; 1998.
19. Lythgoe MFX, Pollak TA, Kalmus M, Haan M, Chong WK. Obsessive, prolific artistic output following subarachnoid hemorrhage. Neurology. 2005;64(2):397–8.
20. Winner E, Von Karolyi C. Artistry and aphasia. In: Taylor Sarno M, editor. Acquired aphasia. 3rd ed. San Diego: Academic Press; 1998. p. 375–411.
21. Bäzner H, Hennerici M. Stroke in painters. Int Rev Neurobiol. 2006;74:165–91. San Diego: Academic Press
22. Nutt AE. Shadows bright as glass: the remarkable story of one man's journey from brain trauma to artistic triumph. New York: Free Press; 2011.
23. ABC Medical Mysteries. Jon Sarkin. ABC News; 2009. https://abcnews.go.com/. Accessed 2 Oct 2023.
24. Robinson J. Adapt-ability: film can help us look disability in the eye. New York Times; 2021. https://www.nytimes.com/. Accessed 30 Sept 2023.
25. Robinson J. Film: how life looks through my 'Whale Eyes'. New York: New York Times; 2021. https://www.nytimes.com/. Accessed 30 Sept 202
26. Engle EC. Genetic basis of congenital strabismus. Arch Ophthalmol. 2007;125(2):189.
27. Cleveland Clinic. Strabismus (crossed eyes). Cleveland, OH: Cleveland Clinic; 2019. https://my.clevelandclinic.org/. Accessed 20 Jan 2023
28. Yale Medicine. Adult strabismus: Yale Medicine; 2023. https://www.yalemedicine.org. Accessed 20 Jan 2023.
29. von Graefe A. Über das Doppelsehen nach Schiel Operationen und Incongruenz der Netzhäute. Arch Ophthalmol. 1954;1:109.
30. Rohrbach JM. Albrecht von Graefe in the present, the past, and the future. Graefes Arch Clin Exp Ophthalmol. 2020;258(6):1141–7.

31. Tan SY, Zia JK. Albrecht von Graefe (1828–1870): founder of scientific ophthalmology. Singapore Med J. 2007;48(9):797–8.
32. Hirschberg J. Geschichte der Augenheilkunde v. 6, 1915. Wilhelm Engelmann: Leipzig and Berlin; 1914.
33. Scholtz S, MacMorris L, Krogmann F, Auffarth GU. Lights and darks of a picture. The life of Giovanni Francesco Barbieri, "il Guercino"—the squinter. Strabismus. 2019;27(1):39–42.
34. Livingstone MS, Conway BR. Was Rembrandt stereoblind? N Engl J Med. 2004;351:1264–5.
35. Livingstone MS, Lafer-Sousa R, Conway BR. Stereopsis and artistic talent: poor stereopsis among art students and established artists. Psychol Sci. 2011;22(3):336–8.
36. Livingstone MS, Conway BR. Reply to Marmor and Shaikh: was Rembrandt stereoblind? N Engl J Med. 2005;352(6):632.
37. Conway BR, Livingstone MS. Perspectives on science and art. Curr Opin Neurobiol. 2007;17(4):476–82.
38. Marmor MF, Shaikh S. Was Rembrandt stereoblind? N Engl J Med. 2005;352(6):631–2. author reply -2
39. Tyler CW. Evidence that Leonardo da Vinci had strabismus. JAMA Ophthalmol. 2019;137(1):82–6.
40. Shakarchi AF, Guyton DL. A geometric analysis of eye dominance suggesting that Rembrandt and Leonardo da Vinci had straight eyes after all. JAMA Ophthalmol. 2020;138(1):101–2.
41. Wu F, Selvadurai C, Smithwick Q, Cain J, Cavallerano J, Silver P, Goldring E. The seeing machine camera: an artistic tool for the visually challenged conceived by a visually challenged artist. Leonardo. 2012;45(2):141–7.
42. Stevenson LJ. Neither contradictory nor coincidental: finding strength in cognitive abnormalities. Santa Cruz: University of California; 2019.
43. Cechanowicz L. Art, world building and design thinking: a conversation with Laura Cechanowicz. In: Luke RG, editor. Alumni profile. USC School of Design: Echos; 2018.
44. Wolff U, Lundberg I. The prevalence of dyslexia among art students. Dyslexia. 2002;8(1):34–42.
45. Alden S, Pollock VL. Dyslexia and the studio: bridging the gap between theory and practice. Int J Art Des Educ. 2011;30(1):81–9.
46. Symonds H. Teaching and professional fellowship report 2004/5: designing effective curriculum for dyslexic students within art and design in higher education; 2005.
47. Winner E, Casey MB, Dasilva D, Hayes R. Spatial abilities and reading deficits in visual art students. Empirical Studies Arts. 1991;9(1):51–63.
48. Shaywitz SE. Overcoming dyslexia: a new and complete science-based program for reading problems at any level. New York: Alfred A. Knopf; 2003.
49. Critchley M. The dyslexic child. London: Heinemann Medical; 1970.
50. Naidoo S. Specific dyslexia: the research report of the ICAA Word Blind Centre for dyslexic children. London: Pitman; 1972.
51. Georgiewa P, Rzanny R, Hopf JM, Knab R, Glauche V, Kaiser WA, Blanz B. fMRI during word processing in dyslexic and normal reading children. Neuroreport. 1999;10(16):3459–65.
52. Stevenson LJ. Creative coding: neither contradictory nor coincidental: interactive video installation; 2019. http://laurastevensondesign.com/creative-coding/. Accessed 5 Aug 2023.
53. Cechanowicz L. Worldbuilding design portfolio: Laura Cechanowicz; 2017 http://laurastevensondesign.com/creative-coding/. Accessed 5 Aug 2023.
54. Grandin T. The way I see it: a personal look at autism and Asperger's. Arlington: Future Horizons Inc.; 2011.
55. White R. Interview with Temple Grandin. Medscape Psychiatry. 2005;10:1.
56. Guggenheim D. Still: A Michael J. Fox movie. Apple TV+; 2023.
57. Kisida B, Bowen DH. New evidence of the benefits of arts education; 2019. https://www.brookings.edu/. Accessed 23 July 2023.
58. Hetland L, Winner E, Veenema S, Sheridan KM. Studio Thinking 2: the real benefits of visual arts education. New York: Teachers College Press; 2015.
59. Lloyd K. Benefits of art education: a review of the literature. Scholarship Engage Educ. 2017;1(1):6.

Faces and Face Blindness (Prosopagnosia)

11

Even before written documentation, faces were recorded on material objects, capturing both human behaviors and medical conditions. Some were generic faces and others were historical portraits. Now, photography and digital technologies allow us to capture faces mechanically. Within this, imaging tools have given us access to how a brain may conceptualize faces, although we still do not know precisely how an artist paints a face nor do we understand why some people fail to recognize faces. This chapter turns to neurotypical and neuroatypical faces in research, artistic practices, and as a medium of human communication. Topics include a summary of the face historically, research surrounding the inability to decipher faces (prosopagnosia), and the array of ways human faces influence neurocultural perspectives.

11.1 A Brief Chronology of Faces and Face Blindness

The world's oldest known portrait was found in 2006 in the Vilhonneur grotto near Angoulême and is thought to be 27,000 years old. Drawn with calcium carbonate and using the bumps in the wall to give form to the face, it features two horizontal lines for the eyes, another for the mouth and a vertical line for the nose [1]. There are also artifacts suggestive of portraits conceived between 9000 and 6000 BCE in the Pre-Pottery Neolithic B period. The Moche culture of Peru was one of the few ancient civilizations to produce portraits.

Face blindness is harder to identify in ancient history, although it is possible that the Greek general Thucydides offers a description of cases acquired during the Peloponnesian War (431–404 BCE):

> For the disorder which had originally settled in the head passed gradually through the whole body, and, if a person got over the worst, would often seize the extremities and leave its mark… Some again had no sooner recovered than they were seized with a forgetfulness of all things and knew neither themselves nor their friends ((2.49), p. 117, in [2]).

A. Ione, *Neuroscience and Art*, Neurocultural Health and Wellbeing,
https://doi.org/10.1007/978-3-031-62336-3_11

Some of the earliest surviving painted portraits of people who were not rulers are the Greco-Roman funeral portraits found in the dry climate of Egypt's Fayum district, dating from the second to fourth century CE. These famous death masks or mummy portraits were painted during the Roman occupation of the area. Despite the Roman preference for cremation, the Egyptians who continued to bury their dead painted the masks on wood in a pigmented wax technique called encaustic [3]. Figure 11.1 shows a portrait of a youth with a surgical cut on the lower lid of one eye from this period [3, 4], possibly a treatment for trachoma, a disease identified in Egypt as early as 15 BCE [5]. The portrait also suggests strabismus, or heterotropia,

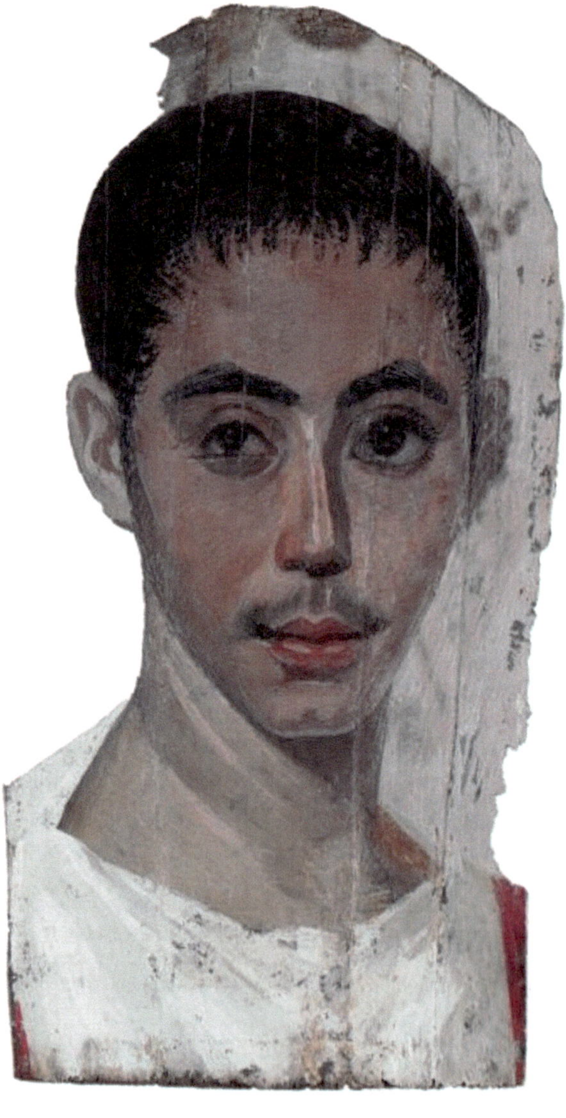

Fig. 11.1 Portrait of a Youth with a Surgical Cut in One Eye. Roman Period (Egypt), 190–210 CE. Medium: encaustic on limewood. Dimensions: H. 35 × W. 17.2 cm (13 3/4 × 6 3/4 in.). Courtesy of The Metropolitan Museum of Art, Rogers Fund, 1909

but it is unclear. His left eye seems to be looking straight at us, but the right eye appears to be looking up a little bit, which would make sense if that eye had been damaged by the infection.

The history of faces further includes donor portraits, and many faces that have become iconic. For example, to this day, there is an ongoing interest in Leonardo da Vinci's portrait of Lisa Gherardini, known as the *Mona Lisa,* painted between 1503 and 1506. Since 2005, this painting has been enclosed and housed in the Louvre's largest room, the *Salle des États*, which is also home to other remarkable Venetian paintings such as *The Wedding Feast at Cana* by Veronese. The face is fascinating in terms of its mysterious expression, the ongoing debates about the identity of the sitter, and the number of research analyses on why the painting looks the way it does [6–11]. These range from the strabismus studies discussed in the last chapter [12–14] to one by a dentist who argues that she is not smiling because she is wearing an expression common to people who have lost their front teeth [6].

With the development of oil painting and improved printmaking technologies, the ability to render facial expressions and different emotions improved, as work by the versatile Dutch painter Rembrandt Harmenszoon van Rijn (1606–1669) demonstrates. In the last chapter, Fig. 10.2 presented a self-portrait crafted when he was 54. Here, he has unsparingly depicted his own aging face, including his hypothesized strabismus, building up the paint in high relief to convey his furrowed brow, the heavy pouches beneath his eyes, and his double chin. Comparing that portrait with the more gestural composition he did as a younger man, as shown in Fig. 11.2,

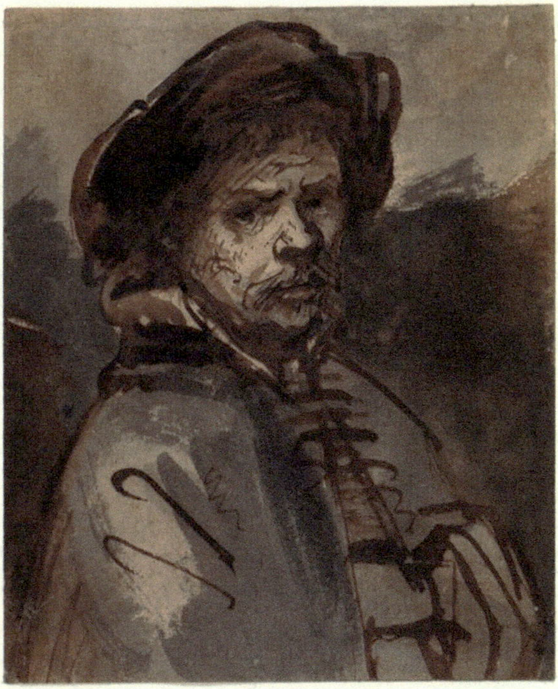

Fig. 11.2 Rembrandt: Self-Portrait (reworked by another hand). Artist: Rembrandt (Rembrandt van Rijn) (Dutch, Leiden 1606–1669 Amsterdam). Date: 1635–1640. Medium: Pen and brown ink, brush and brown and gray ink, brown and gray washes. Dimensions: 5 3/8 × 4 1/2 in. (13.6 × 11.4 cm). Courtesy of the New York Metropolitan Museum of Art, Robert Lehman Collection, 1975

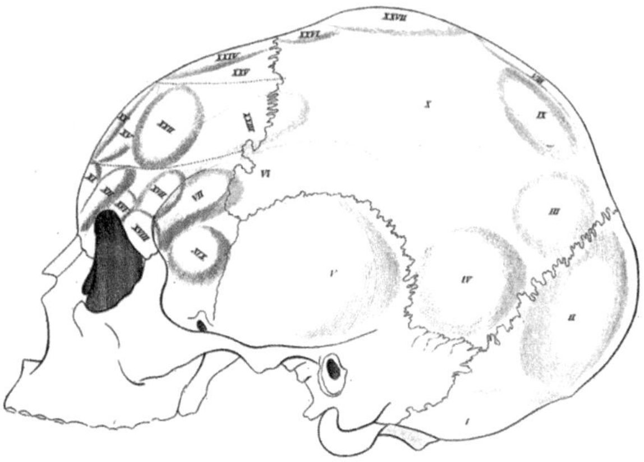

Fig. 11.3 Gall's physiological map. Plate from *Anatomie et physiologie du système nerveux en général, et du cerveau en particulier, avec des observations sur la possibilité de reconnoître plusieurs dispositions intellectuelles et morales de l'homme et des animaux par la configuration de leurs têtes*/Par F. J. Gall et G. Spurzheim [15]. Courtesy of the Wellcome Collection

reminds us that we recognize faces by facial features. The two renderings additionally show how much our faces change as we age, even as they stay remarkably the same.

Physiognomy speaks to the practice of assessing a person's character or personality from their outer appearance—especially the face. This, too, speaks to the constancy of our ever-changing facial features. The practice was documented even in ancient times, and in the century following Rembrandt's work, phrenology became a popular approach to it. Now associated with the German physician Franz Josef Gall (1758–1828), phrenology involves the measurement of bumps on the skull to predict mental traits. It is based on the concept that the brain is the organ of the mind and that certain brain areas have localized specific functions or modules (Fig. 11.3).

While Gall's phrenology is now discredited, he is credited for arguing that mental functions arose from the brain, rather than from the soul, heart, or liver [16], and he is also associated with the idea of brain modularity. More specifically, the idea that the brain is the major controlling center for the body goes back to Hippocrates and Galen. Gall's contribution was to link the brain to the mind, and his formal principles constituted one of the earliest attempts to localize specific functions to anatomical regions of the brain [17].

It is not surprising that Gall and others focused on faces in their medical and scientific research. Faces are usually the first thing we notice in people, and facial expressions are a part of human exchange. Face recognition is now considered a highly complex cognitive process involving a dedicated network of brain regions. It is largely associated with the fusiform gyrus and the less studied inferior occipital gyrus [18–20]. When one is face blind, this network is not working typically. Given this, it is noteworthy that Arthur Wigan (1785–1847), in 1844, was among the first to write a case study that explicitly described the inability to remember faces [21]. This face blind case is additionally noteworthy retrospectively because we now know that about 2.5% of the population has some form of congenital or developmental prosopagnosia. Estimated prevalence rates range from 0.64% to 5.42% [22]. Wigan wrote:

> A gentleman of middle age, or a little past that period, lamented to me his utter inability to remember faces. He would converse with a person for an hour, but after an interval of a day could not recognize him again. Even friends, with whom he had been engaged in business transactions, he was unconscious of ever having seen. Being in an occupation in which it was essential to cultivate the good-will of the public, his life was made perfectly miserable by this unfortunate defect, and his time was passed in offending and apologizing. He was quite incapable of making a mental picture of anything, and it was not till he heard the voice, that he could recognize men with whom he had constant intercourse. He was not at all fanciful in the matter, but was always in a state of alarm lest he should lose the sense of sight—imagining that the disorder or disease lay in that organ alone. When I inquired more fully into the matter, I found that there was no defect in vision, except that his eyes were weak, and that any long continued employment of them gave him pain. There was no appearance of that morbid vanity which induces a person to affect peculiarities of constitution, but on the contrary a strong desire to conceal his defect from the world. I endeavored in vain to convince him that an acknowledgment of the defect would be the best means of removing the unfortunate effect it had produced in alienating friends. He was quite determined to conceal it, if possible, and it was impossible to convince him that it did not depend solely on the eyes [21].

Another face blind case was described in 1867 by Antonio Quaglino, a professor of ophthalmology at the University of Pavia. His patient, LL, developed a sudden inability to recognize a person by his or her face, after he suffered a right hemisphere stroke. While the stroke obliterated his ability to recognize familiar individuals from their faces, he was able to read even the smallest characters very well, with excellent near and far vision [23].

In the 1870s, John Hughlings Jackson (1835–1911) also described clinical cases of visual agnosia for faces and places; often, the two go together. Hughlings Jackson, in 1872, noted a left hemiplegia case in which a person was unable to recognize people and places and "at one time he did not know his own wife." Hughlings Jackson also notes that having wandered far from home this patient "was unable to find his way back" (p. 32, [24, 25]). Hughlings Jackson related these to what he called imperceptions, or the lack or failure of perception. Another Hughlings Jackson case, in 1875, was a patient, Eliza T, age 59, who "developed fluctuating drowsiness, confusion, misidentified time, letters, and faces, and developed left sided hemiparesis" (p. 736, [17]).

Charles Darwin's 1872 classic study *The Expression of the Emotions in Man and Animals* [26] offers additional commentary on faces. Prior to *Expression*, the face was of scientific interest primarily to those who claimed they could read character, personality, or intelligence from the facial features. Darwin ignored the features and focused on the visible but temporary changes in appearance. He treated emotions such as anger, fear, disgust, and so forth as separate and discrete entities or modules [26, 27]. His study, still cited today, took for granted that it is the form or morphology of facial expression that conveys information about which emotion is occurring, and he used photographs and engravings to convey his understanding of emotional expressions. Darwin also believed that facial expressions of emotion were universal and not unique to humans [26–28]. Still, his studies omitted many features found in other analyses: He did not provide methods for measuring faces. He did not distinguish between voluntary and involuntary expressions. He did not consider deceptive expressions [27] nor did he ponder how the inability to distinguish faces might translate into an individual's life and/or facial expressions.

Like Darwin, the nineteenth century neurologist Jean-Martin Charcot (1825–1893) embraced photography and drawings as a tool in understanding human mental states, seeing psychiatric hospitals as an excellent place to study them. *The Iconographie photographique de la Salpêtrière* (1876–1880) [29], a collection of texts and photographs that represent female patients during his tenure at the hospital and asylum, records patients diagnosed primarily with hysteria or epilepsy and shows a range of facial expressions. Charcot is also among those who appear to have identified a type of face blindness in an otherwise healthy person who had difficulty remembering a close friend by his face [30]. Charcot's 1883 Monsieur X case is sometimes regarded as one of the earliest descriptions of prosopagnosia [31, 32].

How different these expressions on and about faces are from Claude Monet's (1840–1926) poignant and yet clinical portrait of his first wife Camille Doncieux (1847–1879) on her death bed. From 1872 through 1878, Camille was one of his favorite models. He often depicted her in the family garden and other natural spots, frequently with their son Jean. Toward the end of her life, Camille suffered greatly from dyspepsia and other medical complications that followed the birth of her children. In 1877, shortly before the birth of her second son, she had become ill. Surgery was recommended for "uterine ulceration", but she refused the surgical procedure (presumably for Camille to have a hysterectomy) [33]. A contemporaneous Monet letter to the homeopathic physician Dr. Georges de Bellio is virtually the only surviving documentation about Camille's illness. The artist wrote: "I am very frightened because [the doctors] are not hiding the seriousness of [Mme. Monet's] illness from me.... [T]hey are talking about an operation, which horrifies my wife terribly" (p. 928, [33]).

Their second son Michel was born in March 1878, and Camille died at the age of 32, on September 5, 1879. This loss left Monet in a state of immense grief [34]. Decades later, Monet, then an old man, recalled the complex emotions he had experienced while creating the posthumous portrait of his wife [33]. While we tend to associate Monet with *plein-air* (outdoor) landscape painting and impressionistic light, his clinical appraisal of how to reproduce the colors of his beloved's face even as he grieved conveys that our complex emotions and practices frequently operate in tandem:

> I found myself staring at the tragic countenance, automatically trying to identify the sequence, the proportion of light and shade in the colors that death had imposed on [her] immobile face. Shades of blue, yellow, gray…Even before the thought occurred to memorize the face that meant so much to me, my first involuntary reflex was to tremble at the shock of the colors. In spite of myself, my reflexes drew me into the unconscious operation that is but the daily order of my life. Pity me, my friend (p. 13 [33]).

In 1890, face blindness as a condition surfaced again in terms of visual agnosia in Heinrich Lissauer's *Ein Fall von Seelenblindheit nebst einem Beitrag zur Theorie derselben* [35, 36]. Called "mental blindness" at the time, Lissauer's research was published 3 years after Wilbrand's formulation of the Seelenblindheit concept. Lissauer wrote that he found no detailed clinical contributions to the specific question of mental blindness since Wilbrand's work and proposed a distinction between apperceptive and associative agnosia [35, 36]. He used apperceptive agnosia to identify patients whose impairment occurs at the stage where the physical characteristic of the stimulus is processed and integrated in a structured percept. By contrast, he asserted that associative agnosia was more accurate in cases where a fully specified structural description of the stimulus is achieved but fails to produce the associations stored in semantic memory that gives it its meaning [35, 36].

More in the cultural than the pathological sphere was the impetus that led Dr. David Inglis to report on his own face blindness in the 1899 *Transactions of the Michigan State Medical Society for the Year* [37]. This short paper relates that after he had been called upon to give testimony regarding face recognition in a murder trial, he decided that a paper on the variations in natural capabilities would influence public discussions. Using Daltonism to explain that aspects of our biological makeup vary,[1] Inglis noted that he himself had an analogous difficulty in seeing faces. His key point was that the inability to develop face recognition skills happens despite having unimpaired vision and memory. As Dr. Inglis described it:

> I am face-blind. I can by no means recall to my mind the memory of a face. I have no mental picture of the faces of those nearest and dearest to me. I do not see faces in my dreams. Necessarily such a defect is a great drawback for I am continually in difficulties from failure to recognize acquaintances. No training, no voluntary effort has apparently any power to correct the defect. It is again not a faculty untrained but a faculty lacking [37].

Much like color-blindness, people who are face blind are so accustomed to seeing the world this way that they often do not learn it is a "condition" or learn that they have the condition until later in life. Moreover, as Inglis noted, "It is again not a faculty untrained but a faculty lacking" [37]. No doubt this explains why the neurological literature contained few references.

At the beginning of the twentieth century, Picasso's Gertrude Stein portrait presents another fascinating face. Marianne Teuber has argued that the philosopher/psychologist William James' (1842–1910) ideas, discussed in Chap. 7, may have influenced Picasso's development of Cubism through James' student, the writer Gertrude Stein (1874–1946) [38]. Stein studied with James while attending Radcliffe, before moving to Paris and becoming a force in defining modern literature and art [38]. James, who had trained as a physician, urged Stein to attend

---

[1] As discussed in Chap. 4, John Dalton (1766–1844) is best known as an English chemist, physicist, and meteorologist who also did research into color-blindness, sometimes called Daltonism. Dalton's research came about because both he and his brother shared this visual deficiency. While his explanation of the trait was incorrect, Dalton is credited for recognizing that the condition was hereditary.

medical school, and she subsequently attended Johns Hopkins Medical School for 2 years. She went to Paris after losing interest and dropping out of the program [38]. Tauber says that when Picasso painted her now famous portrait there, between 1905 and 1906 and at the end of his so-called "Rose Period," they talked about William James' ideas [39]. According to Teuber's analysis, these conversations influenced Picasso's pictorial approach and the development of the Cubist style [38]. Given Stein's history, perhaps we can also say that Gertrude Stein's face, albeit indirectly, not only influenced the development of modern art, but also had a little known link to medicine for she enjoyed giving medical advice throughout her life [38].

Between 1932 and 1950, about 25 prosopagnosia cases were reported. This includes the work of Joachim Bodamer (1910–1985), a German neurologist, who in 1947 coined the term "prosopagnosia" to describe three patients with difficulty recognizing faces, but with no other visual processing difficulties. These were the result of head injuries. Bodamer proposed that face blindness constituted a relatively specific form of visual agnosia [40]. His rationale was that a patient with prosopagnosia was typically able to distinguish the individual elements of a human face but could not appreciate its unique character nor could he discriminate facial expression. Bodamer's cases included a 24-year-old man who sustained a bullet wound to the head and lost his ability to recognize his friends, family, and even his own face. This patient was, however, able to recognize and identify them through other sensory modalities such as auditory, tactile, and even other visual stimuli patterns (such as gait and other physical mannerisms) [40].

In 1955, Christopher Pallis (1923–2005), an Anglo-Greek neurologist, documented a patient's experience (recorded in a diary) of the onset of prosopagnosia following a stroke [41]. In 1962, when Beyn and Knyazeva presented another case, CH., aged 39, they noted that between 1950 and 1962 there were 15 descriptions of prosopagnosia that had appeared in the literature [42]. Their patient was admitted to their neurology department with complaints of inability to recognize colors or faces, a defect of spatial orientation, right-sided visual field defect, impairment of memory, a feeling of tension in the left half of the body, mild headaches, and increased irritability. Although unable to identify familiar faces, rooms, or buildings, spatial perception as such was not disturbed, and there was no constructional disability. The patient could even distinguish between different types of buildings shown in pictures but was unable to recognize actual buildings or places that had previously been familiar to him. When shown objects, he based his identification on individual attributes or details much in the way that he recognized a particular person by his voice rather than by his face. The authors concluded that the defect of visual recognition could be ascribed to a failure to react to the combination of those specific features or properties of an object which endows it with uniqueness. As in other cases of prosopagnosia, the disorder in this case was restricted to visual recognition [42].

Throughout the twentieth century, as discussed in Chap. 7, film changed our perception of reality, including the way faces and portraits entered our lives. A mid-twentieth century example of this was discussed in that chapter through Ingmar Bergman's (1918–2007) film *Persona* (1966). Bergman used many close-ups to bring the audience into the movie, including doubling the faces and dramatic effects created by drawing on the facial similarities of the two primary characters. As noted during that discussion, the film's most iconic shot shows Elisabet and Nurse Alma facing each other in the middle of the night in front of what appears to be the mirror of a bathroom cabinet. As the two of them discover (or imagine?) an uncanny resemblance, they move as if to kiss one another, and their faces then overlap cinematically. Watching this affective sequence, we wonder who the face is and what it is looking at so intently as it looks at us and we look at it. In effect, this cinematic mapping that obscures each face is creating an effect like associative face-blindness. This is because we can ascertain there is a face but have some measure of difficulty identifying it.

In the 1980s, with the advent of computed tomography (CT) and magnetic resonance imaging (MRI), more information about face processing in the brain became available. For example, in 1982, Antonio R. Damasio, Hanna Damasio, and Gary W. Van Hoesen used critical analyses of postmortem and CT scan data in their analyses of prosopagnosia. They concluded it is associated with bilateral lesions of the central visual system and that prosopagnosia is not specific to human faces [43].

Soon after MRI scans were introduced, they were supplemented with fMRI studies capable of detecting changes associated with blood flow. This technique relies on the fact that cerebral blood flow and neuronal activation are coupled. The difference between the MRI and the fMRI is noteworthy in terms of prosopagnosia. MRIs, which reveal the anatomical structure, show the brains of prosopagnosiacs as normal, although some have a thicker than average fusiform face area. Deviations from a normal brain are more evident with an fMRI, which shows brain activity, or the damage, most noticeably in the fusiform face area [44, 45].

Also noteworthy is that imaging studies of individuals with acquired prosopagnosia have at times backed autopsies and/or altered earlier diagnoses. Oliver Sacks' case of Dr. P, for example, recorded in his 1986 book *The Man Who Mistook His Wife for a Hat* [46], was based on an examination that did not include brain scans of this patient. He updated the original diagnosis of visual agnosias (including prosopagnosia) after the autopsy showed that Mr. P.'s prosopagnosia was a symptom of his Alzheimer's [47].

Another avenue opened by imaging technologies was the ability to study facial perception through portrait painting. For example, in 2000, portrait painter Humphrey Ocean was among the first artists to have his brain scanned while creating art. In this case, the scanning was done while he was drawing a series of faces. Ocean's brain activity was contrasted with a subject who had no training in art. As noted, it is generally said that facial perception is associated with increased blood flow to the fusiform face area. With this expert, however, efforts to depict facial form showed less blood flow as compared with novice painters. In other words, the results appear to show that a novice requires greater cerebral "visual effort," as shown by the increased regional cerebral brain flow in the fusiform face area.

These differences may appear counterintuitive at first but are likely due to the fact that Ocean, an experienced portrait painter, spent hours each day looking at faces and analyzing them, so he was well practiced at face perception. Perhaps it is because Ocean is so skilled in facial perception that what these researchers found is that he is less likely to need to ponder the features and gestalt of faces? Another possibility is that he has developed some level of efficiency in processing faces, leaving him to allocate more cerebral effort to capturing the deeper qualities of the person? In any case, the results showed that Ocean had greater activation in the right frontal area than did the novice painter. Perhaps this suggests that the practiced painter used "higher-order" cognitive processing [30]? Was he "thinking" the face as well as "seeing" it?

This showing that he was not activating the visual face area to the degree of the novice painter resonates with a study discussed in Chap. 2 in regard to the use of Sign language. As pointed out then, the deaf were unable to "propositionize" until Sign gave them language. Once Sign became a complex and thriving natural language, one with an independent syntax and grammar as functional and powerful as any found in the oral languages of the world, it was processed in the left hemisphere of the brain, as are spoken languages. This key discovery revealed that what is critical to the control of language by the left hemisphere is linguistic structure—not auditory processing or vocal articulation—and that Sign had a linguistic structure [48–50]. Ocean, as a portrait painting expert, may have similarly adapted functionally as he developed his expertise in translating faces as an artist.

The Ocean study was more about capturing the form of a face than its emotional qualities or an identity. However, the impetus to create faces or portraits are often more complex than the form itself. Sharon Sprung, who was commissioned to paint Michelle Obama's official portrait, expressed this when she was asked why she painted portraits. After a few silent minutes, as tears welled up in her eyes, she said: "[M]y father had died when I was 6, and all the photographs were destroyed. So I had nothing of him that was left, and remembering his face, and other people's faces, was so embedded in my being, that that's why I'm a portrait painter" [51]. How differently she might have responded if she were among those who have difficulty seeing faces, like Bill Choisser (1947–2016) who is often credited with moving the condition prosopagnosia into the mainstream cultural sphere in the 1990s.

It was in the 1990s that congenital or developmental prosopagnosia was first recognized as a specific neurological deficit present from childhood and often occurring in a number of members of the same family. Choisser, baffled by his inability to recognize people throughout his life, had consulted with several doctors and found them no help at all. They concluded his was an emotional problem and recommended that he see a psychiatrist [52]. After posting a message about his condition to a Usenet group in 1996, Choisser was contacted by many others who shared that they had the same problem [52]. After this, the prosopagnosia population began to meet one another online and share what it is like to live with the inability to recognize faces. As noted, we now know that about 2.5% of the population has some form of congenital prosopagnosia. Estimated prevalence rates range from 0.64% to 5.42% [22].

In 1999, cognitive neuroscientists Brad Duchaine and Ken Nakayama, a prosopagnosiac, likewise benefitted from the reach of the Internet. They used it to seek out individuals with long-term face recognition difficulties and initiated quantitative studies. From this point, research has continued to grow and spread. Since about 2000, researchers have become more aware of congenital and inherited cases. It is also known that congenital prosopagnosia is a lifelong deficit [53–56].

Also in 1999, primate researcher Jane Goodall wrote about her inability to distinguish faces. Her experience encapsulates the experience of many who only discover later in life that they have lived with this condition their entire lives:

> In the course of my travels, one thing detracts from my enjoyment of meeting people. I suffer from an embarrassing, curiously humbling neurological condition called prosopagnosia, which, translated, means I have problems in face recognition. I used to think it was due to some mental laziness, and I tried desperately to memorize the faces of people I met so that, if I saw them the next day, I would recognize them. …Sometimes I knew that people were upset when I did not immediately recognize them—certainly I was. And because I was embarrassed, I kept it to myself. Quite by chance, when talking to a friend recently, I found that he suffered from the same problem. I could not believe it. Then I discovered my own sister, Judy, knew similar embarrassment. Perhaps others did, also. I wrote to the well-known neurologist Dr. Oliver Sacks. Had he ever heard of such an unusual condition? Not only had he heard of it—he suffered from it himself! And his situation was far more extreme than mine (p. 1, [57]).

In 2015, the British National Health Service added prosopagnosia or face blindness to its list of long-term conditions that can be coded onto patient's notes [58]. This recent recognition no doubt explains why some health professionals retain only limited awareness of the number of people impacted by this condition, particularly congenitally and developmentally.

11.2 The Complexity of Face Blindness/Prosopagnosia

As the above demonstrates, faces have been interwoven with human life in a multitude of cross-disciplinary ways, and the inability to recognize faces was for the most part outside of the discussion until the nineteenth century. When visual agnosias and reports of face blindness began to emerge, the assumption was that the condition was largely rare and usually an acquired one.

Since Bodamer coined the term prosopagnosia in 1947, it has been used to characterize a person unable to recognize faces. The word is a combination of the Greek word for face, *prosopon*, and *agnosia*. *Agnosia* itself is a combination of two Greek words. It derives its meaning from *a-* (without) and *gnosis* (knowledge). As this suggests, prosopagnosia is one of a subset of agnosias, by which people lose their knowledge. Whereas there are several types of agnosias based on the type of sensation involved (visual, auditory, tactile, and so forth), the agnosia of interest here is the inability to recognize familiar faces or faces more generally—prosopagnosia or face blindness.

Potential causes of acquired prosopagnosia include Alzheimer's disease, brain tumors, dementia, head injuries, cerebral hypoxia (from lack of oxygen), seizures, epilepsy, stroke, and toxins. Incidence is difficult to establish because most accounts rely on a patient's self-report. In addition, it is unlikely, for example, that a stroke would produce a single cognitive impairment [59]. Many of the people who are labeled with prosopagnosia in this literature have visual field defects and achromatopsia too. Substantial lesions in bilateral posterior regions can lead to shape and object agnosia as well as prosopagnosia [59]. Acquired prosopagnosia is most often found in adults.

As Chap. 4 noted, some, but not all, painters acquire prosopagnosia after a stroke. Analyses of their self-portraits show that independent of the premorbid style of the affected painter, neglect-related signs or style elements can be found in impacted painters [60–64]. For example, Anton Räderscheidt, one painter often discussed in relation to acquired prosopagnosia, had to relearn how to paint after his 1967 stroke. His poststroke portraits are constructed on the basis of what he saw in his right visual field [60, 61, 65, 66].

Otto Dix's (1891–1969) case is more complicated due to the multiple injuries he sustained over the course of his life. Profoundly affected by the brutality of World War I, both physical injuries and recurring nightmares transformed his painting style long before his first stroke in 1967. Many of the early stylistic changes gave visual form to shell shock. Comparing his self-portraits before and after this stroke, we can see that his sense of self changed and that his last self-portraits were substantially different formally. Four days after the neurological event, he was capable of constructing simple forms, such as trees, but showed left visual field neglect. He appeared to compensate for neglect within 2 weeks [66]. The works are no longer frontal, and some suggest this points to a certain degree of prosopagnosia although he is still filling up the entire page [65]. While his earlier work offers an almost photographic rendering of his face, the later work has a caricaturistic feel to it. Hans Kinkel's preface to Florian Karsch's catalogue of Dix's graphic art describes the painter's reaction to the stroke and the series of late self-portraits as an "obsessingly urgent self-inquiry" [67], which comports with the urge to communicate discussed in Chaps. 5 and 10 in reference to the cases of Tommy McHugh (b. 1950) and Jon Sarkin (b. 1953). Also of note is that there is a wide chasm between those who claim he recovered his artistic skills and those who hypothesize his style merely changed dramatically [65, 66, 68].

With congenital or developmental prosopagnosia, the kind Jane Goodall and Oliver Sacks experienced throughout their lives, an individual never adequately develops the ability to recognize faces. In other words, these prosopagnosiacs have never recognized faces normally, and they do not overcome the condition as they age. Because they are born with it, the impairment is often not readily apparent to them and thus goes unreported. As a result, many congenital and developmental prosopagnosiacs are unaware of their prosopagnosia even into adulthood, as Jane Goodall's surprise on learning of her condition indicates. Congenital and developmental prosopagnosia are frequently grouped in the literature.[2] When there is a

[2] Duchaine prefers the term developmental to congenital, although here I use both. His point is a valid one and worth noting. He says: "In the literature [developmental prosopagnosia] is also often called congenital prosopagnosia, but here I will refer to it as developmental prosopagnosia because it is unclear whether face recognition abilities in [developmental prosopagnosias] are atypical at birth" [69].

genetic basis, it is sometimes called pure developmental prosopagnosia. It should be noted, however, that there are known cases where prosopagnosia resulted in infancy or early childhood from closed head injury in childhood [18]. As is often the case in closed head injury, none of the patients had evident lesions on the MRI scan [70].

Studies of the brain using functional MRI and other techniques show that numerous brain areas respond much more strongly to faces than to other visual categories. Although the right hemisphere is more important for face recognition than the left hemisphere, a fairly similar set of areas is present in the left hemisphere. These areas are interconnected, and all become active when faces are viewed [71]. Recent studies of acquired prosopagnosia show that it results from damage to the network, whereas developmental prosopagnosia is caused by atypical development of the network. Given that many areas as well as connections between areas contribute to face processing, there are many ways for the network to malfunction.

Functional locality is worth pondering with regard to prosopagnosia. Despite recent developments showing neuroplasticity and brain adaptation with many conditions to date, it does not apply to congenital and developmental face blindness. This suggests that there are substantial limits on plasticity. Recovery from face blindness is rare, and the role of the fusiform and the inferior occipital gyrus in the network implementing face recognition do not appear to be taken over by functionally or neuroanatomically related areas [18]. In other words, this demonstrates yet again that recognizing faces is complex and that the brain treats faces differently from other objects.

Face complexity also shows up in other ways. For example, we are sensitive to inverted faces [72], but not inverted objects; therefore, identifying an inverted (upside-down) face is much more difficult than doing the same for non-facial objects [70]. Curiously, people with prosopagnosia are better at identifying inverted faces than upright ones, which suggests that upright normally viewed faces are processed by a different cluster of neurons than are inverted faces [30].

With associative prosopagnosia, people cannot recognize the face of someone who should be instantly recognizable, such as a friend or a family member. They do indeed see a face but are unable to associate it with a particular person, as was the case with the *Persona* film discussed earlier. They can still recognize them by other means like the sound of their voice or how they walk. Apperceptive prosopagnosia is a more severe form. People with this variant cannot clearly see the form of the face and therefore find it difficult to distinguish one face from another and are unable to recognize a person's facial expressions or other nonverbal cues.

What is significant here from a cultural perspective is that the prevailing assumption is that the ability to recognize and identify people is an important social skill and one that people normally achieve effortlessly in infancy. Part of this assumption is that during the first year after birth, infants begin to extract a large amount of information from faces and they begin to recognize identities [73]. It is also assumed that we determine how others may be feeling from looking at their faces. But what happens if face recognition is never fully achieved? How do people adapt when they lack effective face recognition skills? Social competence is one issue with impaired face processing because interactions are stifled and inhibited by a confusion over who people are. The result is that people who have prosopagnosia are often withdrawn and may have feelings of inadequacy. For example, Oliver Sacks, the world's most famous neurologist, thought his incredible shyness was an outcome of his prosopagnosia [74].

11.3 Congenital Prosopagnosia and Art

After the success of James Robinson's film on his strabismus, discussed in the last chapter, he expanded his work into a series called "Adapt-Ability" that explores how it feels to live with a disability and the ways in which society can adjust to be more inclusive of people with disabilities [75]. One of these films, *I Have Face Blindness*, features Paul Kram [75]. Kram explains how difficult it is to describe his condition to others. When he tries to tell those he meets that he may not recognize them if they meet again, people are apt to respond, "Oh, I'm bad at names, too," which he finds frustrating.

A severe case of an artist with prosopagnosia is that of Carlotta, introduced in *Lost in Face-Die Welt Mit Carlotta's Augen*, a short film by brain scientist Valentin Riedl and animation artist Frédéric Schuld [45]. Through the film we learn about her condition and the many problems it caused while growing up in Munich, Germany, in the 1960s. She was ridiculed in school by teachers and students alike because everyone looked the same to her. When the recess bell rang and they were supposed to gather with their classmates, she could never identify her class because there wasn't one assigned spot for each class. Waiting for her mother outside of a store, she would often go home with someone who emerged that was not her mother. She only realized this kind of mistake when she repeatedly ended up at strange houses. She was treated as dumb and lazy in school and by her family because of these kinds of behaviors. Yet, after school, she would come home and read everything she could.

Once she left school, Carlotta looked for a job that wouldn't involve identifying people. Although she didn't know there was a name for her problem, she understood that it was a part of who she was. In the years that followed, she worked as a horse groomer, a truck driver, a cement mixer, and a film projectionist. She even bought a boat, left Germany, and went to live off the Australian coast for about a year. After turning 40 and back in Munich, she discovered in a printed article that she likely had a genetic condition called prosopagnosia. Soon thereafter, she read in a book that one could draw one's face by touching it. This was the beginning of her self-portraiture and a quest to understand what she looked like.

Exhibitions of these haptic portraits in Germany followed. One led a neuroscientist, Valentin Riedl, to get in touch with her [45, 76]. He subsequently made the film about her life with prosopagnosia and scanned her brain. In the film, he tells her that the MRI scan showed no damage. According to Dr. Riedl, whatever makes her brain different from others is too small to be visible [45].

Carlotta's haptic self-portraits are quite unlike the portraits painted by the American artist Chuck Close, discussed next, and her life was as well. Whereas she isolated herself and eventually found her way back to the world and her own face through art, Chuck Close, another prosopagnosiac, adapted from a young age, although he, too, was unaware that he was face blind until adulthood. It is also unclear if she had other neurological problems, whereas we know Close had several.

11.4 Chuck Close: Portrait Painting with Prosopagnosia

Chuck Close (1940–2021) is largely remembered for the large-scale photorealist and abstract portraits he crafted of himself and others. Intriguingly, his oeuvre was defined by faces, or heads to use his term, although he was face blind. Being unable to perceive the identity of a face was what first inspired him to do the faces [77]. He could recognize people have faces, but not to whom the face belongs, even if it was a family member. In other words, for most of us, putting faces together is a holistic operation; it is like listening to a symphony. We do not perceive the task step-by-step; rather, we pull it all together without a thought. Chuck Close, however, could not pull the elements of a face into focus three dimensionally due to his prosopagnosia. Therefore, the artist approached his heads analytically, using photographs instead of actual faces to visualize this live, three-dimensional part of human beings. Close said that once he changed a face into a two-dimensional object, he could commit it to memory, and he had a photographic memory for things that are two-dimensional.

His prosopagnosia was a part of his unusual medical profiles. As he put it: "I'm a rolling neurologic clinic, you name it, I've got it" [77]. He also believed he flourished as an artist not in spite of his neurologic conditions but because of them—especially his dyslexia and face blindness. In part, this was due to his family's realization that art nourished him, rather than concluding his troubles indicated he was stupid, as was the case with Carlotta. His own determination to tackle his disabilities was also evident from a young age:

> When I was 5, I asked my father for an easel and he made me an easel. Then I asked both of my parents, who were very poor, to buy me a set of genuine artist oil paints in a wooden box that I saw in a catalogue, which they did. And then when I was 8, long before I was physically disabled, I realized I had a learning disability. In the '40s and '50s nobody knew about dyslexia, they just assumed that you were dumb or lazy, so I had terrible academic problems. I couldn't memorize anything. I still literally cannot add or subtract. I don't know the multiplication tables [78].

Growing up, he used a slow, step-by-step process to achieve his goals and learned how to manage his learning process. He later said that everything in his work was directly related to his learning disabilities, and this step-by-step approach later became the backbone to his head construction, as discussed shortly. It is also important to note that when he says "physically disabled" in the quote above he is referring to a 1988 spinal artery collapse that left him in a wheelchair, also discussed shortly. His framing above somewhat obscures that he also had many physical challenges as a child. One was a neuromuscular condition that made it difficult for Close to lift his feet. A bout with nephritis, an inflammation of the kidneys, kept him out of school for most of sixth grade. In school, his dyslexia, which was not diagnosed at the time, led to poor grades. Along with his undiagnosed prosopagnosia, Close also had vision problems including astigmatism, nearsightedness, and what he calls "some lazy eye stuff."

Due to his muscle weakness, Close was clumsy as a kid and took a long time to learn how to walk. "I was uncoordinated and often had muscle fatigue," Close says. "My legs would lock up and I would fall down." Several doctors thought the vision problems and muscle weakness were symptoms of myasthenia gravis, a neurologic disorder caused by faulty biochemical transmissions between the nerve and muscular systems, but this was never confirmed. It was clear that Close never could use his arms very well when holding them over his head. For example, he would have to switch hands four or five times when changing a light bulb [77].

In school, Close was able to surmount some of his difficulties with reading and language through devising art-related projects to fulfill assignments. For example, when one teacher assigned a paper on the travels of Lewis and Clark, he drew an extensive map of the exploration instead [79]. Eventually, he completed a Master's at Yale and began teaching. While teaching, Close was unable to perceive the faces of his students due to his face blindness, and so he devised a method for keeping track of them and their work. He would do small sketches of each piece they did in class next to their name in his roll book. If they later needed a letter of recommendation, he could recall their work through these recorded sketches [80–82].

It was in the 1970s, in his thirties, that he learned about dyslexia. This came about at the grade school his young daughter attended. After hearing a lecture on the condition, he suspected that he had this learning disability himself. As explained in the last chapter, dyslexia is a learning disorder marked by difficulty reading due to problems identifying speech sounds and learning how they relate to letters and words (decoding) that affects one out of every five men and women. Since there was so little awareness of the condition when Close was young, it is not surprising that his case was missed. As also noted in the last chapter, the cause of dyslexia is not known. As with prosopagnosia, there is no physical evidence of it (e.g., such as damaged tissue) that highlights it in the brain, although fMRI studies do suggest that the brains of people with dyslexia work differently than the brains of other people. Essentially, when people read, three areas in the brain, one in the front and two in the back, are generally activated. In people with dyslexia, there are significant differences in Broca's area and the left inferior temporal region for nonword reading and a phonological transformation task [83].[3]

Close's career unfolded in three distinct stages. The first began in the 1970s when his distinctive heads were admired by both the cognoscenti and the public. At this point, his work was celebrated, and most found his vibrant friendly personality appealing. The imposing presence of his large images, like his over six-foot frame, gave him a larger-than-life persona. Photography was a key element in developing

[3] People with phonological dyslexia have trouble isolating individual phonemes, which leads to difficulty sounding out and pronouncing words.

his personal style at this point. His heads had a photorealistic quality to them, to use a term he hated, and were totally out of step with the art world styles of the time.

As noted, most of us see faces three dimensionally, and it is the sculptural shading that catches the three-dimensional structure of the face. This is what seemed to be missing in Chuck Close's world due to his prosopagnosia; it left him unable to recognize people in the flesh, in three dimensions, but he could copy the detail and capture the three-dimensional quality of faces from a two-dimensional photograph. Therefore, he developed an analytical approach that used photography to translate each face into a flat two-dimensional image that would help him to figure out the three-dimensional aspect of faces even if it did not significantly enhance his in-the-world perception of them [77].

This translation approach is not at all like the kind of translation discussed in the last chapter in terms of strabismus. Looking at the two side by side reinforces that brain functioning is complex. As noted in Chap. 10, strabismus is characterized as the misalignment of one eye in relation to the other, resulting in the failure of the two eyes to simultaneously focus on the same image and this results in the loss of binocular vision [84]. In other words, most of us see binocularly because our eyes fuse the images from each eye in the occipital cortex of the brain. With strabismus, this fusion does not work properly, and as a result, there is limited stereoscopic depth perception, and people with this condition usually have difficulty seeing three-dimensional images.

By contrast, Chuck Close and others with prosopagnosia have a difficulty with face recognition. As noted earlier in this chapter, face recognition is now considered a highly complex cognitive process involving a dedicated network of brain regions and is largely associated with the fusiform gyrus and the less studied inferior occipital gyrus [18]. Obviously, when one is face blind, this network is not working typically. So, while both conditions involve visual perception, the functional problems are dissimilar. It is also important to note that even though this book has introduced these conditions with a focus on artists, they are distributed throughout the general population and not confined to artists.

Returning to Chuck Close's biography and his prosopagnosia, whereas his heads helped him see faces to some degree, they did not cure his condition, especially when seeing people in an unexpected context. This procedure also explains why many early pieces (e.g., Nancy, 1968, or Phil, 1969) so fully resemble black-and-white photographs. As a result of their striking resemblance to photography, some critics were initially critical in their reaction. Hilton Kramer of *The New York Times* and *The New Criterion* called him a lunatic and characterized his work as trash. John Canaday, *The New York Times* critic before Kramer, lambasted this early work as well, only to tell Close in the mid-1980s, just before this critic died, that he was wrong [78]. The esteemed critic Clement Greenberg was also unkind. He commented that it was no longer possible in the twentieth century to even make a portrait [79].

The second phase of Close's career began in 1988, after a seizure left him paralyzed from the neck down. The cause was diagnosed as a spinal artery collapse. He called it "The Event," and the event led to several months of his undergoing rehabilitation therapy. Although the cause was unknown, its seriousness was never in question. Physical therapy helped in strengthening his muscles and he soon had slight movement in his arms. He could walk but only for a few steps. He relied on a wheelchair thereafter, although he did retain the use of his limbs to some degree [77].

After The Event, Close's commitment to continuing his art never wavered. "In the hospital, when the painter had movement only in his head and neck, he told a friend he'd work with a brush clenched in his teeth—even spit paint on the canvas if he had to" [77]. As he began to work again, he felt his near paralysis didn't affect his art all that much because he had begun to loosen his style before The Event. After it, he needed to change his materials to some degree, but his materials had changed throughout his career. Moreover, he was able to continue to use the grid pattern, a part of his standard process. When working on a portrait painting, as noted, he would first take a photograph. Then he would pencil a grid onto a blank canvas. Working from the top left corner down to the bottom right, he would fill in each individual square with the appropriate color or colors. Using this methodical process day by day would eventually result in the large-scale head.

Close ultimately regained the use of his arms and was able to paint with a brush strapped to his hand and forearm [85]. He replaced his elevator chair with a motorized easel that allowed him to use foot pedals to raise, lower, and rotate his canvas. This setup put the whole painting within reach. Still using photographs to create his heads, changes in his work were frequently described as adding an exciting element not evident before. The inability to precisely render was accompanied by larger grids, with luscious dabs of bright color making each grid an almost abstract painting with the result having a pixelated look that nonetheless retained the concept of the photographic image when viewed at a distance.

According to Close, art saved him twice. The first time was when his father died. At that time, he was 11. "It was the absolute worst thing that's ever happened to me in my life," he says of that first loss, "but there was an odd gift in it as well." Experiencing overwhelming grief at such a young age made it easier when he became quadriplegic, he says, as he knew he could get through it [86]. This was the second time art saved him. At this point, Close was 48. Assistants and technologies enabled him to accommodate his diminished motor control, and he continued to produce artworks that matched his signature style. His moveable easel and technological accommodations for the loss of motor control bring to mind how Renoir adapted to his rheumatoid arthritis, discussed in Chap. 7, although the conditions and the details surrounding the two cases differ significantly.

The third stage of Chuck Close's career is generally dated from about 2016–2017 to his death in 2021. Several women accused him of sexually inappropriate verbal behavior when he was alone with them in his studio between 2005 and 2013. Although none accused him of physical assault and Close denied the accusations, two museums canceled exhibitions and others removed his work from display. A few years before his death, he apologized, saying: "I never reduced anyone to tears, no one ever ran out of the place. If I embarrassed anyone or made them feel uncomfortable, I am truly sorry, I didn't mean to. I acknowledge having a dirty mouth, but we're all adults" [87]. During this third period, he divorced his wife, Leslie, in 2017, after 43 years of marriage. A second marriage also ended in divorce. He changed his living style, and his behavior became erratic. At this point, he was older, so it was a time of life when bodies degenerate unevenly, and Close was no exception; he also began to look run down.

Medically, after 30 years of muscular atrophy, he had only the slightest motor control at this time. As an artist, the changes resulted in a conversation with himself through self-portraiture [79]. On the medical side, Close had received a diagnosis of Alzheimer's disease in December 2013. This was amended to frontotemporal dementia in 2015. Frontotemporal dementia is caused by a group of disorders that gradually damage the brain's frontal and temporal lobes. Close died at 81, on August 19, 2021, due to cardiopulmonary failure [88]. His neurologist suggested that his degenerating health at the end of his life could have possibly influenced his behavior toward the women who accused him of sexual mistreatment. "He was very disinhibited and did inappropriate things, which were part of his underlying medical condition … Frontotemporal dementia affects executive function. It's like a patient having a lobotomy—it destroys that part of the brain that governs behavior and inhibits base instincts" [88]. Because the first reported instance was in 2003, others suspect that even if the biological factors were to blame, it is likely the situations are more complex than a neurodegeneration assumption allows.

The jury is still out on his legacy. Many, including women, have argued that it is healthier for us to see his works [85] and it is wrong to create a litmus test for what goes up on the wall, particularly since he was never charged for anything. Another view is that some male artists with similar behavior are shown with wall text noting the controversy.

11.5 David Hockney: Portrait Painting Without Prosopagnosia

Like Chuck Close, British-American artist David Hockney (b. 1937) is known for his portraits, although Hockney's oeuvre is significantly more diverse. One fascinating aspects of Hockney's portraiture is the series of self-portraits he has created since he was a teenager. Still, he has also reached out in many directions, experimenting with technologies like photography, iPads, and video. In addition, he frequently returned to traditional genres like landscapes and still lifes. Unlike Chuck Close who was unable to identify faces, Hockney does portraits because, to him, "Faces are the most interesting things we see; other people fascinate me, and the most interesting aspect of other people—the point where we go inside them—is the face. It tells all" (p. 5, [89]).

Although both Hockney and Close paint portraits, their background are quite different. Initially a good student, in school Hockney found it was better to be a mediocre student because the school curriculum didn't include many art courses for the better students. In other words, problems like dyslexia and prosopagnosia were not creating side challenges. That said, he is neuroatypical in his own ways. A gay synesthete, he began to lose his hearing at a young age. Indeed, Hockney began using hearing aids in 1979, at the age of 42, although he was aware he was going deaf before this. In the 1980s, he famously wore one red and one blue hearing aid [90]. It was also at this point that he began to explore the works of other deaf artists, for example, Francisco de Goya (1746–1828) [91]. Hockney claimed in his biography that as he became deaf he tended to retreat more into himself [92] and saw it as an opportunity to reinvent himself [92]. In 1985, after testing a new hearing aid, he wrote to the painter R. B. Kitaj (1932–2007) that "Music is more alive again and sound seems spatial, and made me think that over the last years to compensate for my muffled ears I developed a strong visual space sense" (p. 232, [89]). Later in life, he frequently claimed that his hearing loss had sharpened his other senses.

A mild heart attack, in 1990, left Hockney hospitalized for a week. Upon release his work continued apace [89]. By 2004, Hockney was "deaf as a post" [93] and continued to work creatively and incessantly, depending heavily on two hearing aids, and he still needed to lean in for conversations. Additional health problems emerged in 2012:

> I had had a very minor stroke that had kept me in London, and the first drawing afterwards took me two days to do.... The stroke only manifested itself in my speech. I found I couldn't finish sentences and although it came back after about a month I find now I talk a lot less (p. 472, [93]).

Even though the stroke had affected his speech initially, he stopped worrying when he realized he was still able to paint. To his mind, if he could draw and paint then he was fine; he could speak and communicate in another way. Hockney's recovery included a project focused on painting faces of people close to him [94]. In part, these portraits were a part of his recovery from an overall bad year [95]. Aside from the stroke, one of his studio assistants in East Yorkshire, part of a tight-knit circle of aides, collaborators, and former lovers, died in Mr. Hockney's home after drinking drain cleaner on the heels of a drug binge.

All in all, Hockney believed that the hearing loss affected him more deeply than the stroke [95]. In an interview with curator Edith Devaney while preparing for the "A Bigger Picture" exhibition, he told her he had become exhausted while preparing for the show, with the mental and physical exhaustion intensified by the tragic death of his assistant Dominic Elliott and his stroke. He didn't work for a bit, which is unusual for him. After the stroke, he also focused on charcoal works for a time. Then, as he put it:

> I came into the studio and eventually I did start painting, and I saw J-P (Jean-Pierre Gonçalves de Lima) with his head in his hands, and I said, 'Well. we both feel like that, don't we: I'll paint it.' So, I painted him and then I remembered Van Gogh's Sorrowing Old Man [94].

This is a work Van Gogh had painted while he was convalescing from a severe relapse in his health some two months before his death [96]. Devaney believed the poignant portrait of J-P was really a Hockney self-portrait. As it turned out, and to bring this back to what Hockney said about faces being the most interesting thing about people, this portrait was the start of the "81 Portraits and a Still-Life Series" exhibited in 2018 [94]. The still life came about when a sitter had to cancel at the last moment. As Devaney noted, after the J-P portrait, Hockney began an exploration of the individual, and this lifted a weight that he had been carrying around.

Close to his 18th birthday, Hockney was asked if he remained as prolific as always. He replied, "Yes, really. I mean, I like painting, I like making pictures, I now see there's new ways to make pictures. And I find that very exciting" (p. 498, [93]).

11.6 Conclusion

Neurobiologist Semir Zeki once commented that "artists are in some sense neurologists, studying the brain with techniques that are unique to them, but studying unknowingly the brain and its organization nevertheless" (p. 80, [97]). The distinctive approaches of the painters discussed in this chapter underscore that each adapted to their own biological condition in ways that were unique to their circumstances. As such, as each embraced living and communicating, the outcomes reflected their own aspirations and biological framework. Artistic works like those presented here add visual information to the quantitative research on faces and face blindness. As the artists often convey a range of emotions in faces of others as well as self-portraits, those with and without prosopagnosia have also given us insight into who they are, how they see, and the exchange between neuroscience and art in neurocultural terms.

References

1. Sage A. Cave face 'the oldest portrait on record'. June 5, 2006. https://www.thetimes.co.uk/. Accessed 5 Sept 2023.
2. Hughes JD. Responses to natural disasters in the Greek and Roman world. In: Pfeifer, K., Pfeifer, N. (eds) Forces of nature and cultural responses. Dordrecht: Springer; 2013. doi: https://doi.org/10.1007/978-94-007-5000-5_7. Accessed 15 May 2023.
3. Walker S. Ancient faces: mummy portraits in Roman Egypt. London: Taylor & Francis; 2020.
4. Roman Period. Portrait of a youth with a surgical cut in one eye, 190–210 CE. New York: Metropolitan Museum of Art.
5. Taylor HR. Trachoma: a blinding scourge from the Bronze Age to the twenty-first century. East Melbourne: Centre for Eye Research Australia; 2008.
6. Borkowski JE. Mona Lisa: the enigma of the smile. J Forensic Sci. 1992;37(6):1706–11.
7. Chakravarty A. Mona Lisa's smile: a hypothesis based on a new principle of art neuroscience. Med Hypotheses. 2010;75(1):69–72.
8. Kontsevich LL, Tyler CW. What makes Mona Lisa smile? Vision Res. 2004;44(13):1493–8.
9. Livingstone MS. Is it warm? Is it real? Or just low spatial frequency? Science. 2000;290(5495):1299.
10. Marsili L, Ricciardi L, Bologna M. Unraveling the asymmetry of Mona Lisa smile. Cortex. 2019;120:607–10.
11. Santos-Bueso E, Vico-Ruiz E, García-Sánchez J. Eye pathology in the paintings by Leonardo da Vinci (I). The Mona Lisa. Arch Soc Esp Oftalmol. 2014;89(2):e12–3.
12. Marmor MF. Leonardo da Vinci probably did not have strabismus. JAMA Ophthalmol. 2019;137(11):1331–2.

13. Tyler CW. Evidence that Leonardo da Vinci had strabismus. JAMA Ophthalmol. 2019;137(1):82–6.
14. Tyler CW. Leonardo da Vinci probably did not have strabismus-reply. JAMA Ophthalmol. 2019;137(11):1332–3.
15. Gall FJ, Spurzheim J. Anatomie et Physiologie du Système Nerveux en Général, et du Cerveau en Particulier (4 volumes with an accompanying Atlas). Paris: Schoell; 1810–19.
16. Sacks O. A Neurologist's notebook: face-blind: why are some of us terrible at recognizing faces? The New Yorker. August 30, 2010:36–43.
17. Pearce JMS. Hughlings Jackson's "imperception" and anosognosia. J Neurol Neurosurg Psychiatry. 2002;72(6):736.
18. Hadjikhani N, De Gelder B. Neural basis of prosopagnosia: an fMRI study. Hum Brain Mapping. 2002;16(3):176–82.
19. Chen C-C, Chen CM, Tyler CW. Depth structure from asymmetric shading supports face discrimination. PLoS One. 2013;8(2):e55865.
20. Chen C-C, Kao K-LC, Tyler CW. Face configuration processing in the human brain: the role of symmetry. Cereb Cortex. 2007;17(6):1423–32.
21. Wigan AL. The duality of the mind. London: Longman, Brown, Green, and Longmans; 1844.
22. DeGutis J, Bahierathan K, Barahona K, Lee E, Evans TC, Shin HM, et al. What is the prevalence of developmental prosopagnosia? An empirical assessment of different diagnostic cutoffs. Cortex. 2023;161:51–64.
23. Rossion B. Damasio's error—prosopagnosia with intact within-category object recognition. J Neuropsychol. 2018;12(3):357–88.
24. Code C, Joanette Y, Lecours AR, Wallesch C-W. Classic cases in neuropsychology. Taylor & Francis; 2004.
25. Jackson JH. Case of disease of the brain—left hemiplegia—mental affection. Medical Times Gazette. 1872;1:513.
26. Darwin C, Ekman P, Prodger P. The expression of the emotions in man and animals. London: Harper Collins. 1872/1998
27. Ekman P. Darwin's contributions to our understanding of emotional expressions. Philos Trans R Soc Lond B Biol Sci. 2009;364(1535):3449–51.
28. Prodger P. Darwin's camera: art and photography in the theory of evolution. Oxford and New York: Oxford University Press; 2009.
29. Charcot J-M, Bourneville DM, Régnard P. Iconographie photographique de la Salpêtrière (3 volumes): Progrès médical; 1876–1880.
30. Solso RL. About faces in art and in the brain. Cerebrum. Dana Foundation; 2004. https://dana.org. Accessed 25 Sept 2022.
31. Zago S, Allegri N, Cristoffanini M, Ferrucci R, Porta M, Priori A. Is the Charcot and Bernard case (1883) of loss of visual imagery really based on neurological impairment? Cogn Neuropsychiatry. 2011;16(6):481–504.
32. Charcot J-M, Bernard DAF. Un cas de suppression brusque et isolée de la vision mentale des signes et des objets (formes et couleurs). Progrès Médicale. 1883;11:568–71.
33. Gedo MM. Mme Monet on her deathbed. JAMA. 2002;288(8):928.
34. Harris JC. Camille on her deathbed. Arch Gen Psychiatry. 2003;60(1):13.
35. Lissauer H. Ein Fall von Seelenblindheit nebst einem Beitrage zur Theorie derselben. Arch Psychiatrie Nervenkrankheiten. 1890;21(2):222–70.
36. Lissauer H, Jackson M. A case of visual agnosia with a contribution to theory. Cogn Neuropsychol. 1988;5(2):157–92.
37. Inglis D. Moral imbecility. In: Society MSM, editor. Transactions of the Michigan State Medical Society for the year 1899, vol. XXIII. Grand Rapids: The Society; 1899. p. 377–87.
38. Teuber M. Gertrude Stein, William James, and Pablo Picasso's Cubism. In: Bringmann WG, Lück HE, Miller R, Early CE, editors. A pictorial history of psychology. Chicago: Quintessence Publishing Co.; 1997. p. 256–64.
39. Giroud V. Picasso and Gertrude Stein. New York and New Haven: Yale University Press; 2007.
40. Bodamer J. Die prosopagnosie. Arch Psychiatrie Nervenkrankheiten. 1947;179:6–53.

41. Pallis CA. Impaired identification of faces and places with agnosia for colours. Report of a case due to cerebral embolism. J Neurol Neurosurg Psychiatry. 1955;18:218–24.
42. Beyn ES, Knyazeva GR. The problem of prosopagnosia. J Neurol Neurosurg Psychiatry. 1962;25(2):154–8.
43. Damasio AR, Damasio H, Hoesen GWV. Prosopagnosia. Anatomic basis and behavioral mechanisms. Neurology. 1982;32(4):331.
44. Dingfelder S. My life with face blindness. The Washington Post Magazine. August 21, 2019.
45. Riedl V. Lost in face-die welt Mit Carlotta's Augen. Corso Film; 2021.
46. Sacks O. The man who mistook his wife for a hat and other clinical tales. New York: Perennial Library; 1987.
47. Rawlence C, Morris M, Nyman M. Michael Nyman: The man who mistook his wife for a hat (1987 TV Opera); 1987.
48. Stokoe WC. Sign and culture: a reader for students of American Sign Language. Silver Spring: Linstok Press; 1980.
49. Stokoe WC. Language in hand: why sign came before speech. Washington, DC: Gallaudet University Press; 2002.
50. Stokoe WC, Armstrong DF, Karchmer MA, Van Cleve JV. The study of signed languages: essays in honor of William C. Stokoe. Washington, DC: Gallaudet University Press; 2002.
51. Heinrich W. Portrait artist didn't expect to keep secret for so long. The New York Times. September 8, 2022; Sect. A.
52. Davis J. Face blind. Wired. November 1, 2006.
53. Duchaine B, Nakayama K. Dissociations of face and object recognition in developmental prosopagnosia. J Cogn Neurosci. 2005;17(2):249–61.
54. Duchaine BC, Nakayama K. Developmental prosopagnosia: a window to content-specific face processing. Curr Opin Neurobiol. 2006;16(2):166–73.
55. Wilmer JB, Germine L, Chabris CF, Chatterjee G, Williams M, Loken E, et al. Human face recognition ability is specific and highly heritable. Proc Natl Acad Sci. 2010;107(11):5238–41.
56. Kennerknecht I, Grueter T, Welling B, Wentzek S, Horst J, Edwards S, Grueter M. First report of prevalence of non-syndromic hereditary prosopagnosia (HPA). Am J Med Genet Part A. 2006;140A(15):1617–22.
57. Goodall J, Berman PL. Reason for hope: a spiritual journey. New York: Warner Books; 1999.
58. Face Blind UK. Face Blind UK: living with prosopagnosia: https://www.faceblind.org.uk/. Accessed 29 Dec 2022.
59. Cousins R. Prosopagnosia after stroke: potentials for impairment and treatment. Top Stroke Rehabil. 2013;20(6):471–7.
60. Bäzner H, Hennerici M. Stroke in painters. Int Rev Neurobiol. 2006;74:165–91. San Diego: Academic Press
61. Chatterjee A. The neuropsychology of visual artistic production. Neuropsychologia. 2004;42(11):1568–83.
62. Blanke O. Visuospatial neglect in Lovis Corinth's self-portraits. Int Rev Neurobiol. 2006;74:193–214. San Diego: Academic Press
63. Blanke O. I and me: self-portraiture in brain damage. Basel: Karger; 2007. p. 14–29.
64. Blanke O, Ortigue S, Landis T. Colour neglect in an artist. Lancet. 2003;361(9353):264.
65. Bäzner H, Hennerici MG. Painting after right-hemisphere stroke—case studies of professional artists. Front Neurol Neurosci. 2007;22:1–13.
66. Petcu EB, Sherwood K, Popa-Wagner A, Buga AM, Aceti L, Miroiu RI. Artistic skills recovery and compensation in visual artists after stroke. Front Neurol. 2016;7:76.
67. Dix O, Karsch F, Kinkel H. Otto Dix: das graphische Werk: Fackelträger-Verlag; 1970.
68. Jung R. Neuropsychologie und neurophysiologie des Kontur- und Formsehens in Zeichnung und Malerei. In: Wieck HH, editor. Psychopathologie musischer Gestaltungen. Stuttgart and New-York: Schattauer; 1974. p. 29–88.
69. Duchaine B. Developmental prosopagnosia: cognitive, neural, and developmental investigations. In: Calder AJ, Rhodes G, Johnson MH, Haxby JV, editors. Oxford handbook of face perception. Oxford: Oxford University Press; 2011. p. 821–38.

70. Civile C, McLaren RP, McLaren IPL. The face inversion effect—parts and wholes: individual features and their configuration. Quart J Exp Psychol. 2014;67(4):728–46.
71. Duchaine B, Yovel G. A revised neural framework for face processing. Annu Rev Vis Sci. 2015;1:393–416.
72. Thompson PG. Margaret Thatcher: a new illusion. Perception. 1980;9(4):483–4.
73. Frank MC, Vul E, Johnson SP. Development of infants' attention to faces during the first year. Cognition. 2009;110(2):160–70.
74. Sacks O. On the move: a life. New York: Alfred A. Knopf; 2015.
75. Robinson J. Adapt-ability: film can help us look disability in the eye. New York Times; 2021.
76. Venema V. Prosopagnosia: the artist in search of her face. BBC News. 2020.
77. Farley T. Disabilities are at the heart of Chuck Close's art. Brain Life; 2011. https://www.brainandlife.org/. Accessed 18 Sept 2022.
78. Bui P, Close C. Art in conversation: Chuck Close with Phong Bui. Brooklyn: The Brooklyn Rail; 2008. https://brooklynrail.org. Accessed 18 Sept 2022.
79. Hylton WS. The mysterious metamorphosis of Chuck Close. The New York Times, July 13, 2016.
80. Andersen K. Portraits of perception with Chuck Close. New York: World Science Festival; 2010.
81. Close C, Sacks O. Oliver Sacks and Chuck Close talk about their own brain disorders. New York: Big Think; 2010.
82. Society for Neuroscience. Chuck Close: my life as a rolling neurological clinic. In: Dialogues between neuroscience and society. New Orleans: Society for Neuroscience; 2012.
83. Georgiewa P, Rzanny R, Hopf JM, Knab R, Glauche V, Kaiser WA, Blanz B. fMRI during word processing in dyslexic and normal reading children. Neuroreport. 1999;10(16):3459–65.
84. Engle EC. Genetic basis of congenital strabismus. Arch Ophthalmol. 2007;125(2):189.
85. Smith R. Chuck Close's uneasy legacy. New York: New York Times; August 21, 2021; Sect. C.
86. Hawkins K. How painting saved a man with paralysis. BBC News; July 8, 2015. https://www.bbc.com/news. Accessed 30 May 2023.
87. Freeman N. Chuck Close, acclaimed artist with a complicated legacy, Dies at 81. Vanity Fair; 2021.
88. Johnson K, Pogrebin R. Chuck Close, a painter of outsize reality, is dead at 81. New York; New York Times; 2021 August 20, 2021; Sect. A.
89. Hockney D. David Hockney photographs. London and New York: Petersburg Press; 1982.
90. Hearwear MC. The future of hearing [Art]. Br Med J. 2005;331(7514):459.
91. Sonnenstrahl DM. Deaf artists in America: Colonial to contemporary. San Diego: Dawn Sign Press; 2002.
92. Hockney D, Stangos N. That's the way I see it. London: Thames and Hudson; 1993.
93. Hockney D, Holzwarth HW. David Hockney: a chronology: Taschen; 2020.
94. Marlow T, Barringer TJ. David Hockney: 82 portraits and 1 still-life. London: Royal Academy of Arts; 2016.
95. Finkel J. He's back, in a Defiant Blaze of Color. New York: The New York Times. October 27, 2013; Sect. AR.
96. Hulsker J. The complete Van Gogh: paintings, drawings, sketches. New York: H.N. Abrams; 1980.
97. Zeki S. Art and the brain. Daedalus. 1998;127(2):71.

Synesthesia and the Senses

12

Many years ago, while giving a few talks on synesthesia in London, I stopped for tea with a colleague's mother, Active, and a musician in her nineties; she asked what I was speaking about while in the city. When I said synesthesia, she responded "What's that?". I explained that it is a perceptual phenomenon in which stimulation of one sensory or cognitive pathway leads to involuntary experiences in a second sensory or cognitive pathway. To my surprise, she then said, "I didn't know there was a name for it" and went on to explain that she had experienced color-music synesthesia her entire life. People like her who report a lifelong history of such experiences are known as synesthetes. Not all synesthetes experience color-music—a color sensation in response to music—for there are a range of types, as this chapter delineates.

For example, Richard Cytowic related a quite different kind of case in his book *The Man Who Tasted Shapes* [1]. While a young medical resident, Richard Cytowic was intrigued by a host's comment at a dinner party. The host, Michael Watson, while checking on the meal, reported that the chicken needed to cook longer because it didn't have enough points. He then described that for him tactile sensations always accompanied the taste of food. "I feel it—its weight, its texture, whether it's warm or cold, everything. I feel it like I'm actually grasping something" (p. 4 [1]).

Watson, like my colleague's mother, had never heard of synesthesia. Like her, when told about the condition, he responded: "You mean there's a name for it?" He then added, "people have always said I'm just crazy" (p. 5, [1]), which is a common response when individuals casually report their experiences to friends and family, particularly at a young age. Equally noteworthy, when Cytowic's fellow medical residents heard about Watson's condition, they suspected a brain injury or malformation. Cytowic was advised to avoid the topic; he was told it would ruin his career.

Cytowic decided to pursue the topic nonetheless and has since become an expert in the field. Part of the reason is that he met Watson in 1979 shortly after reading

A. Ione, *Neuroscience and Art*, Neurocultural Health and Wellbeing,
https://doi.org/10.1007/978-3-031-62336-3_12

Russian neuropsychologist A. R. Luria's *The Mind of a Mnemonist: A Little Book about a Vast Memory* (1968) [2].[1] At that time, B. F. Skinner's behaviorism dominated psychology, and Luria's book introduced a radically different kind of approach to neuropsychology. In the book, Solomon Shereshevsky, called "S," is the mnemonist. He was a Russian journalist with a seemingly unlimited memory that functioned in tandem with his fivefold synesthesia. As Luria explained, when a bell was rung, Shereshevsky not only heard it but also tasted saltwater; saw "a small, round object" rolling in his field of vision; and felt "something rough like a rope" with his fingers (in [4]). "What a crumbly, yellow voice you have," he once told L. S. Vygotsky" (p.24, [2]).

Both Luria and Cytowic have suggested that synesthesia has a strong relationship with memory, or hyperthymesia, for synesthetic percepts are vividly remembered fields [2, 4, 5]. The reason the behaviorist paradigm of that era had no room for this condition is that reports were considered subjective and unverifiable in a lab. When Cytowic met Watson, the literature on the topic was mostly limited to Luria's book, some studies by Lawrence Marks [6, 7] and a small body of work that dated back a century or more and was largely unknown.

All of these examples suggest the range and challenges of synesthesia. Moreover, as researchers continue to probe synesthesia's parameters, it has become less clear where the boundaries are [8–17]. This complex landscape was captured by Bosley and Eagleman in reporting on twin studies in 2015, where their data suggested that synesthesia is a heritable condition with incomplete penetrance and therefore is substantially influenced by epigenetic and environmental factors [10].

In an interview about 30 years after Cytowic began his research, he, too, noted that despite the recent explosion of research, we still understand synesthesia incompletely:

> For decades, one of the ways the establishment has denied synesthesia's reality has been by insisting that individuals were "just remembering" childhood associations from coloring books or refrigerator magnets. Much effort was spent refuting this… I do not claim that memory isn't involved in synesthesia—of course it is, just as an elevated memory is one of the skills that the trait often bestows. My dismay has to do with the "just." It is too easy to forget the enormous nuance inherent in the large, heterogeneous field.… An individual inherits (or has a spontaneous mutation for) a biological propensity to hyper-connect brain neurons, but then during the long period of human development must be exposed to cultural artifacts such as calendars, food names, and alphabets. How do the two interact? (pp. 242–3, [18]).

The complexity of synesthesia makes it a perfect topic for closing this book since the trait includes quantitative research; studies focused on imprinting, epigenetic, and environmental factors; a range of people who have offered anecdotal reports; artistic projects that communicate the experience; and both clinical and nonclinical populations. Some recent findings overlap with historical investigations. Others

[1] *The Mind of a Mnemonist: A Little Book about a Vast Memory* (1968) [2] also had a major impact on Oliver Sacks' commitment to the value of case studies. Sacks has said the book read so much like a novel that at first he did not realize it was a case study [3].

revisit long-standing debates. Cases of lost or acquired synesthesia add extensive information to the synesthesia equation. Furthermore, metaphorical and cultural examples span the ages.

12.1 Synesthesia's Definitional Elasticity

Defining the many forms of synesthesia is challenging due to the variations in names used to characterize multisensory experience. In film and cinema studies, for example, the multisensory aspect is frequently referred to as *cinesthesia*. Historically, terms like color audition, pseudochromesthesia, psychochromesthesia, and photisms were among those used for what we now call synesthesia. The data complicates the puzzle more. Recent quantitative studies have confirmed that individual synesthetes offer consistent reports over time. Yet, and despite their reliably even when the individual cases are studied longitudinally, the trait is nonetheless idiosyncratic overall. In other words, it is more likely than not that any individual's synesthetic experiences will differ from those of other synesthetes. In addition, acquired synesthesia can often be related to an identifiable physiological event, but this is not always the case. Similarly, the loss of synesthesia may or may not accompany a known event.

As noted, synesthesia is generally defined as a phenomenon in which the stimulation of one sensory or cognitive pathway leads to involuntary experiences in a second sensory area. In this respect, it offers an example of a deviation from the rubric that we have five distinct senses. Detailed studies, however, have shown that synesthesia is much broader than this generic definition implies. For example, synesthesia happens in two ways: The first is "sensorial synesthesia," where one sense, such as smell, is involuntarily and simultaneously also perceived by one or more additional senses, such as sight and/or hearing. This type of synesthesia speaks to the better-known definition of the word. It also conforms best with the etymology of the word itself, which comes from the Greek *syn* (together) and *aisthēsis* (sensation or perception) [19].

A second form of synesthesia is called "ordinal sequence" or cognitive category synesthesia. While this form includes sensations, it also encompasses ways of knowing that are put together or serialized:

> [With "ordinal sequence" synesthesia there are] certain sets of things which individual cultures teach us to put together and categorize (and also usually serialize) in some specific way—such as letters, numbers, or people's names. [These] also get some kind of sensory addition, such as a smell, color, or favor. The most common forms of ordinal sequence synesthesia involve colored written letter characters (graphemes), numbers, time units, and musical notes or keys. For example, the synesthete might see about a foot or two before her, different colors for different spoken vowel and consonant sounds, or perceive numbers and letters, whether conceptualized or before her in print, as colored; e.g., the letter "a" as pink, "b" as blue, and "c" as green, no matter what color of ink they are printed with (p. 17, [12]).

The ordinal sequence cohort is particularly important because the majority of synesthesias, about 88% [13, 21], include grapheme-color. Grapheme-color synesthesia is a form of synesthesia in which an individual's perception of numerals, words, and letters is associated with the experience of colors (Fig. 12.1). Another form in this category is spatial sequence synesthesia (SSS), also known as sequence-space synesthesia and number-form synesthesia. As Figure 12.2 shows, this happens when abstract ordinal sequences are experienced as specific geometric forms in space [22–26]. These types of synesthesia could not have existed in infancy since they require learned knowledge (although they may have been present in other forms, as Duffy demonstrates in *Blue Cats and Chartreuse Kittens* [20]).

Another idiosyncratic element of the synesthetic experience is that some synesthetes are projectors and other associators. Projectors see colors, forms, or shapes

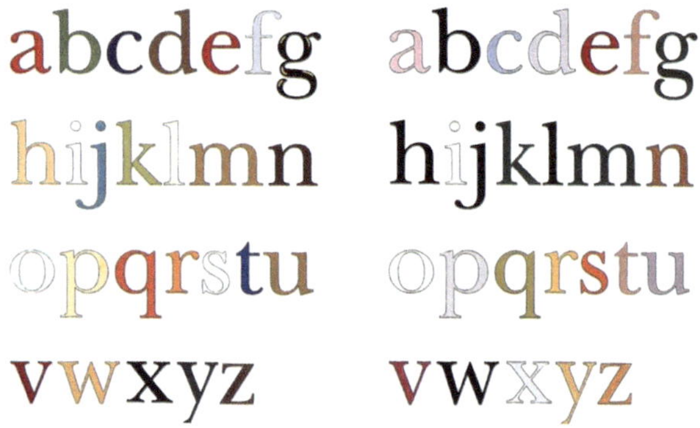

Fig. 12.1 Two synesthetic colored alphabets. Patricia Lynne Duffy's alphabet is on the left and Carol Steen's alphabet is on the right. Published in *Blue Cats and Chartreuse Kittens* by Patricia Lynne Duffy [20]. Painted by Carol Steen. Courtesy of Patricia Lynne Duffy and Carol Steen

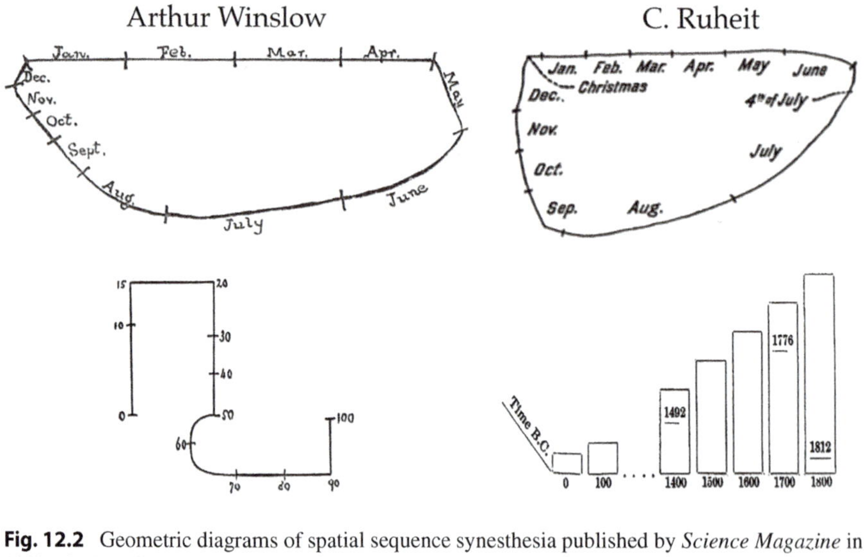

Fig. 12.2 Geometric diagrams of spatial sequence synesthesia published by *Science Magazine* in 1883. C. Ruheit (right) shows how he perceives the months and important dates [27]. Arthur Winslow's conception of the month and his number diagram are shown on the left [28]

when stimulated. This is the widely understood version of synesthesia. Those who are associators feel a very strong and involuntary connection between the stimulus and the sense that it triggers [16]. Furthermore, it has been proposed that all human infants start out as "synesthetes" [29] and the senses then differentiate as a part of a gradual pruning of connections that begins at about four to six months – and can last well into the teens.

As is often the case when discussing synesthetic parameters, some have raised questions about presenting the condition in terms of pruning. This is because synesthesia is generally thought of as a physiological "combining of the senses." The objection is that pruning suggests an actual "coming together" of sensory percepts as if the senses were separated initially. Yet, the brain has not yet differentiated at this stage of an infant's development. "Combining" does not capture that at this point the brain does not respond separately to visual, tactile, auditory, olfactory, and gustatory stimuli. Therefore, the issue is more one of why a small percentage of the population develops atypically [12].

Another definitional complication is that even non-synesthetes deal with perception and sensation relationally (e.g., when we speak about a warm blue, a loud color, a sharp cheese, or compare the taste of red and green apples). All of this raises yet another question: If synesthesia is not fully genetic and has developmental components, why do some younger synesthetes construct a sustained cross-sensory awareness, even as in others it recedes or is lost? [30]

It is also important to note that although synesthesia is characterized as a neurological condition those who have it throughout their lives usually do not experience it as a pathological condition or as a difficulty. In other words, most individuals do not see their experience as a disability, a handicap, or something imagined; rather for them, it is simply how their perception is experienced [5, 31–34]. Moreover, synesthetes describe their experiences in largely positive terms, seeing it more as a tool, a trait, or a special gift. As synesthete Sean Day puts it: "Synesthesia is not a disease, nor is it a deficit, in most cases. I tend to refer to synesthesia as a 'trait,' like left-handedness, or having red hair or green eyes" (p. 18, [35]). Synesthesia researcher and synesthete Patricia Lynne Duffy discussed this trait with many she interviewed about their involuntary synesthesia. She explains: "Neither I nor any of the synesthetes I have interviewed have experienced their synesthesia as a 'pathology'—although if it can be accurately described as such it is unique in being one of the very few that virtually none of its hosts wish to be cured of!" (p. 42, [20]).

Artist Carol Steen also believes this trait enhances her life (Figs. 12.3 and 12.4; also, see Fig. 6.1). She has pointed out that sometimes it is hard, even for a synesthete, to understand the experience before one has a concept for it [36]. In her case, she didn't recognize that synesthesia was an element of her neurological profile until she was an adult. According to Steen, this was partially due to her not having a name for what she experienced. Another factor was how others responded when she mentioned it. Her silence began at about the age of 7. Like Michael Watson, one of her schoolmates acted as if she was strange after she mentioned that the letter "A" was a beautiful pink (Fig. 12.1). In an interview conducted in 2019, Steen comments on this history and offers a familial connection:

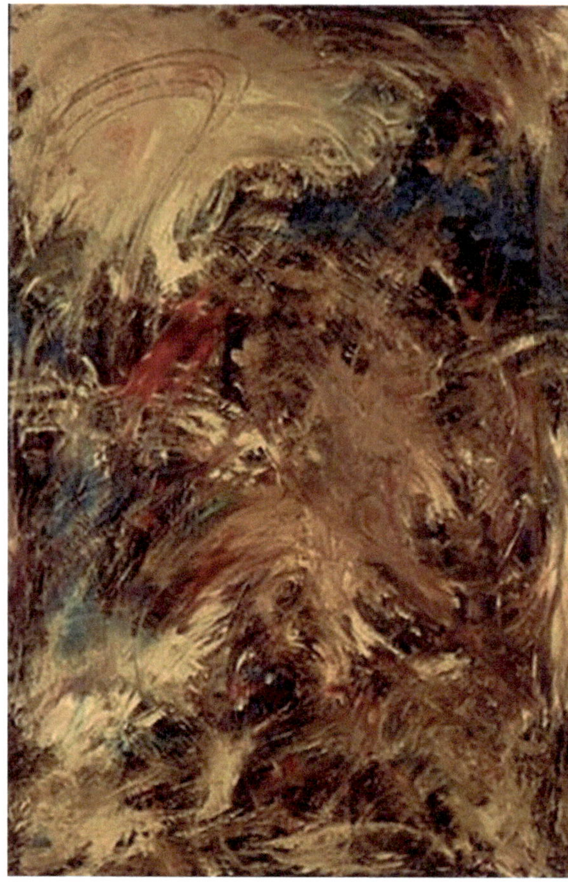

Fig. 12.3 "Runs off in Gold," 2003. Artist: Carol Steen. Courtesy of the artist

> My father was a reluctant synesthete. I'll always remember the day I came home from college on a semester break. As was usual when my family was together, we ate dinner in the dining room. My father was an artist, and we were used to talking about all sorts of topics… I announced, "The number five is yellow." My father looked at me and immediately said, "No, it's yellow ochre" … I quickly asked him what were the colors for the numbers 2 and 6. To me, both were the exact same value but one was green, the other blue, I just wasn't sure which number was which color. My father said, "Two is green." … I tried to ask him additional questions, but he changed the topic. For the next thirty years he refused to talk about his colored numbers and letters. Finally, when there was some published information about synesthesia, he began to speak about his colors. But he always spoke about them with restraint, as if he wasn't quite sure it was okay to discuss synesthesia. [37]

This vignette further illuminates the idiosyncratic pairings confirmed in color-number research and also introduces the familial connection borne out over studies of synesthesia in families [11, 14]. Later in this chapter, verification tools designed to test individual associations are discussed. At this point, suffice it to say that the disagreement between Carol Steen and her father over the number 5 is much like the difference shown in the visual example of color-letter synesthesia, presented in

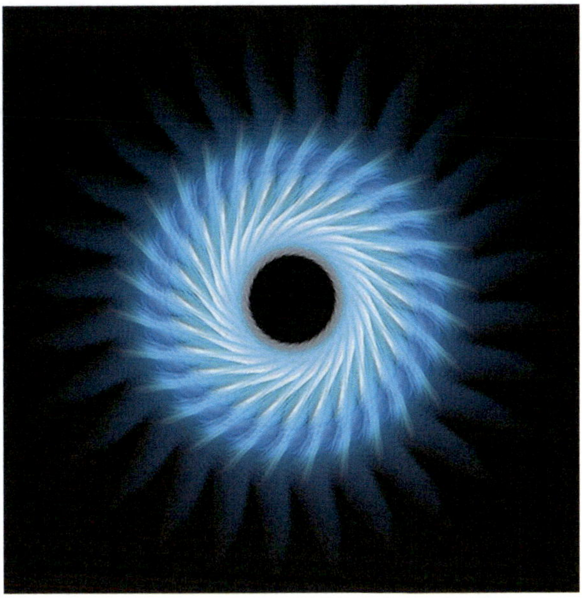

Fig. 12.4 Blue Hypnagogic Mandala, 2014. Artist: Carol Steen. Courtesy of the artist

Fig. 12.1, where Duffy's colored alphabet, on the left, is paired with the colored alphabet of artist/synesthete Carol Steen, on the right.

Sometimes, synesthesia is acquired after identifiable physiological changes and may also appear as a symptom of other conditions, e.g., epilepsy or autism [11, 38–40]. Case studies of lost or acquired synesthesia are particularly informative physiologically because they sometimes accompany specific events, although this is not always the case. The color-blind artist Jonathan I (Chap. 4) lost his synesthesia after a head injury that took away his ability to see color. While it may seem like the loss of his color synesthesia was insignificant as compared to his other losses, he profoundly missed it, for he had always felt that the music he heard was enriched by the colors [41, 42]. It was this case that convinced neurologist Oliver Sacks that synesthesia is a physiological phenomenon [42].

By contrast, contemporary philosopher David Chalmers (b. 1966) reports the music-color synesthesia he experienced as a boy simply went away in his twenties. Here, no specific event or health change was involved. Chalmers thinks this loss helped him become "fascinated by the world of sensory experience, especially the experience of color and the experience of music" [43, 44]:

> As a child I was a music-color synesthete — songs had color for me. Somewhat disappointingly most songs were murky shades of brown or olive green, but every now and then there was something distinctive. I remember that "Here, There, and Everywhere" by the Beatles was bright red. The synesthesia went away around the age of 20. I have a journal entry from around then, saying "Songs don't have colors anymore. What happened?" As a child, I'm not sure I realized there was anything special about this, but I think it helped get me fascinated by the world of sensory experience, especially the experience of color and the experience of music. [43, 44]

It is now known that there are at least sixty and possibly more than hundred subtypes of synesthesia. These occur in about 4–6% of the population [45]. Moreover, we all increasingly speak about and partake in multisensory experience within our technological world, an environment that encourages multisensory behavior.

12.2 Synesthesia in History

Synesthetic references are evident throughout history, with number-sound-music relationships and formulations going back at least as far as Pythagoras (c. 570–495 BCE). Written cross-sensory references are also evident historically, although until the nineteenth century these were typically characterized as rhetorical, metaphoric, or stylistic. We also find philosophical questions that incorporate sensory cross-modality in imaginative ways. For example, as noted in Chap. 4, John Locke (1632–1704) mentioned a blind man "who fancied that the idea of scarlet was like the sound of a trumpet" (p. 136, [46]).

As noted in Chap. 1, Aristotle's (384–322 BCE) conclusion was that we have five distinct perceptual capacities, which are now characterized as five distinct senses – touch, sight, hearing, smell, and taste. In his view, if a sensory faculty is lost, knowledge is lost as well. This characterization of five distinct senses has echoed down through time, particularly his valuations of how the senses should work in a well-functioning being. Yet, Aristotle's theory of perception also emphasizes sensory integration, a process that is itself not sensory. For Aristotle, in addition to the classical five senses, there is a *common sense* (*koine aisthesis; sensus communis*) by which experiences are forged out of the separate contributions from the different sense organs [47].

Wade is among those who have pointed out that the Aristotelian characterization of the five senses – again, touch, taste, smell, vision, and hearing – is too limited. Wade has proposed a number of candidates for a sixth sense (a muscle sense, a temperature sense, a movement sense) [48]. Oliver Sacks, too, was fascinated by the idea of a "sixth sense," seeing proprioception as a logical candidate, defining it as "unconscious, invisible, but arguably more vital than any or all of the other five senses put together" (p. 227, [3]). He wrote:

> One could be blind and deaf, like Helen Keller, and still lead a fairly rich life, but proprioception [is] crucial for the perception of one's own body, the position and movement of one's limbs in space, crucial indeed for the perception of their existence. How, if proprioception were extinguished, would a human being survive? (pp. 227–228, [3])

This "sense" that the five distinct senses do not cover the territory is also evident throughout history. As this book has detailed, often discrete historical characterizations came from an underdeveloped knowledge of physiology. For example, Aristotle hypothesized that the sense organs were located in the head (smell in the middle, sight above, and hearing to the side) while also concluding that the source of sensation is the heart [49].

Medieval scholars in both Europe and the Islamic world reconfigured Aristotle's sensory model. In addition to the five "outer" or "external" senses (i.e., touch, taste, smell, hearing, and sight), they postulated a set of "inner" or "internal" senses. These inner senses were widely believed to be housed in the ventricles of the brain. While the naming conventions differed from place to place and publication to publication, they generally were said to include *fantasia* and *imagination* in the front ventricle, then through the vermis to *cogitiva*, *estimativa* in the middle ventricle, and finally *memorativa* in the rear ventricle [50–52]. The meaning of these terms is self-evident except for *estimative*. The standard example of *estimativa* is that what it represented is articulated by thinking of sheep who were said to know innately that wolves are dangerous. This knowledge was said to be the result of the sheep's estimative faculty. Estimation was believed, in effect, to be able to read the "meaning" or "intention" of the wolf off of the image of it transferred from the front ventricle [52]. These refinements to Aristotle's model show that at this point some proposals were already postulating a functional separation between an inner (subjective) modality and an external reality.

Gregor Reisch's 1503 *Margarita Philosophica*, now considered the first "modern" encyclopedia on brain anatomy, offers a version of this. He depicts four outer senses, with touch left out. The four are connected to the *sensus communis* in the front ventricle. As was common in earlier publications, this theory was only one of the topics Reisch discussed. The book also offered theories about geography, arithmetic, music, geometry, astronomy, and optics.

In terms of the sensorium, sound was often linked with motion. In the seventeenth century, Thomas Willis (1621–1675) had postulated that the external ear gathers together the particles constituting sounds [50, 53]. Descartes (1596–1650) spoke of the sensory experience in his writing on sound and music. He proposed a musical theory intended to identify the affections or properties of sound, as far as they may be known mathematically, an inclination that dates back to Pythagoras. As a part of this, Descartes wondered which sounds are capable of pleasing and arousing the passions [54].

In the following century, in *Opticks*, Isaac Newton (1642–1727) assumed that sound and motion were connected to musical tones and color tones as well. He also assumed they have frequencies in common [55]:

> [In the rays of light] there is nothing else than a certain power and disposition to stir up a sensation of that Color. For as sound in a Bell or musical String, or other sounding Body, is nothing but a trembling Motion, and in Air nothing but that Motion propagated from the Object, and in the Sensorium 'tis a sense of that Motion under the form of sound; so Colors in the Object are nothing but a disposition to reflect this or that sort of rays more copiously than the rest; in the rays they [the colors] are nothing but their dispositions to propagate this or that Motion into the Sensorium and in the Sensorium they are sensations of those Motions under the forms of Colors (p. 13, in [56], p. 90, [57]).

Correlation of color, sound, and motion appears in a variety of ways in early references, with the language used often attempting to explain them in terms of universal correspondences. This inference in effect masks and obscures that deviations from the universalistic types of assumptions are repeatedly noted in individual and medical reports, as discussed shortly. Culturally, manifestations in artistic forms had their own texture and allure. One, based on Newton's idea that musical and color harmonies are related by frequencies of light waves and sound waves, was developed by the French Jesuit Louis Bertrand Castel, a topic examined shortly in the color organ section of this chapter.

Jonathan Swift's *Gulliver's Travels* (1726, part 3, "A Voyage to Laputa") includes a cultural description of synesthesia in literature. His fictitious ship surgeon Lemuel Gulliver learns of what appears to allude to color-touch synesthete when visiting the Academy of Lagado (a parody of the Royal Society):

> [A] man born blind … had several apprentices in his own condition. Their employment was to mix colors for painters, which their master taught them to distinguish by feeling and smelling. It was indeed my misfortune to find them at that time not very perfect in their lessons…." (p. 207, [58]).

It appears that Swift got his ideas from Robert Boyle's *Experiments and Considerations Touching Colours* published in 1664 [59], which offered detailed notes for studying colored surfaces and liquids. Boyle proposed that colors were sensory images rather than genuine properties of material objects, and he discussed a case he learned of from Sir John Finch (1626–1682), a Padua-trained physician and Fellow of the Royal Society. Boyle, through Finch, describes John Vermassen who was about 33 at the time of Finch's report. He had become blind from smallpox around the age of 2 and was working as an organist:

> … the doctor discoursing with him over night, the blind man affirm'd, that he could distinguish colors by the touch, but that he could not do it, unless he were fasting; any quantity of drink taking from him that exquisiteness of touch, which is requisite to so nice a sensation. That hereupon the doctor provided again the next morning seven pieces of ribbon, of these seven colors, black, white, red, blew, green, yellow, and gray, but as for mingled colors, this Vermaasen would not undertake to discern them, though if offer'd, he would tell that they were mix'd [59].

We do not know if this is an early case of synesthesia or if he simply had a more refined sense of touch than that possessed by sighted individuals [60]. As Jewanski and his colleagues point out, recent investigations of touch-color synesthesia similarly report colors from touch (e.g., Braille characters), but the subjects do not see the true color of the text by touching it. Since there are no accounts that the dyes have known tactile differences, it is possible that if different dyes were used on different fabrics the colors did not evoke synesthesia.

Shortly after Boyle's report, the philosopher John Locke (1632–1704) offered another reference to synesthesia in *An Essay Concerning Human Understanding.* Published in 1690 [61], it contains a question now known as Molyneux's problem (see Chap. 2). Locke asked if a person born blind might immediately identify a shape previously familiar to them only by touch if they were made to see. In considering the question – and if the loss of one sensation might enhance others – Locke mentions a blind man "who fancied that the idea of scarlet was like the sound of a trumpet" (p. 136, [46]). A medical report by Thomas Woolhouse, in 1710, mentions another case of a blind man. This man could see sound-induced color visions.

As these cases indicate, associations of color with blindness come up repeatedly in the historical literature. There are contemporary reports as well. For example, Richard Cytowic described a blind synesthete who said: "I find it *easy* and *satisfying* to picture street maps and am a good navigator. The maps that I see are colorful, and perhaps that's what makes map visualization both easy and *pleasurable*" (p. 50 [5]).

Another contemporary report is found in *My Path Leads to Tibet*, written by Sabriye Tenberken (b. 1970) [62]. In her book, Tenberken tells us that in Tibet the blind were frequently treated as less than human and denied education, work, respect, and a role in the community. Tenberken was totally blind when she travelled there, although she had not always been blind. Until she was 12, Tenberken was able to make out faces and landscapes. In addition, she had loved painting and colors. Even when she could no longer decipher shapes and forms, she was still able to use colors to identify objects (also see [63]).

Key here is that Tenberken had experienced synesthesia with the visual construction of form and color as a child, and her synesthesia intensified once she went blind. It appears that her early vision, although impaired, served as an aid as she lived her life as a blind woman. She described this as follows:

> As far back as I can remember, numbers and words have instantly triggered colors in me.…The number 4, for example, [is] gold. Five is light green. Nine is vermilion.…Days of the week as well as months have their colors, too. I have them arranged in geometrical formations, in circular sections, a little like a pie. When I need to recall on which day a particular event happened, the first thing that pops up on my inner screen is the day's color, then its position in the pie (in [63], p. 212–213).

Another case of acquired synesthesia coupled with the loss of vision was Jacques Lusseyran (1924–1971). This case is somewhat opposite to that of the color-blind painter. As noted in Chap. 4, after a small stroke, the painter Jonathan Isaacson lost his synesthesia and even his imagination of color when he became color-blind, having acquired a form of cerebral achromatopsia. When Lusseyran, by contrast, completely lost his ability to see, his synesthesia intensified. He began to identify himself as "visually blind" and to visualize his world internally. Indeed, after becoming blind, Lusseyran lost his grasp of visual imagery, even forgetting the faces of his parents and people he had loved.[2] This had a tremendous impact on his experience of music because Lusseyran, who was blinded at the age of 7, claimed his synesthesia replaced his actual perception of music and prevented him from becoming a musician. He wrote:

> [S]ome blind people are not made for music; I was one of the visual blind. I did not become a musician, and the reason was a strange one. I had no sooner made a sound on the A string, or D or G or C, than I could no longer hear it. I looked at it. Tones, chords, melodies,

[2] While Lusseyran calls himself one of the "visual blind" he is better known for having become a leader in the French resistance against Nazi Germany's occupation of France in 1941. He was 17 years old at the time [63].

> rhythms, each was immediately transformed into pictures, curves, lines, shapes, landscapes, and most of all colors.…At concerts, for me, the orchestra was like a painter. It flooded me with all the colors of the rainbow. If the violin came in by itself, I was suddenly filled with gold and fire, and with red so bright that I could not remember having seen it on any object. When it was the oboe's turn, a clear green ran all through me, so cool that I seemed to feel the breath of night.…I saw music too much to be able to speak its language (pp. 81–82, [64]).

As noted in Chap. 2, vision research since the twentieth century has shown that representations are not given to us fully formed but constructed in perception by the brain through the seamless integration of a huge number of components. Through this research, it was clear that vision was an analytic process involving the differing sensitives within the cerebral and retinal systems [65–67]. This research also relates to the question of how someone who becomes blind might nonetheless retain color awareness. No doubt pre-blindness exposure comes into play because blind synesthetes who had learned colors before becoming blind are unlike the congenital blind.

This difference was noted even before twentieth century research establishing color construction. Frederick Starr mentioned this in his 1887 paper "Note on Colored-Hearing," where he emphasizes the importance in distinguishing the congenital blind from synesthete subjects exposed to color before totally losing their vision [68]:

> In closing we desire to add a few notes made on blind subjects at the State College for the Blind, Vinton, Iowa, in 1887. We seemed to find among the congenital blind absolutely no conception of color, and no tendency to imagine it in terms of other sensations. On the other hand, we believe that those who have become blind in childhood, after learning colors, very soon come to perceive color sensations in their hearing. We examined perhaps a dozen. Of these, three were born blind, or were blind from infancy. Of the rest, at least three had color hearing to a marked degree. The most interesting case was Miss N. This lady had pseudo-chromesthesia before she became blind. Musical notes are colored to her [68].

Historical cases also show awareness of other forms of synesthesia. A literary work, *The Memoirs of Martinus Scriblerus* [69] appeared in 1713–1714, a few years after the 1710 Woolhouse study reported on the sound-induced visions of a blind man. *Martinus Scriblerus* is an incomplete satirical work supposedly written by the members of the Scriblerus Club. Rather than presenting a case of physiological blindness, it mocks blind adherence to new cultural trends and outdated beliefs. Each chapter satirizes a different fad or fashion of the time through the story of Martinus Scriblerus' upbringing by his parents. Alexander Pope and Dr. Arbuthnot are credited on its title page, although the book is sometimes associated with Jonathan Swift as well. Color-touch synesthesia comes up in the work when we are told that Martinus Scriblerus "first found out the *Palpability of Colors*; and by the delicacy of his Touch could distinguish the different Vibrations of the heterogeneous Rays of Light" (p. 123, [69]).

Later, E. T. A. Hoffmann's (1776–1822) character Johannes Kreisler offers a compelling case for asking if Hoffmann was a synesthete. In *Kreisleriana*, Hoffmann writes: "I happened to be wearing a coat … Its color was in C sharp minor, so in order to give those seeing it some peace of mind I had a collar made for it in the color of E major" (p. 130, [70]). Given how specific the language is, it seems plausible that Hoffmann either was a synesthete or was well acquainted with the phenomenon.

12.3 Early Scientific Investigations of Synesthesia

Research and cultural references quickened in the nineteenth century. An early publication by Johann Wolfgang von Goethe, *Zur Farbenlehre* (Theory of Colors) (1810), referenced color-hearing. While his theories are no longer considered valid overall, his proposals stand out because of Goethe's efforts to offer a subjective theory of color that was not based on a mystical type of analysis [71]. He took issue with Newton's theories and pointed out that color, for an individual, depends not only on light but also on our senses and how we perceive them [45, 71]. What is particularly compelling is that, despite this book's inaccuracies, Goethe astutely realized that colors are the product of the visual system and do not depend only on the light that reaches the eye [45]. That said, there is no inference of a color area in the brain, but because the presentation does offer a hands-on method for experientially observing some natural phenomenon (e.g., aftereffects), it remains of interest today. He also references Giuseppe Arcimboldo's (1526–1593) "colorific music," discussed in the color organ section [71].

Many of the early nineteenth century investigations included a range of self-described reports. Georg Tobias Ludwig Sachs' 1812 doctoral thesis is often cited as the first convincing account of synesthesia in the medical literature. He experienced several forms of synesthesia and placed his observations on vision within Goethe's theoretical framework, offering examples of his colored letters, tones of the musical scale, his spatial numbers, and spatial days of the week [60, 72]. Sachs also described his own albinism and that of his sister. Few have read it since the original was published in Latin, although it was widely cited in the medical literature of the nineteenth century (see [60]), perhaps through the 1824 German translation [72].

Another self-reported case was that of a medical doctor, Perroud, in 1863 [73]. It was when he was about 15 years of age that he began to see certain letters, especially vowels, intimately bound with colors [74]. The doctor discusses several eye-based explanations, only to reject them in favor of a brain-based explanation. Given that his synesthesia didn't manifest until he was a teenager, it is not surprising that Perroud did not believe that the trait was always congenital. He also cited a case of a 30-year-old man who had reportedly experienced his own synesthesia for only 12–15 years. William O. Krohn claimed Perroud was the first to recognize the physiological character of these phenomena and to attempt an explanation [74].

J. A. Nussbaumer reported the case of the two Nussbaumer brothers in 1873 [75]. Included with their own observations about their synesthesia was a 19-section questionnaire, reported to be the first questionnaire in the history of this phenomenon [76]. The author published these observations when he was a 23-year-old student in Vienna, with a letter from a medical professor, Prof. Brühl, affirming the report scientifically [74, 76]. Nussbaumer probed his sensations in relation to Johannes Müller's law of the specific energy of senses (1826), a doctrine that stated each sensory organ is linked with its own separate nerve center in the brain: the eye can only see, but not hear; the ear can only hear, but not see [75].

Also, in 1873, Jakob Hock (1831–1890) encouraged physiologists and acousticians to pay attention to this phenomenon. One outstanding feature of his research was his analyses of the eyes of synesthetes, from which he concluded: "While sounding the tuning fork, I could not, under such conditions, perceive the slightest change. Hence no perceptible change takes place at the retina and the optic nerves during the distinctly specified color perception" (p. 161, in [75]).

In the 1870s and the 1880s, Eugen Bleuler (1857–1939) and Karl Bernhard Lehmann (1858–1940) in Switzerland and Sir Francis Galton (1822–1911) in Britain began to interview large groups of people to determine whether they experienced color associations [22, 23, 77, 78]. In 1881, Bleuler and Lehmann reported on 76 cases [78]. The tendency of synesthesia to cluster in families was first reported in 1883 by Sir Francis Galton [23], who had already reported on visualized numbers with colors in 1880 [22].

One of the outstanding features of Galton's 1883 book, *Inquiries into Human Faculty and Its Development*, is that it includes notations, references to familial sharing of a trait, and graphics derived from communications he received about mental imagery after his earlier publications. Plate IV from Galton's book offers an abridged summary (Fig. 12.5). As this graphic shows, Galton investigated color association, spatial sequence synesthesia, and so forth. Whereas Galton had initially concluded that these were no more than associations, he, too, was soon convinced that this was a physiological phenomenon despite its kinship to mental imagery. By 1883, he had concluded synesthesia is fixed and automatic in nature [23].

Among the internal graphics are examples of spatial sequence imagery. Figures 12.6 and 12.7 translate how two subjects described their spatial sequence synesthesia, with D. A's verbal description describing his own sketch. An engraver's translation of it appears in the book:

> From the very first I have seen numerals up to nearly 200, range themselves always in a particular manner, and in thinking of a number it always takes its place in the figure. The more attention I give to the properties of numbers and their interpretations, the less I am troubled with this clumsy framework for them, but it is indelible in my mind's eye even when for a long time less consciously so. The higher numbers are to me quite abstract and unconnected with a shape. This rough and untidy production is the best I can do towards representing what I see. There was a little difficulty in the performance, because it is only by catching oneself at unawares, so to speak, that one is quite sure that what one sees is not affected by temporary imagination....the mental picture never seems [flat] but in a thick, dark gray atmosphere deepening in certain parts, especially where 1 emerges, and about 20. How I get from 100 to 120 I hardly know, though if I could require these figures a few times without thinking of them on purpose, I should soon notice. About 200 I lose all framework. I do not see the actual figures very distinctly, but what there is of them is distinguished from it at the age of seven or eight. I need hardly say that the impression is not that of perfectly straight lines, I have therefore used no ruler in drawing it (pp. 119–120, [23]).

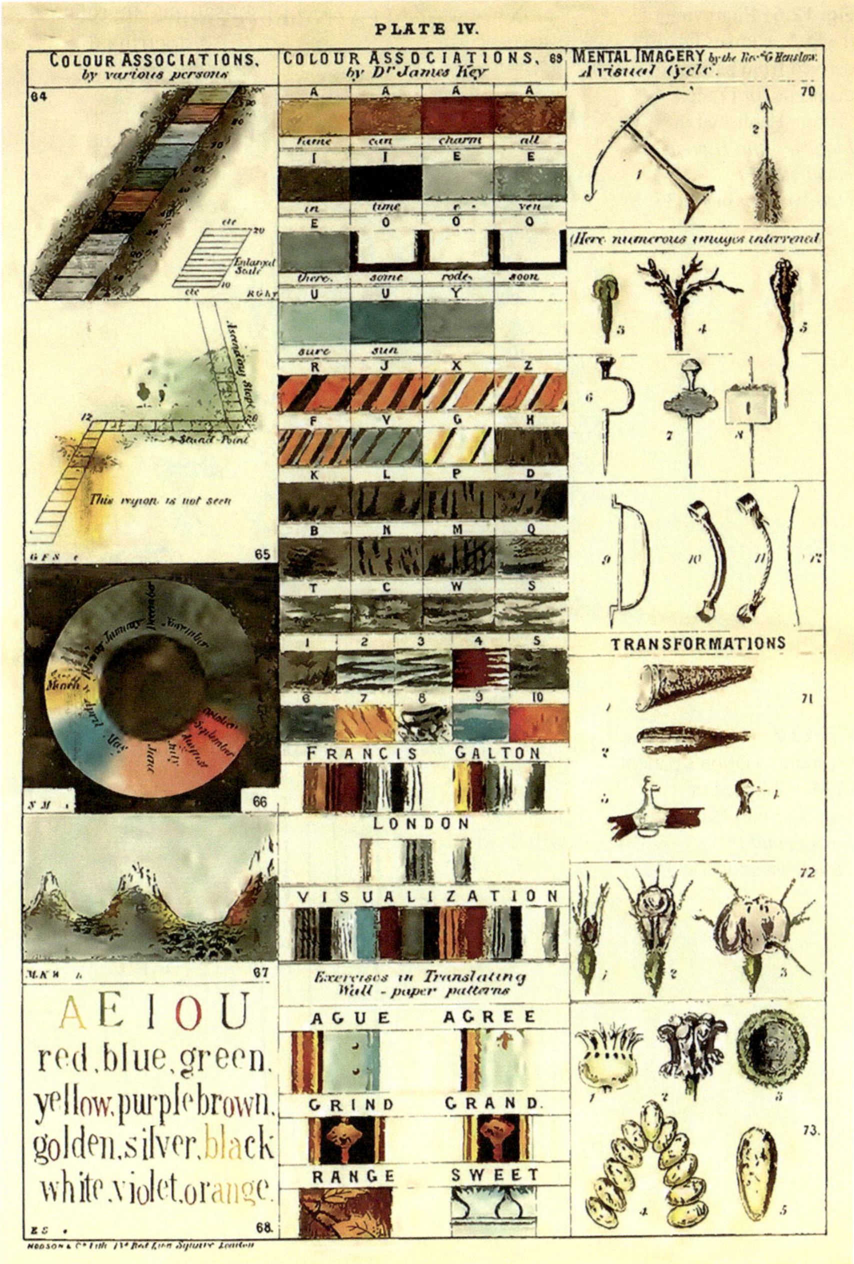

Fig. 12.5 Plate IV from Francis Galton's *Inquiries into Human Faculty and Its Development* (1883). This compilation is an abridged summary of Galton observations on mental imagery. This book brought what we now call synesthesia to the attention of the scientific community

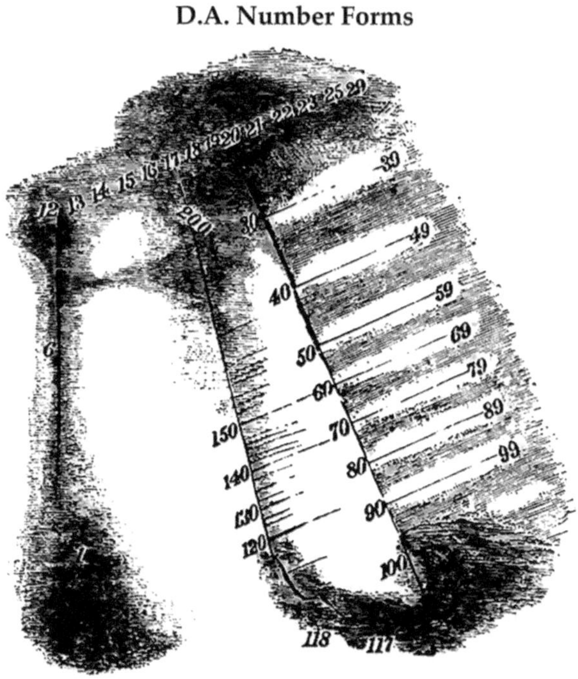

Fig. 12.6 Engraving of the D.A. number forms derived from patient diagrams for Francis Galton. Published in *Inquiries into Human Faculty and Its Development* in 1883 [23]

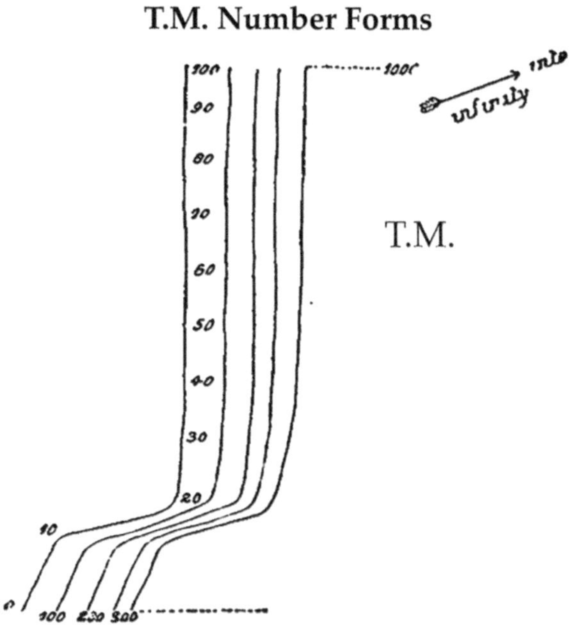

Fig. 12.7 Number forms of Francis Galton's patient T. M. Published in *Inquiries into Human Faculty and Its Development* in 1883 [23]

While these kinds of diagrams are informative, early researchers did not have a mechanism for quantifying or verifying subjectively described or diagrammed experiences [79]. This was a limitation that confronted Théodore Flournoy (1854–1920) whose extensive and pioneering investigations began in 1882 after reading Bleuler and Lehmann's book [80, 81]. He had studied with Wilhelm Wundt, devoted himself to physiological psychology, and founded one of the first laboratories of experimental psychology. His was the most extensive treatment of synesthesia in the nineteenth century and came primarily in the form of an 1883 monograph published under the title *Des Phénomènes de Synopsie* (*Synopsia Phenomena*) [81]. Once dismissed, Flournoy is now being re-evaluated due to his discussion of a number of aspects highly recognizable to modern neuroscientists and psychologists such as attributions of gender and/or personality to triggering stimuli [80, 82].

Interest at that time is also evident in events devoted entirely to the neurodevelopmental condition of synesthesia such as a roundtable included in the first International Conference of Physiological Psychology (Paris, 1889). This meeting also included a sub-meeting in which a committee on synesthesia presented its conclusions [79]:

> We do not know who suggested devoting a separate section to "Colored Audition" (at that time, a synonym for synesthesia), but we can speculate that discussions must have taken place during the conference about creating a new section to deal with it. Two important researchers at that time, Francis Galton and Henry Beaunis (1830–1921), who also had published about synesthesia before 1889 were members of the patronage committee and may have supported the idea of including synesthesia within the conference, the focus of which was otherwise on hallucinations, hypnotism, heredity, and muscular sensations. These topics were themselves related to contemporary views of synesthesia (p. 360, in [79]).

In 1890, the term synesthesia was coined to standardize the many ways people referred to sensory perceptions [20], although in an 1893 issue of *The American Journal of Psychology* we still find several articles on synesthesia that use an array of terms. William O. Krohn, in "Pseudo-chromesthesia, or the association of colors with words, letters, and sounds," makes the point that it is difficult to track the research because more than one word was used at that time. Other articles included "Note on Color-Hearing" by Frederick Starr and Mary Whiton Calkins' article titled "A statistical study of pseudo-chromesthesia and of mental forms" [83].

Starr's article begins by saying that in 1887 he met a remarkable woman with color-hearing, and thinking he had a new field for study, he went hunting for others (Fig. 12.8). After discovering Galton's study [22], he dropped the project for a time. Upon discovering another case, he decided to record it without offering a theory:

> G. L. is a young lady of unusually quick and bright mind; she has some artistic power and is a writer of ability. From early childhood she has had color-hearing. When a very little girl she was laughed at because she said that names were colored. That in her case it is truly sound that is colored is shown by the fact that names were colored before she could spell or read at all. At the same time, she has mental imagery of great clearness and now sees the colored words when she hears them. Although all words are colored, letters, numbers and

NOTE ON COLOR-HEARING.

PROF. FREDERICK STARR.

Reprinted from the American Journal of Psychology, Vol. V, No. 3.

In 1887 I met a remarkable case of color-hearing and made quite full notes. The matter was one of which, at that time, I had never heard, and thinking I had a new field for study, I went hunting for victims, subject to pseudo-chromesthesia. I studied somo blind cases and had a theory. Just then I discovered Galton's Study upon the same subject and dropped the matter. I made a second observation upon my original subject in 1891, and this spring made a third one, and believe the case of sufficient interest to make a record of it.

I have no theory to offer. I do not agree with Mr. Galton's, however, but consider one of the physiological explanations preferable. When I first studied the subject, I believed that the phenomenon was due to crossing or mingling of fibres of the auditory and visual nerves—a theory which I see has been held by some writers. I am not qualified to urge the theory or reject it; but I am heartily thankful to Mr. Krohn for his late valuable synopsis of the subject, and its literature.

G. L. is a young lady of unusually quick and bright mind; she has some artistic power and is a writer of ability. From early childhood she has had color-hearing. When a very little girl she was laughed at because she said that names were colored. That in her case it is truly sound that is colored is shown by the fact that names were colored before she could spell or read at all. At the same time she has mental imagery of great clearness and now *sees* the colored words when she hears them. Although all words are colored, letters, numbers and personal names are the most clearly and definitely affected. In my first test I gave many words like snow, sky, ink, grass, and the like, to see if natural quality of objects had influenced the result, but could not see that it had. Unfortunately different lists were used on the three occasions, and only letters, numbers and a few personal names occur in all these.

The letters of the alphabet bear the following colors for G. L.:

A. Dirty-white.
B. Radish-color.
C. Corn-color.
D. Dark-brown.
E. Cloudy-white or light-gray.
F. Light-gray.
G. Grayish-yellowish-white.
H. Dark-gray, greenish tinged.
I. Dirty-white.
J. Dark-greenish-bluish-gray.
K. Gray.
L. Yellow.
M. Geranium-red.
N. Red, with white tinge.
O. Pearl-white.
P. Deep-bright-green.
Q. Light-gray-green.
R. Rich-reddish-brown.
S. Sunshine-yellow.
T. Blue.
U. Gray-white.
V. Radish-color and brown.
W. Red, with white cloud.
X. Brownish-beet-color.
Y. Dirty-white.
Z. Deep-beet-color.

Fig. 12.8 Frederick Starr (1858–1933) on "Note on Color-Hearing" (pseudo-chromesthesia), 1893 [68]

> personal names are the most clearly and definitely affected. In my first test I gave many words like snow, sky, ink, grass, and the like, to see if the natural quality of objects had influenced the result, but could not see that it had. Unfortunately different lists were used on the three occasions, and only letters, numbers and a few personal names occur in all these (p. 1, [68]).

Starr closes with the case of a woman who became blind in childhood. She acquired synesthesia later in life and had learned colors before going blind. This patient experienced color sensation upon hearing and with musical notes. Exploring a dozen blind subject, he found three others that matched her experience. As noted earlier, his report further establishes the importance in distinguishing the congenital blind from subjects exposed to color before totally losing their vision.

Edward Holden's reports on his daughter Mildred's color associations and his other investigations are contemporaneous with Starr. The most complete one was published is his 1891 *Nature* article [74, 84]. He also published two articles in *Science* (1885 and 1895) [85, 86]. In the 1885 report, Holden writes:

> In 1880, while I was in Washington, I read Mr. F. Galton's note on "Visualized numerals," in *Nature* of Jan. 15 of that year. After I came to Madison, probably late in 1881, or early in 1882, I mentioned my own entire inability to visualize numerals, or anything else of the kind, to a member of the university faculty, Professor O. I was interested to learn, that, when a boy, he had always conceived the vowel sounds as having color, and that he still retained some traces of this early habit. I spoke of this subject at my dinner-table shortly after; and my daughter Mildred, then about eight years old, said she also had colors for the days of the week (p. 242, [85]).

All of the publications mentioned above stimulated dialogue [22, 23, 27, 28, 87–89]. For example, one series of letters in *Science Magazine* describes colored number synesthesia and offers diagrams for months and numbers much like those Galton presented in 1883 as shown in Figs. 12.5, 12.6, and 12.7. Some of the *Science* diagrams (Fig. 12.2) resemble the description in the Tenberken report quoted above. Briefly, Arthur Winslow, a geologist, wrote that he perceived his months similarly to C. Ruheit of New Haven who had submitted a drawing of his geometric associations earlier. To show the striking similarity, Winslow presented his own diagram and added: "A diagram of numbers which also forms itself in my mind, is of peculiar form, and is equally difficult to explain … I cannot think of any number less than a hundred, that does not place Itself immediately in its appropriate place in the diagram" (p. 243, [28]).

At this point, researchers had already concluded that while in rare instances associations changed spontaneously, they were generally not altered by voluntary efforts. Ruheit of New Haven ended his 1883 letter with the comment that the information suggested an interesting field for study and "I, for one, would be glad to hear other facts bearing on the subject" (p. 142, [27]).

Recent investigations into spatial sequence synesthesia and other forms do continue to broaden this discussion. Eagleman wrote:

> [Our] method of quantifying 3D coordinates introduces at least two advantages. First, it can open the door to quantifiable verification of SSS [Spatial Sequence Synesthesia] … A second advantage of quantifying in 3D space is the ability to identify patterns across the spatial coordinates of different subjects. In the same way that color-grapheme synesthetes may display some amount of imprinting from exposure to patterns seen in everyday life [90], there may be the possibility of similar influences on spatial forms as seen, for example, in the frequent placement of numerals 1–12 in a clocklike pattern [23, 91–93]. [This paper identifies] several common features of month forms. Finally, the flip side of identifying patterns is quantifying individual differences. … The fact that one quarter of the ellipses ran counterclockwise casts at least some doubt on the possibility that subjects are imprinting on clock faces (p. 1269, [17]).

By 1898, the term chromesthesia was included in the *American Pocket Medical Dictionary*, a book advertised as containing the pronunciation and definition of over 26,000 of the terms used in medicine and the kindred sciences. Chromesthesia, however, did not have a stand-alone listing. Rather, it was referenced under the term audition, with reference to color [94].

L. B. Alford reported on two cases of synesthesia in a 1918 issue of *The Journal of Abnormal Psychology*. A pathologist at the Boston State Hospital, Alford notes the degree of correspondence of associated color sensations in a pair of male twins, aged 27 years old, and very similar in mental makeup [95]. His general summary mentions that the secondary sensations are remarkably constant. A related case he cites is the self-described report of a Dr. Harris who said his synesthesia remained practically unaltered since childhood. Alford also cites Edward S. Holden's 1891 report on his daughter Mildred's color associations discussed above. At the end, he concludes that synesthesia is a common occurrence and is not a mental abnormality:

> Synesthesia is evidently of common occurrence, although it is difficult to estimate even approximately its frequency, because the figures of different observers and even those of the same observer on different material vary within such wide limits. Galton [23] gives the percentage of occurrence as 3.3 of the general population, while Mercante [96] estimates that it exists in 80%, of all grade school children. Calkins found it present in 6.6%, of certain classes of Wellesley College in 1891, and in 15.7 in 1892. Some reason for the difference in these figures is to be found in the fact that the condition has been described under a number of different names—color audition, psychochromesthesia, photisms, etc., some of which include many more manifestations than others, and again in the fact that it is not possible definitely to separate synesthesia from visions, and the so-called "forms" and "personifications." Accepting the lowest estimates as correct, however, it remains that synesthesia is a common phenomenon (pp. 1–2, [95]).

Recent studies vary on the percentage of synesthetes within the population. Many suggest that a little over 4% of the adult population have some form of synesthesia involving one of the five classical senses [34]. This number is worth thinking about in light of the invisible nature of synesthesia. As pointed out above, many people do not even realize they are synesthetes until it is pointed out to them. Moreover, although the first scientific documentation is generally dated to 1812, as the examples cited above suggest, it appears the trait was known and documented earlier.

Before turning to art and synesthesia, it is important to add that the earlier literature does include synesthesia-like descriptions that reference pathological experiences such as epilepsy and/or migraine. William Richard Gowers (1845–1915), for example, presents an intelligent man, a patient of 26 years of age, who experienced a variety of sensations: sounds, taste, smell, and visual colors during his epilepsy attacks. Gowers also notes that his auditory cases of epilepsy included visual auras:

> The most frequent associations of the auditory aura were with a visual aura (in seven cases), and an olfactory aura (in two cases). … In several cases in which a visual and auditory aura were combined, the two corresponded in character—*i.e.*, in their degree of elaboration. … in a case of Dr. Hughlings Jackson's of cerebral tumor… a sound of bells ringing referred to the left ear, preceded left-sided fits in an early period of the disease. After death we found a tumor outside the opposite optic thalamus. It was extensive, but had evidently commenced between the thalamus and the upper temporo-sphenoidal convolution (pp. 69–70, [97]).

12.4 Art and Synesthesia

In the nineteenth century, in tandem with the scientific cases entering the literature, we also see a number of intersections among synesthesia, culture, and the arts. In 1855, the composer Joachim Raff reportedly stated that color impressions were elicited from various instruments. Raff reported that for him, "the sound of the flute produced the sensation of intense azure blue; of the hautboy, yellow; cornet green; trumpet, scarlet; the French horn, purple; and the flageolet, gray" (p. 22, [74]). One wonders if he was aware of Locke's correlation of scarlet with the sound of a trumpet.

The nineteenth century Symbolist movement is often evoked in reference to synesthesia because Symbolist artists regarded music as the highest step on the ladder of the arts [45, 98–100]. Their references to its universality and ethereal quality are striking given how idiosyncratic the synesthetic experience is. To reiterate, even individuals with verifiable synesthesia fail to agree on correspondences. In any case, their philosophical positioning largely followed Richard Wagner in assuming that visual, auditory and other sensory elements were properly attuned in one Gestalt experience [101].

Literary figures associated with the movement include poets such as Stéphane Mallarmé (1842–1898) and Arthur Rimbaud (1854–1891) who expressed interest in color-sound affinities. Rimbaud's 1871 poem *Voyelles (Vowels)* begins: "Black A, white E, red I, green U, blue O: you vowels, / Some day I'll tell the tale of where your mystery lies" ([102–104]; also, see pp. 94, 191, [103]). This kind of example illustrates why synesthesia is frequently aligned with metaphor. That said, whereas the references to the imagery and metaphors of literary works are harder to evaluate, some examples in the literature are so specific that they evoke the sense they are written in tune with an actual experience rather than as an attempted correlation that implies the metaphoric conclusion, even if we cannot know for certain.

Although not a Symbolist, Vincent van Gogh (1853–1890) is among the nineteenth century visual artists currently assumed to be a synesthete. He equated colors with sound and said that he took piano lessons to become acquainted with the subtleties of color tone. This quest didn't sit well with his piano teacher. Thinking his pupil was mad because of his constant comparisons of the piano keys to colors, the teacher discontinued the lessons. Later, in his writings, Van Gogh compared painting with pencils to playing the violin with a bow [55].

As Cytowic points out, the first photograph of a presumably synesthetic individual was taken in 1872. The portrait is of 8-year-old Ellen Emerson, the daughter of American Transcendentalist and philosopher Ralph Waldo Emerson. According to fellow philosopher Henry David Thoreau, a close family friend, Ellen said to him in 1845 that she could tell the color of a great many words and she amused the children at school by doing so [105]. Apparently, this was the first time Thoreau had heard of such an experience. As Cytowic correctly asks, where were all the synesthetic children historically? This is a question that has more potency now that we can identify reported cases in the literature and have research devoted to childhood synesthesia as well [13, 15, 30].

In the twentieth century, artists of the *Der Blaue Reiter* (The Blue Rider) group conducted cross-modal experiments. It is harder to ascertain those that represented genuine synesthesia, those that were metaphorical, and those that were focused on enhancing multisensory perception for the artist as well as the audience. For example, as discussed in a later section, Wassily Kandinsky characterized himself as a synesthete, while Paul Klee did not. Klee's artistic goals centered around merging percepts in his visual presentations. Yet many confuse and conflate the works the two produced.

Also in the twentieth century, the writer Vladimir Nabokov (1899–1977) wrote about his colored alphabet in his autobiography, *Speak, Memory*. Here, he explained that as a child he saw all the letters of the alphabet as having distinct colors and was deeply disturbed when someone gave him a box of colored letters because all of the letters had the wrong colors. Much like Carol Steen and her father differed on their numbers, Nabokov and his mother agreed that the colors were wrong, but they were unable to agree on what the "right" colors should be [106–108]. Nabokov also said that for him the tint of vowels depended on the sound [109].

A contemporary pairing of synesthesia and art chronicled by Richard Cytowic is the painter David Hockney (b. 1937). After reading an interview with Hockney in *Architectural Digest* (September 1980) that suggested the painter was a synesthete, Cytowic contacted him and later tested him to learn more. Hockney, who was unable to read music, said that despite his increasing deafness he played music over and over when working on his set designs for musical productions [110]. Cytowic said it was unclear if Hockney has projected photisms, but the artist did discuss that his synesthesia involves an association of music, shape, color, and space. There is also a kinesthetic sense while he is painting to music that triggers colors and shapes [110].

12.5 Syntax: Color Organs

At the beginning of the recent explosion of synesthesia research, in 1974, Lawrence E. Marks produced a chronological summary of the literature with 110 names for source materials [6]. Many have come up in this chapter and show the interplays among psychology, philosophy, physics, and physiology in early interpretations. One early project Marks includes was proposed by Giuseppe Arcimboldo

(1526–1593), an Italian painter who created imaginative portraits filled with fruits, vegetables, flowers, and so forth. The painter was also interested in color/music/sound relationships and attempted to devise instruments that matched color to tonal scales. For this, Arcimboldo consulted with a musician at the court of Rudolph II about a keyboard instrument that would use colored strips of paper, much as the French Jesuit Louis Bertrand Castel did later (Fig. 12.9) [111].

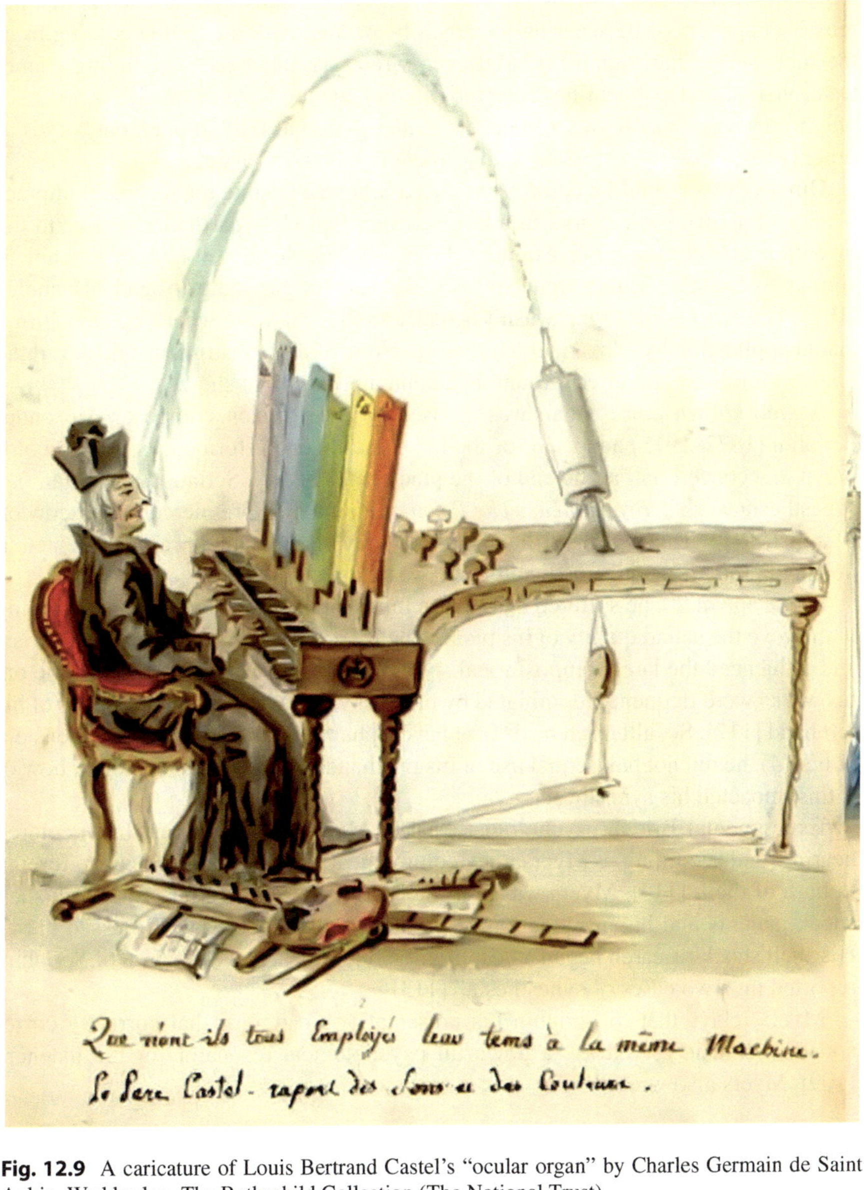

Fig. 12.9 A caricature of Louis Bertrand Castel's "ocular organ" by Charles Germain de Saint Aubin. Waddesdon, The Rothschild Collection (The National Trust)

Castel based his color organ on Newton's idea that musical and color harmonies are related by frequencies of light waves and sound waves. He built an "ocular organ"—a *clavecin oculaire* – around 1720 and exhibited the instrument in 1735 [6]. Working with the instrument maker Rondet, Castel's design had colored paper strips on top of a harpsichord. When a key was pressed, the paper strips were lit by candlelight, and the colors would come up [55, 101]. Figure 12.9, a caricature of Castel's instrument by Charles Germain de Saint Aubin, shows that he chose colors for each note. This image is also of interest in multisensory terms because it includes a painter's palette at the organist's feet. Another multisensory reference is that the musician appears to be scenting the air as he pedals, with the aroma coming from the tube on the harpsichord. Castel even wrote synesthetically, including sound, taste, vision, and color in his description: "For we are born in music, & we have only to open our ears in order to taste it ... and...one has only to open one's eyes in order to taste a Music of colors & to judge it" (p. 304, in [6]).

Other efforts to build a color harpsichord followed despite the technical impracticality of showing the colors in a large concert hall [55], particularly prior to the invention of electricity. For example, Frederick Kastner developed an instrument similar to Castel's ocular organ in 1870 that used a gas lamp instead of candles [101]. The term color organ wasn't actually used until 1893 where we find it in a patent application by Alexander Rimington. None of these instruments achieved the kinds of multisensory effects found in a contemporary light show.

A well-known color organ was devised by the Russian composer Alexander Scriabin (1872–1915) and used for the first synesthetic performances using a color organ in a concert hall at the end of the nineteenth century. Scriabin's most famous synesthesia work is *Prometheus: The Poem of Fire*, which premiered in Moscow on March 2, 1911. This work is still performed today. As with many cases presented in this book, the musician's medical profile was more complicated.

At the age of 20, he suffered an overuse injury of his right hand when attempting to improve the sound quality of his piano touch. This injury caused a profound crisis and influenced the later compositional style of his piano works. From this time on, his works were frequently dominated by unusual virtuosic use and wide spans of his left hand [112]. So, although he did not have a phantom limb, like Paul Wittgenstein (Chap. 4), he did not have equal use of his two hands either. We do not know how or if this impacted his synesthesia.

Experimental British psychologist Charles Myers examined Scriabin's colored hearing, and according to Myers, they communicated in French, a language foreign to both of them [113]. Myers was presented in Chap. 8 where his work with shell shock patients and his introduction of the term shell shock were discussed [114]. His shell shock research began a year after he outlined his investigation of Scriabin, reported in "Two cases of synesthesia" [113].

Myers relays that to Scriabin his presentation of a color that correctly corresponds to music served as "a powerful psychological resonator for the listener" [109]. Myers also wrote:

> Scriabin's attention was first seriously drawn to his colored hearing owing to an experience at a concert in Paris, where, sitting next to his fellow-countryman and composer Rimsky-Korsakov, he remarked that the piece to which they were listening (in D major) seemed to

> him yellow; whereupon his neighbor replied that to him, too, the color seemed golden. Scriabin has since compared with his compatriot and with other musicians the color effects of other keys, especially B, C major and F# major, and believes a general agreement to exist in this respect. He admits, however, that whereas to him the key of F# major appears violet, to Rimsky-Korsakov it appears green; but this deviation he attributes to an accidental association with the color of leaves and grass arising from the frequent use of this key for pastoral music. He also allows that there is some disagreement as to the color-effect of the key of G major. Nevertheless, as is so universally the case with the subjects of synesthesia, he believes that the particular colors which he obtains must be shared by all who are endowed with colored hearing (pp. 112–113, [113]).

Myers' paper also relays that "Scriabin's *chromaesthesia*" refers to the tonality of the music. For him, as the tonality changes in a piece, the color changes because "the color *underlines* the tonality" (p. 113, [113]):

> In the case of colored hearing which I have previously published [115], the color is dependent on the pitch of the note; in many other cases, already described, it is dependent on the name of the note. But for Scriabin a single note has in itself no color; it has the color of its tonality. Indeed, he insists that whether in or apart from musical compositions, a single tone cannot occur alone; even outside music it is accompanied by overtones, which, in many cases, especially in the case of the sounds of nature, include inharmonic overtones, in addition to the harmonic series (p. 113, [113]).

12.6 Syntax: Klüver Constants

Unlike color organs that reference music, sound, and color, form constants are conjoined with visual frames of reference. Heinrich Klüver (1897–1979), a German-American physiologist, psychologist, and philosopher planted the seed in 1926. The medley of reports on nineteenth century synesthesia had mostly receded by this time, and there was a considerable reduction in the literature on synesthesia [6]. Moreover, in some ways, Klüver's work was orthogonal to both mainstream behaviorism and synesthesia because he wasn't concerned with synesthesia directly even as his work is now frequently cited in conjunction with it.

Oliver Sacks presented a fulsome description of the form constants when writing about hallucinations in 2012 [116]:

> [Heinrich Klüver in *Mescal*] reviewed the world literature on the effects of mescal and described his own experiences with it. Keeping his eyes closed, as Weir Mitchell had done, he saw complex geometrical patterns: "Transparent oriental rugs, but infinitely small ... plastic filigreed spherical objects d'art [like] radiolaria ... wallpaper designs ... cobweb-like figures or concentric circles and squares ... architectural forms, buttresses, rosettes, leaf-work, fretwork." For Klüver these hallucinations represented an abnormal activation in the visual system, and he observed that similar hallucinations could occur in a variety of other conditions — migraine, sensory deprivation, hypoglycenlia, fever, delirium, or the hypnagogic and hypnopompic states that come immediately before and after sleep. In *Mechanisms of Hallucination*, published in 1942, Klüver spoke of the tendency to "geometrization" in the brain's visual system, and he regarded all such geometrical hallucinations as permutations of four fundamental "form constants" (he identified these as lattices, spirals, cobwebs, and tunnels). He implied that such constants must reflect something about the organization, the functional architecture, of the visual cortex—but there was little more to be said about this in the 1940s (p. 81, [116]).

Often called *Klüver constants* now, these patterns are frequently connected with synesthesia although they are also found in non-synesthetic art [117]. Given this, I would propose that what Klüver showed is that there are a limited number of perceptual frameworks built into our perceptual system and that these are probably a part of our genetic endowment. His work does, however, provide a taxonomy for the somewhat vague descriptions that frequently accompanied reports [1, 5, 42, 118, 119]. Also noteworthy is that Klüver wasn't classifying the forms as resulting from synesthesia but rather noting that sometimes synesthesia happened in conjunction with hallucinations. As Critchley noted, certain superindividual form constants found in mescal visions are known to anthropologists who study the varieties of "visions" in the ornamental art of various tribes [120]. Finally, whereas Klüver points out that "the descriptions of abnormal somatosensory experiences by mescalinized persons are often as fantastic and unintelligible as those of schizophrenics," he is only speaking about drug-induced synesthesia.

12.7 Testing and Verification

More refined testing and verification techniques launched a new phase in synesthetic research late in the twentieth century. Key benchmarks included a "the Test of Genuineness" (TOG) developed by Simon Baron-Cohen and his colleagues in 1987. This allowed researchers to decipher whether associations were learned or brain-based percepts. It was revised in 2006 (TOG-R) [121, 122].

In 1988, an early study by Maurer and Maurer suggested that all of us begin life as synesthetes, with subsequent neural development reducing the connections among the senses. Richard Cytowic first published his classic *Synesthesia: A Union of the Senses* in 1989. In it, he presented research conducted on 42 synesthetes, including Watson and Hockney [110]. His research was conducted using early imaging technology and laid out criteria for diagnosis.

With the emergence of the Internet at the end of the twentieth century, as was the case with prosopagnosia (Chap. 11), there were additional means for researchers to expand their research pool, online sites, and discussion groups were also fostering communication among synesthetes. As subjects and studies multiplied, they provided corroborative evidence for subjective reports. In addition, structural brain imaging has revealed differences in the organization of white matter [123].

Modified Stroop interference techniques were also developed for testing. These too have shown that synesthesia is automatic and perhaps obligatory. In the standard Stroop paradigm, color names are presented in colored ink, such as the word RED printed in either red or green ink ("congruent" and "incongruent," respectively). One modified version uses responses in the incongruent condition, which are typically much slower than in the congruent condition. Because the task has nothing to do with reading the word, the interference shows if reading the word is automatic [124]. In the "synesthetic Stroop" paradigm, graphemes are presented in either congruent or incongruent ink colors for each synesthete. As Hubbard and Ramachandran point out, for a synesthete who sees 7 as yellow, a 7 presented in yellow would be congruent, and a 7 presented in any other color would be incongruent. The consistent finding is that synesthetes are slower in the incongruent condition than in the congruent condition. Ramachandran and Hubbard also formulated the cross-activation theory of synesthesia that is based on converging evidence from studies of synesthesia showing that sensory areas for processing real and synesthetic information tend to be neighboring brain regions [125–127].

Figure 12.10 demonstrates one test used to demonstrate the reality of synesthesia. The image on the left, in which a triangle composed of 2s is embedded among a field of 5s, appears like the image on the right to synesthetes although their colors probably would not correspond to the red/green shown here. For non-synesthetes, the triangle comprised of the 2s is hard to identify in the black version. For synesthetes, the triangle would jump out. For those who experience 2s as red and 5s as green, the triangle would be perceived like the image on the right. Synesthetes with other color associations would see their colors in a comparable fashion [128].

Cytowic and Eagleman have similarly theorized that synesthesia results from an elevated level of communication or cross talk between brain areas that specialize in different modes of perception [129]. For all of us, cross talk allows us to integrate diverse sense perceptions. A louder cross talk is experienced by synesthetes due to the cross-circuitry of their brains. Cytowic has suggested that the tendency of all of us to move in time with music may not be a genuine form of synesthesia, even if it does suggest that different brain areas are hardwired in all of us [130].

There are also notable overlaps of synesthesia with other conditions. One is with autism spectrum disorders. Future studies might attempt to identify specific groups

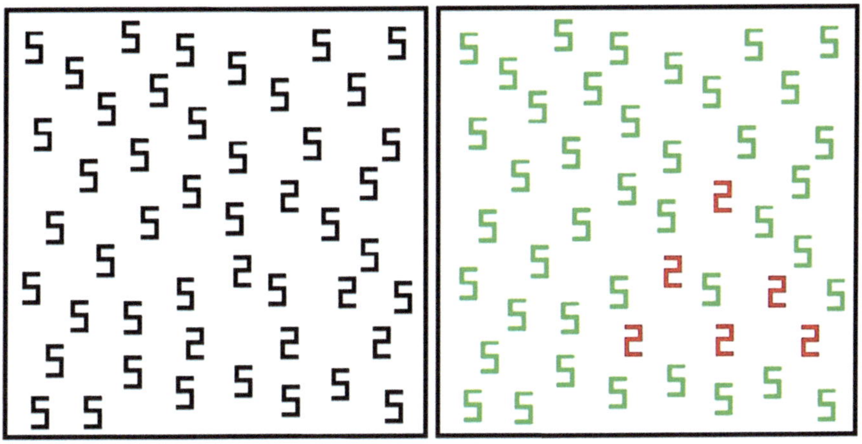

Fig. 12.10 This image demonstrates the automaticity of synesthetic experience. A synesthete might perceive the left panel like the panel on the right. This image first appeared in Ramachandran, V.S., Hubbard, E.M.: Synaesthesia: A window into perception, thought and language. *Journal of Consciousness Studies* 8, 3–34, 2001, Fig. 12.1. The authors and the journal have granted permission for use under the terms of the GNU Free Documentation License [128]

of genes that influence autism, synesthesia, and perception [131]. An open question is the degree to which suppression of synesthetic colors is possible, as measured by negative priming [124].

12.8 Syntax, Synesthesia, Metaphor, and Multisensory Experiences

Characterization of synesthete description as metaphors was common historically. Metaphor is usually defined in terms of speech or language or as a figure of speech by which a characteristic of one object is assigned to another. The idea is that though different it resembles or is analogous to it. In other words, it is comparison by transference of a descriptive word or phrase. In *Poetics*, we find Aristotle's famous quote stating that metaphor implies an intuitive perception of the similarity in dissimilars (*Poetics*, 1459a 3-8] [132]). Literary examples cited above show how easy is to draw a metaphorical inflection from multisensory descriptions.

Chapter 1 pointed out that whereas syntax, like metaphor, tends to be characterized in linguistic terms, all of the arts have their own syntax or grammar. The language painters use to communicate consists of color, line, pattern, form, and so forth. In music, syntax includes pitch, notes, rhythm, melody, harmony, texture, timbre, and so forth. Syntax shares the root *syn-* with synesthesia. Again, the "syn" part of synesthesia brings in the element of combining and of joining together. As the examples above have demonstrated, synesthesia is used in an array of art forms.

Moreover, with the increased use of technology, we are increasingly surrounded by stimuli that encourage multisensory engagement. As a result, there is much discussion of multisensory learning, multisensory art projects, and other manifestations that show the appeal of this kind of blending – even though this kind of blending does not match the perception of a lifelong synesthete who naturally experiences some form of combined sensory awareness.

In the cultural sphere, multisensory experience and integration typically refer to the set of processes by which information arriving from the individual sensory modalities interacts and influences processing in other sensory modalities. While some synesthete's do try to avoid the sensory overload of noisy, busy places, this is not frequently mentioned as a problem. Rather, the difference centers around how vision, audition, touch, and so forth are combined to yield a unified perceptual experience, which is at the core of all experience [133], even if some of the synesthete's experiences are atypical.

As some have noted, there is a benefit in distinguishing between the various kinds of ways in which we do this for the array may have qualitatively different effects on brain functionality, human perception, and human performance [134]. For example, studies have shown that congenital or lifelong synesthesia generally leads to higher intelligence, a better memory for events, vivid imagery, and sensory thresholds. Yet, with this cohort, we also always find exceptions that speak to individual uniqueness. For example, some say synesthetes suffer from a poor sense of direction and in reading maps [130], and others claim their synesthesia helps them remember names and locations, which is one reason testing has advanced our understanding of the condition.

How synesthetes use the trait as a tool comes up frequently in the literature. Duffy has written about how using the personal coding of her synesthesia served as an aid when she was learning Chinese [20]. A professor of neuropathology, with chromatic-phonemic synesthesia and number forms, has explained that synesthesia helps him remember various categories:

> Let me say that this is a *delightful* trait to have. I tend to use it consciously and unconsciously to help me remember correct sequences of numbers, words, phrases, letters, to help me remember names and locations of anatomical structures (especially neuroanatomical structures— you should see the beautiful array of colors in the brain!) and neuropathological classifications. I could go on and on, but those will serve as a few examples (p. 50 [5]).

Research has indicated that understanding the physiology behind how these experiences align with multisensory sensations generally might offer insight into human behavior, sensory congruency and incongruency, the individual synthesis of biological and environmental factors, attention, memory, learning, and so forth. An intriguing cohort largely beyond the scope of this chapter includes studies that show many children outgrow synesthesia in adolescence. Suffice it to say that whether this speaks to hormonal changes, cerebral reorganization, or other factors is unknown. For example, using a randomly sampled population of child synesthetes, Simner and Bain's longitudinal studies of grapheme-color synesthesia showed cases where synesthesia was not developing in the same timeframe for all and for some it died out [30, 135].

12.9 The Larger Picture

Perception has traditionally been viewed modularly, as have the arts, although relational characterizations repeatedly enter into descriptions. Therefore, cross-modality has a long, documented history as well. In recent times, we can identify many non-synesthetic artists who have pursued producing multisensorial work. Jack Ox, who has worked with music and visual art for decades, claims that now it is easy and natural for her to see sonic forms (p. 7, [136]). After Victor Acevedo shifted from digital art to electronic visual music in 2007, his affective works deliberately included a synesthetic component that encourages audiences to engage on multiple sensory levels [137].

There are also artists who offer relational information that allows us to glimpse the kinds of goals an artist may bring to a multisensory approach. We see this well through the work of Wassily Kandinsky (1866–1944) and Paul Klee (1879–1940). Both painters were trained musicians, and both were drawn to the ways one can manipulate abstract possibilities in combining art and music. Yet, although each effectively brought musicality to his painted work, when looking at the motivations of these two colleagues, we find significant points of divergence. Kandinsky, who most now believe was a synesthete, brought more of a top-down, philosophical approach to creating visual music in his artistic statements. Klee, by contrast, was more process-oriented and relied on perceptual elements to bring a dynamic design element into his process-oriented compositions.

The two men first met, briefly, in Franz Stuck's painting class in the Munich Academy in the summer of 1900 [138]. They met again in 1911, and their professional friendship further strengthened after Klee joined the Blue Rider group (founded by Kandinsky and the painter Franz Marc) in 1912. Later, the bonds between Klee and Kandinsky deepened when Klee accepted an appointment at the Bauhaus in 1921 and Kandinsky joined him in 1922.

Working side by side for many years, both painters articulated their projects in terms of the Bauhaus aspiration to unify all of the arts, a non-modular goal that they shared even before their appointments. Indeed, some claim that the resemblances among their works in the early 1920s are so close that an untrained eye might well confuse the two [139]. According to Roskill, "Klee and Kandinsky … [were] like a musical partnership … even while their 'styles' of performance and commentary remained entirely different in cast" (p. xvi, [138]).

As noted, Klee's approach was process-oriented. His hope was to "one day…be able to improvise freely on the keyboard of colors: the rows of watercolors in my paint box" (p. 17, [140]). We see this in Fig. 12.11, his watercolor/drawing titled *Carnival in the Snow* (1923). In this work, acrobats and circus jugglers emerge from a complex pattern of shapes and colors. The composition also shows two jugglers who share a face and torso. Like others in *Der Blaue Reiter* group, Paul Klee believed that the elements of color and shape alone could carry connotations of emotions, moods, and subjective feelings. His method was to begin painting without

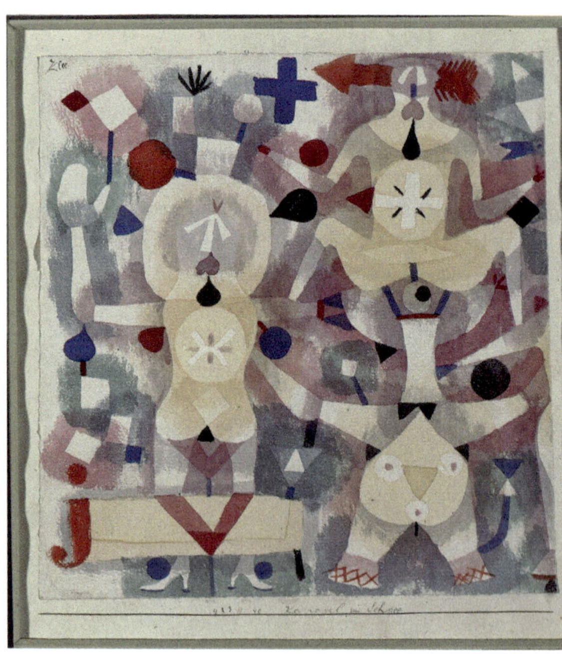

Fig. 12.11 *Carnival in the Snow*. Paul Klee, 1923. Drawing sheet: 24 × 23.5 cm (9 7/16 × 9 1/4 in). In this watercolor, acrobats and circus jugglers emerge from a complex pattern of shapes and colors. [141]. Published in collaboration with the Cleveland Museum of Art and their Open Access initiative

a subject, only settling upon something that could be recognized after layering colored form upon colored form [141].

Kandinsky, by contrast, aspired to develop a vocabulary using his synesthesia that would point toward universals, and it was in this context that he referenced his color-music. As he put it: "Color is the keyboard, the eyes are the hammers, the soul is the piano with many strings. The artist is the hand which plays, touching one key or another, to cause vibrations in the soul" (p. 25, [142]). His theories were published in *Über das geistige in der kunst* (*Concerning the Spiritual in Art*, 1912) [143–146]. Haftmann succinctly summarizes the psychological premises that defined each process, writing "Kandinsky took hold of the world but remained outside it. Klee sank himself in the world" [139].

Kandinsky spoke of "my colors," and his aim was to enhance sensory exchange much like one might develop relative pitch. While we can't test him, penciled notes in his books that spoke of exercises one could do to enhance the experiences do offer some support of this idea. His 1896 description of a performance of Richard Wagner's *Lohengrin* in Moscow reads much like the one visually blind Lusseynan offered of his perceptions of music discussed earlier, but Kandinsky's description of the impact of the color-music of the orchestra is what he translates into painting:

> The violins, the deep tones of the basses, and especially the wind instruments at that time embodied for me all the power of that pre-nocturnal hour. I saw all *my colors* in my mind; they stood before my eyes. Wild, almost crazy lines were sketched in front of me [*italics added*] (pp. 364, [147]).

As I have discussed elsewhere [146], even in musical terms, the contrasts between Klee and Kandinsky are striking. Kandinsky's equated his work with Schönberg's 12-tone music. From this starting point, he concluded that the concept of tonal harmony was undergoing a radical change, meaning that dissonance was becoming a means of expression on par with consonance ([148]. This led Kandinsky to endeavor to join music with his own move toward abstraction and transcendence.

Klee, by contrast, was convinced that the modern music of his day, which Kandinsky applauded, was too academic and overly dictated by educational theory. For this reason, Klee focused on developing an abstract, visually based perceptual language. He based it on historical musical models for he was not seeking innovation so much as fulfillment of his desire to blend the arts into an all-inclusive spiritually felt meshing of sensations.

Finally, whereas Kandinsky saw his art as a spiritual science, Klee never aspired to call his approach science. His far-reaching experiments with color and form were experiential. His writings suggest he revised his motifs as he constructed them, continually adjusting elements in order to tease out intense visual or perceptual reactions [139, 149]. The abstract, subtle relationships that resulted, as such, are hard to characterize linguistically and do not effectively translate in reproductions. Originals do, nonetheless, visually evoke complex chords, rhythms, and tonal variations. Although his work appears deceptively simple at first glance, what makes the actual pieces striking is that the imagery is so infused with the delicate rhythms and intricate counterpoint of musical compositions. The symbolic language becomes secondary because he achieved his goal of playing color like a "chromatic keyboard."

As such, Klee's techniques stimulate thoughts about psychophysically designed experiments, along the lines pioneered by J.J. Gibson [150, 151]. Some may argue that Gibson's work is now somewhat peripheral to research highlighting cognitive operations [152]. A counterargument would be that his interest in the world we see is relevant to the visual arts precisely because the artist establishes an environmental relationship with the artwork while constructing it. This environmental relationship is at the core of neuroculture as well.

12.10 Conclusion

Throughout time, humans have grappled with who we are, how we live together, and the nature of the world. Within this, each of us brings an individual genetic profile and adds environmental influences as our lives develop. Thus, we perceive our reality through both our cultures and our brain apparatus. This book has examined the ways our brains construct our reality and ended with synesthesia because synesthesia is an atypical condition that offers in a largely positive light on the degree to which our perceptions vary. Synesthesia research, reports, and projects thus offer a viable means for looking at neuroscience and art broadly and in a cross-disciplinary manner, as a neurocultural lens does. Both neuroculture and the subset provided by synesthesia include all demographics. Investigations also intersect with neuroscience, psychology, musicology, art history, philosophy, anthropology, and linguistics, to name only a few of the many points of entry into functionality and the experience.

Historical reports and current research into synesthesia also emphasize that there is no "typical" human mind, brain, or body. The array of cases introduced throughout this book further supports this. Earlier, in 1883, Sir Francis Galton made this pointed when he published *Inquiries into Human Faculty and Its Development*, where he discussed synesthesia in detail. Indeed, his opening words emphasized that the differences among humans are profound:

> It will be seen in the end how greatly metaphysicians and psychologists may err, who assume their own mental operations, instincts, and axioms to be identical with those of the rest of [humankind], instead of being special to themselves. The differences between [humans] are profound, and we can only be saved from living in blind unconsciousness of our own mental peculiarities by the habit of informing ourselves as well as we can of those of others (p. 32, [23]).

Since Galton wrote, we have learned more. We now know more about how genetic makeup and environmental factors influence brain development. Both help us to understand how the brain functions as well as the unique ways individuals function. Thus, as detailed throughout this book, the neurocultural lens offers a window that broadens our understanding of neuroscience, art, and the human experience. Furthermore, it is an adaptable tool for communicating about human well-being.

References

1. Cytowic RE. The man who tasted shapes: a bizarre medical mystery offers revolutionary insights into emotions, reasoning and consciousness. New York: Putnam; 1993.
2. Luria AR. The mind of a mnemonist: a little book about a vast memory. London: Cape; 1969.
3. Sacks O. On the move: a life. New York: Alfred A. Knopf; 2015.
4. Johnson R. The mystery of S., the man with an impossible memory. The New Yorker; 2017.
5. Cytowic RE. Synesthesia: a union of the senses. 2nd ed. Cambridge: MIT Press; 2002.
6. Marks LE. On colored-hearing synesthesia: cross-modal translations of sensory dimensions. Psychol Bull. 1975;82(3):303–31.
7. Marks LE. The unity of the senses: interrelations among the modalities. New York: Academic Press; 1978.
8. Spector F, Maurer D. Synesthesia: a new approach to understanding the development of perception. Dev Psychol. 2009;45(1):175–89.
9. Smilek D, Moffatt BA, Pasternak J, White BN, Dixon MJD, Merikle PM. Synaesthesia: a case study of discordant monozygotic twins. Neurocase. 2002;8(4):338–42.
10. Bosley HG, Eagleman DM. Synesthesia in twins: incomplete concordance in monozygotes suggests extragenic factors. Behav Brain Res. 2015;286:93–6.
11. Tilot AK, Vino A, Kucera KS, Carmichael DA, van den Heuvel L, den Hoed J, et al. Investigating genetic links between grapheme-colour synaesthesia and neuropsychiatric traits. Philos Trans R Soc Lond B Biol Sci. 2019;374(1787):20190026.
12. Day SA. What is synesthesia? In: Jewanski J, Day SA, Siddiq S, Haverkamp M, Reuter C, editors. Music and synesthesia abstracts from a conference in Vienna, July 3–5, 2020. Reihe. p. 2020.
13. Simner J, Hubbard E. Synesthesia in school-aged children. In: Simner J, Hubbard E, editors. Oxford handbook of synesthesia. Oxford: Oxford University Press; 2013. p. 64–82.
14. Asher JE, Carmichael DA. The genetics and inheritance of synesthesia. In: Simner J, Hubbard E, editors. Oxford handbook of synesthesia. Oxford: Oxford University Press; 2013. p. 23–45.
15. Maurer D, Gibson LC, Spector F. Synesthesia in infants and very young children. In: Simner J, Hubbard E, editors. Oxford handbook of synesthesia. Oxford: Oxford University Press; 2013. p. 46–63.
16. Dixon MJ, Smilek D, Merikle PM. Not all synaesthetes are created equal: projector versus associator synaesthetes. Cogn Affect Behav Neurosci. 2004;4(3):335–43.
17. Eagleman DM. The objectification of overlearned sequences: a new view of spatial sequence synesthesia. Cortex. 2009;45(10):1266–77.
18. Cytowic R. "We are on to something real and fundamental." An interview with Richard E. Cytowic. Theoria et Historia Scientiarum. 2013;X:239–44.
19. Online Etymology Dictionary. Douglas Harper; 2020. synaesthesia (n.). https://www.etymonline.com. Accessed 16 Oct 2023.
20. Duffy PL. Blue cats and chartreuse kittens: how synesthetes color their worlds. New York: Times Books; 2001.
21. Simner J, Mulvenna C, Sagiv N, Tsakanikos E, Witherby SA, Fraser C, et al. Synaesthesia: the prevalence of atypical cross-modal experiences. Perception. 2006;35(8):1024–33.
22. Galton F. Visualised numerals. Nature. 1880;21(533):252–6.
23. Galton F. Inquiries into human faculty and its development. London: Macmillan; 1883.
24. Jonas C, Jarick M. Synesthesia, sequences, and space. In: Simner J, Hubbard E, editors. Oxford handbook of synesthesia. Oxford: Oxford University Press; 2013. p. 123–48.
25. Ward J. Synesthesia. Annu Rev Psychol. 2013;64:49–75.
26. Price MC, Mentzoni RA. Where is January? The month-SNARC effect in sequence-form synaesthetes. Cortex. 2008;44(7):890–907.
27. Ruheit C. Color associations with the months. Science. 1885;6(133):142–3.
28. Winslow A. Color and other associations. Science. 1885;6(137):243.

29. Maurer D, Maurer C. The world of the newborn. New York: Basic Books; 1988.
30. Simner J, Bain A. A longitudinal study of grapheme-color synesthesia in childhood: 6/7 years to 10/11 years. Front Hum Neurosci. 2013;7
31. Maurer D. A rainbow-coloured alphabet: a developmental perspective on synesthesia. The International Association of Synaesthetes, Artists, and Scientists (IASAS); May 27th and 28th 2022: IASAS; 2022.
32. Ramachandran VS, Hubbard EM. Hearing colors, tasting shapes. Sci Am Special Ed. 2006;16(3):76–83.
33. Simner J, Hubbard EM. The Oxford handbook of synesthesia. Oxford and New York: Oxford University Press; 2013.
34. Simner J, Ward J. Synesthesia: the taste of words on the tip of the tongue. Nature. 2006;444(7118):438.
35. Jewanski J, Day SA, Siddiq S, Haverkamp M, Reuter C. Music and synesthesia abstracts from a conference in Vienna, scheduled for July 3–5, 2020. In: Jewanski J, Day SA, Siddiq S, Haverkamp M, Reuter C, editors. Reihe; 2020.
36. Steen C. Personal communication; 2023.
37. Sidoroff-Dorso AV, Jewanski J, Day SA. Interview with Carol Steen. In: Sidoroff-Dorso AV, Jewanski J, Day SA, editors. Synaesthesia: opinions and perspectives: 30 interviews with leading scientists, artists and synaesthetes: Universitäts- und Landesbibliothek Münster, 2020. p. 377-94.
38. Jacome DE. Volitional monocular Lilliputian visual hallucinations and synesthesia. Eur Neurol. 1999;41(1):54–6.
39. Hubbard EM. Neurophysiology of synesthesia. Curr Psychiatry Rep. 2007;9(3):193–9.
40. Neufeld J, Roy M, Zapf A, Sinke C, Emrich H, Prox-Vagedes V, et al. Is synesthesia more common in patients with Asperger syndrome? Front Hum Neurosci. 2013;7
41. Sacks O. The case of the colorblind painter. An anthropologist on Mars: seven paradoxical tales. New York: Vintage; 1995. p. 3–41.
42. Sacks O. Musicophilia: tales of music and the brain. New York: Alfred A. Knopf; 2010.
43. Chalmers DJ. David Chambers interview: what is it like to be a philosopher? In: Sosis C, editor. What is it like to be a philosopher? https://www.whatisitliketobeaphilosopher.com/. Cliff Sosis; 2016.
44. Chalmers DJ. Synesthesia, personal communication. 2023.
45. Lorusso L, Ione A, Franchini AF, Porro A. Synesthesia and emotional sound. In: Colombo B, editor. The musical neurons. Cham: Springer International Publishing; 2022. p. 13–26.
46. Locke J. Locke's essays. An essay concerning human understanding. And a treatise on the conduct of the understanding. With the author's last additions. Philadelphia: T. Ellwood Zell; 1689/1854.
47. Robinson DN. An intellectual history of psychology. Madison: University of Wisconsin Press; 1995.
48. Wade NJ. The science and art of the sixth sense. In: Bacci F, Melcher D, editors. Art and the senses. Oxford and New York: Oxford University Press; 2011. p. 19–58.
49. Sears E. Sensory perception and its metaphors in the time of Richard of Fournival. In: Bynum WF, Porter R, editors. Medicine and the five senses. Cambridge and New York: Cambridge University Press; 1993. p. 17–39.
50. Finger S. Origins of neuroscience: a history of explorations into brain function. New York: Oxford University Press; 1994.
51. Clarke E, Jacyna LS. Nineteenth-century origins of neuroscientific concepts. Berkeley: University of California Press; 1987.
52. Green CD. Where did the ventricular localization of mental faculties come from? J Hist Behav Sci. 2003;39(2):131–42.
53. Willis T. Two Discourses Concerning the Soul of Brutes, Which Is That of the Vital and Sensitive of Man. The first is physiological, shewing the nature, parts, powers, and affections of the same. The other is pathological, which unfolds the diseases which affect it and its

primary seat; to wit, the brain and nervous stock, and treats of their cures: with copper cuts. London: Printed for Thomas Dring etc.; 1672
54. Buzon F. Compendium of music. In: Nolan L, editor. The Cambridge Descartes Lexicon. Cambridge: Cambridge University Press; 2015. p. 143–5.
55. Campen CV. Visual music and music painting. In: Bacci F, Melcher D, editors. Art and the senses. Oxford, NY: Oxford University Press; 2011. p. 495–512.
56. Kassler JC. Newton's sensorium: anatomy of a concept. Cham: Springer International Publishing; 2018.
57. Newton I. Opticks: or, a treatise of the reflections, refractions, inflections and colors of light. London: Smith and Walford; 1704.
58. Swift J. Gulliver's travels: into several remote nations of the world. London: T and J Allman; 1825.
59. Boyle R. Experiments and considerations touching colours. First occasionally written, among some other essays, to a friend; and now suffer'd to come abroad as the beginning of an experimental history of colours. London: H. Herringman; 1664.
60. Jewanski J, Day SA, Ward J. A colorful Albino: the first documented case of synaesthesia, by Georg Tobias Ludwig Sachs in 1812. J Hist Neurosci. 2009;18(3):293–303.
61. Locke J. An essay concerning human understanding. London: Thomas Bassett; 1690.
62. Tenberken S. My path leads to Tibet: the inspiring story of how one young blind woman brought hope to the blind children of Tibet. New York: Arcade Pub; 2003.
63. Sacks O. The mind's eye. New York: Alfred A. Knopf; 2010.
64. Lusseyran J. And there was light: the extraordinary memoir of a blind hero of the French resistance in World War II. Novato, CA: New World Library; 2014.
65. Hubel DH, Wiesel TN. Receptive fields of single neurones in the cat's striate cortex. J Physiol. 1959;148(3):574–91.
66. Hubel DH, Wiesel TN. Receptive fields, binocular interaction and functional architecture in the cat's visual cortex. J Physiol. 1962;160(1):106–54.
67. Sacks O. The river of consciousness. New York: Alfred A. Knopf; 2017.
68. Starr F. Note on colored-hearing. Am J Psychol. 1893;5(4):416–24.
69. Arbuthnot J, Pope A. Memoirs of the extraordinary life, works and discoveries of Martinus Scriblerus. London: Scriblerus Club; 1725.
70. Hoffmann ETA. Hoffmann's musical writings: Kreisleriana; the poet and the composer. Cambridge: Cambridge University Press; 2003.
71. von Goethe JW, Eastlake CL. Theory of colours. Cambridge: M.I.T. Press; 1970.
72. Schlegel JHG. Ein Beitrag zur nähern Kenntniß der Albinos. Meiningen: Keyssner; 1824.
73. Perroud L. De l'hyperchromatopsie. Memoires de la Soci é t é des Sciences M é dicales de Lyon. 1863;2:37–41.
74. Krohn WO. Pseudo-chromesthesia, or the association of colors with words, letters and sounds. Am J Psychol. 1892;5(1):20–41.
75. Jewanski J. Synesthesia in the nineteenth century: scientific origins. In: Simner J, Hubbard E, editors. Oxford handbook of synesthesia. Oxford: Oxford University Press; 2013. p. 369–98.
76. Jewanski J, Day SA, Simner J, Ward J. The beginnings of an interdisciplinary study of synaesthesia: discussions about the Nussbaumer brothers (1873). Theoria et Historia Scientiarum. 2014;10:149.
77. Jewanski J, Simner J, Day SA, Rothen N, Ward J. The "golden age" of synesthesia inquiry in the late nineteenth century (1876–1895). J Hist Neurosci 2020;29(2):175-202.
78. Bleuler E, Lehmann K. Zwangsmässige lichtempfindungen durch Schall und Verwante erscheinungen auf dem Gebiete der andern sinnesempfindungen: Fues's Verlag (R. Reisland); 1881.
79. Jewanski J, Simner J, Day SA, Rothen N, Ward J. Recognizing synesthesia on the international stage: the first scientific symposium on synesthesia (at The International Conference of Physiological Psychology, Paris, 1889). J Hist Neurosci. 2020;29(4):357–84.
80. Plassart A, White RC. Theodore Flournoy on synesthetic personification. J Hist Neurosci. 2017;26(1):1–14.

81. Flournoy T, editor. Des phénomènes de synopsie (audition colorée) Photismes—schèmes visuels—personnifications; 1883.
82. Lee DD. Flournoy, Théodore. In: Rieber R, editor. Encyclopedia of the history of psychological theories. Springer US; 2012. p. 443–4.
83. The American Journal of Psychology. Volume index for 1893. Am J Psychol. 1893;5(4): 552–6.
84. Holden ES. Colour-associations with numerals, etc. Nature. 1891;44(1132):223–4.
85. Holden ES. Color and other associations. Science. 1885;6(137):242–3.
86. Holden ES. Color associations with numerals, etc. Science. 1899;10(255):738.
87. Chipman KA. Color associations. Science. 1885;6(132):125.
88. Minot CS. Color associations with the months. Science. 1885;6(130):82–3.
89. Slack FM. Color and other associations. Science. 1885;6(135):186.
90. Witthoft N, Winawer J. Synesthetic colors determined by having colored refrigerator magnets in childhood. Cortex. 2006;42(2):175–83.
91. Hubbard EM, Arman AC, Ramachandran VS, Boynton GM. Individual differences among grapheme-color synesthetes: brain-behavior correlations. Neuron. 2005;45(6):975–85.
92. Seron X, Pesenti M, Noël M-P, Deloche G, Cornet J-A. Images of numbers, or "when 98 is upper left and 6 sky blue". Cognition. 1992;44(1):159–96.
93. Hubbard EM, Piazza M, Pinel P, Dehaene S. Interactions between number and space in parietal cortex. Nat Rev Neurosci. 2005;6(6):435–48.
94. Dorland WAN. American Pocket Medical Dictionary: containing the pronunciation and definition of over 26,000 of the terms used in medicine and the Kindred Sciences, along with over 60 extensive tables. Philadelphia: W.B. Saunders Company; 1898.
95. Alford LB. A report on two cases of synesthesia. J Abnormal Psychol. 1918;13(1):1–11.
96. Calkins MW. A statistical study of pseudo-chromesthesia and of mental-forms. Am J Psychol. 1893;5(4):439–64.
97. Gowers WR. Epilepsy and other chronic convulsive diseases: their causes, symptoms, & treatment. London: J & A Churchill; 1881.
98. Leonard A. Picturing listening in the late nineteenth century. Art Bullet. 2007;89(2):266–86.
99. Silvers L. Beyond the senses: the Cenesthetic Poetics of French Symbolism. Modern Philology. 2014;112(2):381–404.
100. von Erhardt-Siebold E. Harmony of the senses in English, German, and French Romanticism. PMLA. 1932;47(2):577–92.
101. Campen, C. Artistic and psychological experiments with synesthesia. Leonardo. 1999;32(1):9–14.
102. Rimbaud A. Vowels. New York: Limited Editions Club; 1996.
103. Rimbaud A. Rimbaud complete. New York: Random House; 2013.
104. Rimbaud A, Ella GA. Voyelles. San Francisco: Press in Tuscany Alley; 1993.
105. Cytowic RE. Synesthesia. Cambridge: The MIT Press; 2018.
106. Holabird J, Nabokov VV, Boyd B. Vladimir Nabokov: AlphaBet in Color. Corte Madera, CA: Gingko Press, Inc.; 2005.
107. Nabokov VV. Speak, memory: a memoir. London: Victor Gollancz Ltd; 1951.
108. Baron-Cohen S, Harrison J. Synesthesia: a challenge for developmental cognitive neuroscience. In: Neurodevelopmental disorders. Developmental cognitive neuroscience. Cambridge: The MIT Press; 1999. p. 491–503.
109. Peacock K. Synesthetic perception: Alexander Scriabin's color hearing. Music Percept. 1985;2(4):483–505.
110. Cytowic RE. Synesthesia: a union of the senses. New York: Springer-Verlag; 1989.
111. Gage J. Color and culture: practice and meaning from antiquity to abstraction. Boston: Little, Brown and Company; 1993.
112. Altenmüller E. Alexander Scriabin: his chronic right-hand pain and its impact on his piano compositions. In: Altenmüller E, Finger S, Boller F, editors. Progress in brain research, vol. 216. Amsterdam: Elsevier; 2015. p. 197–215.
113. Myers CS. Two cases of synaesthesia. Br J Psychol. 1914;7:112–7.

114. Myers CS. A contribution to the study of shell shock. Lancet. 1915;185(4772):316–20.
115. Myers CS. A case of synaesthesia. Br J Psychol. 1911;6:228–38.
116. Sacks O. Hallucinations. New York: Alfred A. Knopf; 2012.
117. Ione A. Studio Ione 2003: paintings by Amy Ione. Berkeley: Diatrope Press; 2003.
118. Berman G, Steen C. Synesthesia: art and the mind. Hamilton, ON: McMaster Museum of Art; 2008.
119. Klüver H. Mescal: the 'divine' plant and its psychological effects. London: K. Paul, Trench, Trubner & Company Limited; 1928.
120. Critchley M. Introduction. In: Mescal: the 'divine' plant and its psychological effects. London: K. Paul, Trench, Trubner & Company Limited; 1928. p. 3–7.
121. Baron-Cohen S, Wyke MA, Binnie C. Hearing words and seeing colours: an experimental investigation of a case of synaesthesia. Perception. 1987;16(6):761–7.
122. Johnson D, Allison C, Baron-Cohen S. The prevalence of synesthesia: the consistency revolution. In: Simner J, Hubbard E, editors. Oxford handbook of synesthesia. Oxford: Oxford University Press; 2013. p. 2–22.
123. Rouw R, Scholte HS. Increased structural connectivity in grapheme-color synesthesia. Nat Neurosci. 2007;10(6):792–7.
124. Hubbard EM, Ramachandran VS. Neurocognitive mechanisms of synesthesia. Neuron. 2005;48(3):509–20.
125. Hubbard EM, Brang D, Ramachandran VS. The cross-activation theory at 10. J Neuropsychol. 2011;5(2):152–77.
126. Ramachandran VS, Hubbard EM. Psychophysical investigations into the neural basis of synaesthesia. Proc Biol Sci. 2001;268(1470):979–83.
127. Specht K. Synaesthesia: cross activations, high interconnectivity, and a parietal hub. Transl Neurosci. 2012;3(1):15–21.
128. Ramachandran VS, Hubbard EM. Synaesthesia: a window into perception, thought and language. J Conscious Studies. 2001;8(12):3–34.
129. Cytowic RE, Eagleman D. Wednesday is indigo blue: discovering the brain of synesthesia. Cambridge: MIT Press; 2009.
130. Reich D. Cross talk: the science and art of synesthesia. American University Magazine. https://www.american.edu/magazine/. Accessed 13 Feb 2020.
131. Taylor MJ, van Leeuwen TM, Kuja-Halkola R, Lundström S, Larsson H, Lichtenstein P, et al. Genetic and environmental architecture of synaesthesia and its association with the autism spectrum—a twin study. Proc Royal Soc B: Biol Sci. 2009;2023(290):20231888.
132. Aristotle, Butcher SH, Fergusson F. Aristotle's Poetics: Farrar, Straus and Giroux; 1961.
133. Talsma D, Senkowski D, Soto-Faraco S, Woldorff MG. The multifaceted interplay between attention and multisensory integration. Trends Cogn Sci. 2010;14(9):400–10.
134. Spence C. Crossmodal correspondences: a tutorial review. Attent Percept Psychophys. 2011;73(4):971–95.
135. Simner J, Harrold J, Creed H, Monro L, Foulkes L. Early detection of markers for synaesthesia in childhood populations. Brain. 2009;132(Pt 1):57–64.
136. Ox J. Color me synesthesia. Leonardo. 2017;32:7–8.
137. Frank P, Frost C, Miller T, Masucci M, Prince PD, Acevedo V, Loeb AL. Acevedo in context: analog media 1977–1987, digital media 1983–2020. Los Angeles, CA: Acevedomedia; 2022.
138. Roskill MW. Klee, Kandinsky, and the thought of their time: a critical perspective. Urbana: University of Illinois Press; 1992.
139. Haftmann W. The mind and work of Paul Klee. London: Faber; 1967.
140. Düchting H. Paul Klee: painting music. Munich and New York: Prestel-Verlag; 1997.
141. The Cleveland Museum of Art. Paul Klee: carnival in the snow. Cleveland: The Cleveland Museum of Art; 1923.
142. Kandinsky W. Concerning the spiritual in art. New York: Dover Publications; 1986.
143. Kandinsky W. Über das geistige in der kunst, insbesondere in der malerei; mit acht tafeln und zehn originalholzschnitten. MÜnchen: R. Piper & Co; 1912.

144. Ione A, Tyler CW. Neurohistory and the arts. Was Kandinsky a synesthete? J Hist Neurosci. 2003;12(2):223–6.
145. Just DKC. Was Kandinsky a synaesthete? Examining his writings and other evidence. Multisens Res. 2017;3–5:447–60.
146. Ione A. Klee and Kandinsky polyphonic painting, chromatic chords and synaesthesia. J Conscious Studies. 2004;11(3-4):148–58.
147. Kandinsky W. Vergo P, editor. Kandinsky: complete writings on art. London: Farber & Farber. 1913/1982
148. Maur KV. The sound of painting. Munich, New York: Prestel; 1999.
149. Klee P, Klee F. The diaries of Paul Klee, 1898–1918. Berkeley: University of California Press; 1964.
150. Gibson JJ. The perception of the visual world. Boston: Houghton Mifflin; 1950.
151. Gibson JJ. The ecological approach to visual perception. Mahwah, NJ: Lawrence Erlbaum Associates; 1987.
152. Zeki S. Closet reductionists. Trends Cogn Sci. 2001;5(2):45–6.

Appendix A: The Mute's Lament by John Carlin

The Mute's Lament

I move— a silent exile on this earth;
As in his dreary cell one doomed for life,
My tongue is mute, and closed ear heedeth not;
No gleam of hope this darken'd mind assures
That the blest power of speech shall e'er be known.
Murmuring gaily o'er their pebbly beds
The limpid streamlets as they onward flow
Through verdant meadows and responding woodlands,
Vocal with merry tones — *I hear them not.*
The linnet's dulcet tone; the robin's strain;
The whippowil's; the lightsome mock-bird's cry,
When merrily from branch to branch they skip
Flap their blithe wings, and o'er the tranquil air
Diffuse their melodies — *I hear them not.*
The touches-lyric of the lute divine,
Obedient to the rise, the cadence soft,
And the deep pause of maiden's pensive song,
While swells the heart with love's elated life,
Draw forth its mellow tones — *I hear them not.*
Deep silence over all, and all seems lifeless;
The orator's exciting strains the crowd
Enraptur'd hear, while meteor-like his with
Illuminates the dark abyss of mind—
Alone, left in the dark — *I hear them not.*
While solemn stillness reigns in sacred walls,
Devotion high and awe profound prevail,
The balmy words of God's own messenger
Excite to love, and troubled spirits soothe —
Religion's dew-drops bright — *I feel them not.*
From wearied search through long and cheerless ways
For faithless fortune, I, lorn, homeward turn;
And must this tankless tongue refuse to breathe
The blest word "Mother," when that being dear
I meet with steps elastic, full of joy

A. Ione, *Neuroscience and Art*, Neurocultural Health and Wellbeing, https://doi.org/10.1007/978-3-031-62336-3

And all the fibres of this heart susceptive
Throb with our nature's strongest, purest love?
Oh, that this tongue must still forbear to sing
The hymn sublime, in praise of God on high;
Whilst solemnly the organ peals forth praises,
Inspired and deep with sweetest harmony!
Though sad and heavy is the fate I bear,
And I may sometimes wail my solitude,
Yet oh, how precious the endowments He,
T'alleviate, hath lavished, and shall I
Thankless return his kindness by laments?
O, Hope! How sweetly smileth Heavenly Hope
On the sad, drooping soul and trembling heart!
Bright as the morning star when night recedes,
His genial smile this longing soul assures
That when it leaves this sphere replete with woes
For Paradise replete with purest joys,
My ears shall be unsealed, and I shall hear;
My tongue shall be unbound, and I shall speak,
And happy with the angels sing forever!

The Mute's Lament. In: *The Instructors of the American Asylum.* American Annals of the Deaf and Dumb. 1. Hartford: Convention of American Instructors of the Deaf and Conference of Executives of American Schools for the Deaf; 1848.

Appendix B: Dulce et Decorum Est by Wilfred Owen

Dulce et Decorum Est

Bent double, like old beggars under sacks,
Knock-kneed, coughing like hags, we cursed through sludge,
Till on the haunting flares we turned our backs,
And towards our distant rest began to trudge.
Men marched asleep. Many had lost their boots,
But limped on, blood-shod. All went lame; all blind;
Drunk with fatigue; deaf even to the hoots
Of gas-shells dropping softly behind.
Gas! GAS! Quick, boys!—An ecstasy of fumbling
Fitting the clumsy helmets just in time,
But someone still was yelling out and stumbling
And flound'ring like a man in fire or lime. —
Dim through the misty panes and thick green light,
As under a green sea, I saw him drowning.
In all my dreams before my helpless sight,
He plunges at me, guttering, choking, drowning.
If in some smothering dreams, you too could pace
Behind the wagon that we flung him in,
And watch the white eyes writhing in his face,
His hanging face, like a devil's sick of sin;
If you could hear, at every jolt, the blood
Come gargling from the froth-corrupted lungs,
Obscene as cancer, bitter as the cud
Of vile, incurable sores on innocent tongues, —
My friend, you would not tell with such high zest
To children ardent for some desperate glory,
The old Lie: Dulce et decorum est
Pro patria mori.

By Wilfred Owen

A. Ione, *Neuroscience and Art*, Neurocultural Health and Wellbeing,
https://doi.org/10.1007/978-3-031-62336-3

Index

A. Ione, *Neuroscience and Art*, Neurocultural Health and Wellbeing,
https://doi.org/10.1007/978-3-031-62336-3

GPSR Compliance

The European Union's (EU) General Product Safety Regulation (GPSR) is a set of rules that requires consumer products to be safe and our obligations to ensure this.

If you have any concerns about our products, you can contact us on ProductSafety@springernature.com

In case Publisher is established outside the EU, the EU authorized representative is:

Springer Nature Customer Service Center GmbH
Europaplatz 3
69115 Heidelberg, Germany

Batch number: 10732044

Printed by Printforce, the Netherlands